Lecture Notes in Mathematics

2127

T0236519

More information about this series at
http://www.springer.com/series/304

Benjamin Sambale

Blocks of Finite Groups
and Their Invariants

 Springer

Benjamin Sambale
Institut für Mathematik
Friedrich-Schiller-Universität Jena
Jena, Germany

ISBN 978-3-319-12005-8   ISBN 978-3-319-12006-5 (eBook)
DOI 10.1007/978-3-319-12006-5
Springer Cham Heidelberg New York Dordrecht London

Lecture Notes in Mathematics ISSN print edition: 0075-8434
                             ISSN electronic edition: 1617-9692

Library of Congress Control Number: 2014954808

Mathematics Subject Classification (2010): 20C15, 20C20, 20C40

Printed on acid-free paper

Springer is part of Springer Science+Business Media (www.springer.com)

Introduction

The classification of the finite simple groups is considered as one of the greatest achievements in mathematics of the twentieth century. The result provides the most basic pieces every finite group is composed of, and thus leads to a better understanding of symmetries arising from nature. The extremely long proof of the classification brings together the work of many mathematicians from different fields. One of the main contributors was Richard Brauer who introduced several innovative notions which became research topics on their own.

One of Brauer's ideas was to distribute the indecomposable representations of a finite group into its blocks. These blocks are algebras defined over an algebraically closed field of prime characteristic p. This shifts many problems about finite groups to questions about their blocks which are "smaller" speaking of dimensions. As an example, block theory was essentially used in Glauberman's famous Z^*-Theorem which in turn is a major ingredient in the proof of the classification mentioned above.

The present work focuses on numerical invariants of blocks and how they are determined by means of local data. Thus, we usually consider a block B of an arbitrary finite group G. Then it is a challenging task to determine the number $k(B)$ of irreducible representations of G in B. This global invariant is strongly influenced by a piece of local information called the defect group D of B. Here, D is a p-subgroup of G which is uniquely determined up to conjugation (and thus isomorphism). This raises the following natural question which will be our main theme:

What can be said about $k(B)$ and other invariants if D is given?

Brauer himself conjectured that the inequality $k(B) \leq |D|$ should be true (here $|D|$ is the order of D). This problem, now known as Brauer's $k(B)$-Conjecture, has been unproved for almost 60 years. In this work we will give a proof of this conjecture under different types of additional hypotheses. These hypotheses often take the embedding of D in G into account. Therefore, we make extensive use of the language of fusion systems—a notion originally invented by Puig under the name Frobenius categories. In many instances the combination of old methods by Brauer

and Olsson using decomposition numbers together with new accomplishments from
the theory of fusion systems turns out to be very successful.

Another even stronger conjecture from block theory, proposed by Alperin in
1986, makes a precise statement about the number $l(B)$ of simple modules of B
in terms of so-called weights. We are able to obtain a proof of Alperin's Weight
Conjecture for several infinite families of defect groups. In fact, these are the first
new results of that kind after Brauer [41], Dade [65] and Olsson [212] settled blocks
with finite and tame representation type over 20 years ago. Similarly, we provide
evidence for Robinson's Ordinary Weight Conjecture which predicts the numbers
$k_i(B)$ of irreducible characters of a given height $i \geq 0$. Note that $k(B)$ is the sum
over the $k_i(B)$ $(i = 0, 1, 2, \ldots)$.

In some favorable cases we answer a more subtle question: What are the possible
Morita equivalence classes of a block with a given defect group? If this can be done,
we get an example of Donovan's Conjecture which asserts that there are only finitely
many of these Morita equivalence classes. Here again our work represents the
first advance after Puig's work [221] about nilpotent blocks and Erdmann's results
[80] for the tame cases—both from the eighties. The verification of Donovan's
Conjecture relies on the classification of the finite simple groups and thus fits in
a recent development started by An, Eaton, Kessar, Malle and others (e.g. [7, 152]).
In summary, the present work develops several powerful methods in order to tackle
long-standing open conjectures in modular representation theory. The tools are far
from being complete, but we hope to give a significant contribution which inspires
further research.

We now describe the content of the book in detail. Of course, the first part serves
as an introduction to the fundamentals of block theory of finite groups. In particular,
we state Brauer's three main theorems, and we give a modern account on the notion
of subpairs and subsections via fusion systems. Afterwards we present many open
conjectures which all play a role in the following parts. Part II comprehends more
sophisticated methods. The first section starts by introducing the notion of basic sets
and other features attached to quadratic forms. Afterwards, I present the following
general bound on $k(B)$ in terms of Cartan invariants:

$$k(B) \leq \sum_{i=1}^{l(b_u)} c_{ii} - \sum_{i=1}^{l(b_u)-1} c_{i,i+1}.$$

Here (u, b_u) is a so-called major subsection and (c_{ij}) is the Cartan matrix of b_u (for
a more general version see Theorem 4.2). This bound, proved in [114], together
with a practicable algorithm for computing Cartan matrices amounts to the "Cartan
method"—one of the main tools for the upcoming applications. We also discuss as
special cases Cartan matrices of small dimensions where our results still apply to
arbitrary blocks. As an example, we obtain the implication

$$l(b_u) \leq 2 \implies k(B) \leq |D|$$

where (u, b_u) is again a major subsection for B. This result from [252] generalizes an old theorem by Olsson [216] for the case $u = 1$. For the prime $p = 2$ we also prove Brauer's $k(B)$-Conjecture under the weaker hypothesis $l(b_u) \leq 3$. Now let $p > 2$, and let (u, b_u) be an arbitrary subsection such that $l(b_u) = 1$ and b_u has defect q. Using the structure of the fusion system $\mathscr{F}$ of B we prove

$$k_0(B) \leq \frac{|\langle u \rangle| + p^s(r^2 - 1)}{|\langle u \rangle| r} p^q \leq p^q$$

where $|\mathrm{Aut}_{\mathscr{F}}(\langle u \rangle)| = p^s r$ such that $p \nmid r$ and $s \geq 0$. Here, $k_0(B)$ can be replaced by $k(B)$ whenever (u, b_u) is major. Finally, we take the opportunity to recall a less-known inequality by Brauer using the inverse of the Cartan matrix.

As another topic from this part we state Alperin's Fusion Theorem and deduce important properties of essential subgroups by invoking the classification of strongly p-embedded subgroups. These results are new for $p > 2$ and appeared in [257] in case $p = 2$. Afterwards, we collect material from the literature about the representation theory of finite simple groups. Here we indicate how to replace the arbitrary finite group G by a quasisimple group under suitable circumstances. The second part closes with a survey about p-blocks of p-solvable groups where we update an old structure result by Külshammer [161].

The third part of the present work gives applications to specific defect groups and represents the main contribution to the field. Its content assembles many recent papers of the present author, and also includes new results which have not appeared elsewhere. The content of these articles is strongly connected and we will freely arrange the material in order to improve readability. The chapter starts with the determination of the block invariants for metacyclic defect groups in case $p = 2$. This was mostly done in my dissertation (based on the work by Brauer and Olsson). But as a new result, we add a proof of Donovan's Conjecture for the abelian metacyclic defect groups which illustrates the power of the classification of the finite simple groups. Even more, this leads to infinitely many new examples supporting Broué's Abelian Defect Group Conjecture. Many of the other new results are likewise centered around defect groups which share properties of metacyclic groups. For odd primes p it is essentially harder to obtain the precise block invariants for metacyclic defect groups. However, as a consequence of a new result by Watanabe, Alperin's Weight Conjecture holds for all non-abelian metacyclic defect groups. Moreover, we are able to verify Brauer's Height Zero Conjecture which boils down to the inequality $k_0(B) < k(B)$ for non-abelian defect groups. This extends former results by Gao [88, 89], Hendren [107], Yang [287] and Holloway–Koshitani–Kunugi [120].

An obvious generalization of a metacyclic group is a bicyclic group, i.e. a group which can be written in the form $P = \langle x \rangle \langle y \rangle$ for some $x, y \in P$. It turns out that only for $p = 2$ we get new p-groups. Using a paper by Janko [140], we classify all fusion systems on bicyclic 2-groups. This leads to an interesting new result which states that a finite group is 2-nilpotent (and thus solvable) provided it has a bicyclic Sylow 2-subgroup P such that the commutator subgroup P' is non-cyclic. With the list of all possible fusion systems in hand, we establish Olsson's Conjecture (i.e. $k_0(B) \leq |D : D'|$) for all blocks with bicyclic defect groups.

Another project started in my dissertation focuses on minimal non-abelian defect groups D. Here D is non-abelian, but every proper subgroup of D is abelian. Using Rédei's classification [242] of these groups, we are able to complete the determination of the block invariants at least in case $p = 2$. As a byproduct we also reveal another example of Donovan's Conjecture for an infinite family of 2-groups. The proof of this result relies on the classification of the finite simple groups. For arbitrary primes p we show that Olsson's Conjecture holds for all blocks with minimal non-abelian defect groups, except possibly the extraspecial defect group of order 27 and exponent 3. This is also related to a theorem about controlled blocks with defect groups of p-rank 2 achieved in a different chapter.

Concerning Alperin's Weight Conjecture and Robinson's Ordinary Weight Conjecture, we give further evidence for several classes of 2-groups which are direct or central products of cyclic groups and groups of maximal class. Speaking of representation type these defect groups might be described as "finite times tame". We emphasize that apart from a small case the classification of the finite simple groups is not needed at this point. For the sake of completeness, we carry out computations for small defect groups as far as possible. The main achievement here is a proof of Brauer's $k(B)$-Conjecture and Olsson's Conjecture for the 2-blocks of defect at most 5. The former conjecture also holds for the 3-blocks of defect at most 3.

In Table 1 we collect many cases where the block invariants are known. Here we use the following abbreviations for three classes of bicyclic 2-groups:

$$DC(m,n) \cong \langle v, x, a \mid v^{2^n} = x^2 = a^{2^m} = 1, \, {}^x v = {}^a v = v^{-1}, \, {}^a x = vx \rangle$$

$$\cong D_{2^{n+1}} \rtimes C_{2^m},$$

$$DC^*(m,n) \cong \langle v, x, a \mid v^{2^n} = 1, \, a^{2^m} = x^2 = v^{2^{n-1}}, \, {}^x v = {}^a v = v^{-1}, \, {}^a x = vx \rangle$$

$$\cong D_{2^{n+1}}.C_{2^m} \cong Q_{2^{n+1}}.C_{2^m},$$

$$QC(m,n) \cong \langle v, x, a \mid v^{2^n} = a^{2^m} = 1, \, x^2 = v^{2^{n-1}}, \, {}^x v = {}^a v = v^{-1}, \, {}^a x = vx \rangle$$

$$\cong Q_{2^{n+1}} \rtimes C_{2^m}.$$

Moreover, $I(B) \cong \mathrm{Out}_{\mathcal{F}}(D)$ denotes the inertial quotient of the block B with defect group D.

As it is often the case, the study of these special cases leads to new ideas and general insights. This can be clearly seen in Chap. 14 where we improve the famous Brauer-Feit bound on $k(B)$ for abelian defect groups. The proof makes use of a recent result by Halasi and Podoski [101] about coprime actions. As a consequence, we are able to verify the $k(B)$-Conjecture for abelian defect groups of rank at most 5 (resp. 3) in case $p = 2$ (resp. $p \in \{3, 5\}$). In the same spirit we show that Brauer's Conjecture remains true for arbitrary abelian defect groups whenever the inertial index of the block does not exceed 255. This result depends on perfect

Table 1 Cases where the block invariants are known

p	D	$I(B)$	Classification used?	References
Arbitrary	Cyclic	Arbitrary	No	Theorem 8.6
Arbitrary	Metacyclic, minimal non-abelian	Arbitrary	No	Theorem 8.13
Arbitrary	Abelian	$e(B) \leq 4$	No	[226, 227, 270]
Arbitrary	Abelian	S_3	No	[271]
≥ 7	Abelian	$C_4 \times C_2$	No	[273]
$\notin \{2, 7\}$	Abelian	C_3^2	No	[272]
2	Metacyclic	Arbitrary	No	Theorem 8.1
2	Maximal class $*$ cyclic, incl. $* = \times$	Arbitrary	Only for $D \cong C_2^3$	Theorems 9.7, 9.18, 9.28, 9.37
2	Minimal non-abelian	Arbitrary	Only for one family where $|D| = 2^{2r+1}$	Theorem 12.4
2	Minimal non-metacyclic	Arbitrary	Only for $D \cong C_2^3$	Theorem 13.18
2	$DC(m, n)$ for $m, n \geq 2$	Arbitrary	No	Theorem 10.23
2	$DC^*(m, n)$ for $m, n \geq 2, m \neq n$	Arbitrary	No	Theorem 10.24
2	$QC(m, n)$ for $m, n \geq 2$	Arbitrary	No	Theorem 10.25
2	$C_{2^n} \times C_2^3, n \geq 2$	Arbitrary	Yes	Theorem 13.9
2	$|D| \leq 16$	Arbitrary	Yes	Theorem 13.2
2	$C_4 \wr C_2$	Arbitrary	No	[160]
2	$D_8 * Q_8$	C_5	Yes	[252]
2	SmallGroup(32, 22)	Arbitrary	No	Proposition 13.10
2	SmallGroup(32, 28)	Arbitrary	No	Proposition 13.11
2	SmallGroup(32, 29)	Arbitrary	No	Proposition 13.11
3	C_3^2	$\notin \{C_8, Q_8\}$	No	[154, 282]

isometries constructed by Usami and Puig (e.g. [227, 270]) which reflect Broué's Abelian Defect Group Conjecture on the level of characters.

In the final chapter we address an inverse problem, i.e. we ask what can be said about defect groups D of B if the number $k(B)$ is given. Brauer's Problem 21 claims that there are only finitely many choices for D. An analysis of the situation $k(B) = 3$ leads to an interesting question about fusion systems with few conjugacy classes. We show that $k(B) = 3$ implies $|D| = 3$ provided the Alperin-McKay Conjecture holds. We also classify finite groups G such that all non-trivial p-elements in G are conjugate.

The present book has outgrown my habilitation thesis which was finished in 2013. I would like to thank Prof. Dr. Burkhard Külshammer for his constant support and encouragement. Further thanks go to Charles W. Eaton, Alexander Hulpke, Radha Kessar, Shigeo Koshitani, Jørn B. Olsson, Geoffrey Robinson, Ronald Solomon, and Robert Wilson for answering me specific questions. I am also grateful to Ines Spilling for her assistance in administrative tasks. Last but not least, I thank my mom for picking me up from the train station when I came back from California.

This work was supported by the German Research Foundation (DFG), the German Academic Exchange Service (DAAD), the Carl Zeiss Foundation, and the Daimler and Benz Foundation.

Jena, Germany   Benjamin Sambale

Contents

List of Tables

Part I
Fundamentals

Chapter 1
Definitions and Facts

Most of the material presented in this chapter can be found in standard text books on representation theory of finite groups. We often adapt the notation from Feit's book [81] or from the book of Nagao and Tsushima [196]. However, usually we do not give precise references here. We try to keep this chapter as brief as possible. In particular, we omit technical definitions if they are not explicitly needed.

Unless otherwise stated, groups are always finite and modules are finitely generated left modules. Moreover, every algebra has a unity element. For elements x, y, z of a group G we write $[x, y] := xyx^{-1}y^{-1}$, $[x, y, z] := [x, [y, z]]$ and sometimes ${}^x y = xyx^{-1}$. The members of the lower (resp. upper) central series of G are denoted by $K_i(G)$ (resp. $Z_i(G)$). In particular, $K_2(G) = G'$ is the commutator subgroup of G. For a p-group P, let $\Omega_i(P) := \langle x \in P : x^{p^i} = 1 \rangle$ and $\mho_i(P) := \langle x^{p^i} : x \in P \rangle$ for $i \geq 0$. For convenience, let $\Omega(P) := \Omega_1(P)$ and $\mho(P) := \mho_1(P)$. The *rank* r of P is the minimal number of generators, i.e. $|P : \Phi(P)| = p^r$ where $\Phi(P)$ is the Frattini subgroup of P. The largest rank of an abelian subgroup of P is called the *p-rank* of P. For a finite group G the set of p-elements (resp. p'-elements) is denoted by G_p (resp. $G_{p'}$). For a natural number n let n_p (resp. $n_{p'}$) be the p-part (resp. p'-part) of n.

A cyclic group of order $n \in \mathbb{N}$ is denoted by C_n. Moreover, we set $C_n^k := C_n \times \ldots \times C_n$ (k factors). A *homocyclic* group has the form C_n^2. A dihedral (resp. semidihedral, quaternion) group of order 2^n is denoted by D_{2^n} (resp. SD_{2^n}, Q_{2^n}). A group extension with normal subgroup N is denoted by $N.H$. If the extension splits, we write $N \rtimes H$ for the semidirect product. A central product is denoted by $N * H$ where it will be usually clear which subgroup of $Z(N)$ is merged with a subgroup of $Z(H)$.

© Springer International Publishing Switzerland 2014

B. Sambale, *Blocks of Finite Groups and Their Invariants*, Lecture Notes in Mathematics 2127, DOI 10.1007/978-3-319-12006-5_1

1.1   Group Algebras and Blocks

Let G be a finite group, and let p be a prime number. We fix a so-called *p-modular system* $(K, \mathcal{O}, F)$ consisting of the following three objects:

- a splitting field K for G of characteristic 0,
- a complete discrete valuation ring $\mathcal{O}$ with quotient field K,
- an algebraically closed field F of characteristic p such that $F \cong \mathcal{O}/\operatorname{Rad}\mathcal{O}$.

The group algebra $\mathcal{O}G$ decomposes into a direct sum

$$\mathcal{O}G = B_1 \oplus \ldots \oplus B_n$$

of indecomposable (twosided) ideals $B_1, \ldots, B_n$.

Definition 1.1 The $B_1, \ldots, B_n$ are the (*p*-)*blocks* of $\mathcal{O}G$ (or just G).

An important observation is that every block B of G is itself an algebra. The corresponding unity element e_B is a primitive, central idempotent, i.e. it cannot be written non-trivially as a sum of two idempotents in the center $\operatorname{Z}(\mathcal{O}G)$.

The canonical map from $\mathcal{O}$ to F induces a bijection between the corresponding sets of blocks of G. Hence, most of the time we will identify the blocks of $\mathcal{O}G$ with the blocks of FG. In contrast to that, theorems by Maschke and Wedderburn show that KG splits as direct sum of full matrix algebras over K. Thus, a block decomposition over K would not be very interesting.

Let M be an indecomposable $\mathcal{O}G$-module. Then there is exactly one block B of G such that $B \cdot M = M$. In this case we say that M *belongs* to B. One can also regard M as a B-module in the natural way.

Definition 1.2 The trivial $\mathcal{O}G$-module belongs to the *principal block* of G denoted by $B_0(\mathcal{O}G)$.

The principal block of $\mathcal{O}G$ corresponds to the principal block of FG.

1.2   Defect Groups and Characters

The algebra structure of a block of a finite group is strongly influenced by its defect group which we will define in the following.

Definition 1.3 Let G be a finite group with p-subgroup Q. Then the map

$$\operatorname{Br}_Q : \operatorname{Z}(FG) \to \operatorname{Z}(F\operatorname{C}_G(Q)), \quad \sum_{g \in G} \alpha_g g \mapsto \sum_{g \in \operatorname{C}_G(Q)} \alpha_g g$$

is called the *Brauer homomorphism* with respect to Q.

Definition 1.4 Let B be a p-block of FG with unity element e_B. A maximal p-subgroup $D \le G$ such that $\mathrm{Br}_D(e_B) \ne 0$ is called *defect group* of B.

We list the most important properties of defect groups.

Proposition 1.5 *Let B be a p-block of G with defect group D. Then D is unique up to conjugation in G. Moreover, $\mathrm{O}_p(G) \subseteq D = S \cap T$ for some $S, T \in \mathrm{Syl}_p(G)$. If $|D| = p^d$, then d is called the* defect *of B. In case $D \in \mathrm{Syl}_p(G)$, B has maximal defect. The principal block has maximal defect.*

As a rule of thumb, the defect of a block measures the simplicity of the block algebra. In particular, the block is a simple algebra if and only if the defect is 0. The defect of a block can also be determined by certain character degrees as we will see in the following.

In order to distribute the irreducible characters of G into blocks, we introduce the central characters. We denote the set of irreducible characters of G over K (i.e. the ordinary characters) by $\mathrm{Irr}(G)$. Note that $k(G) := |\mathrm{Irr}(G)|$ is the number of conjugacy classes of G.

Definition 1.6 Let $\chi \in \mathrm{Irr}(G)$. Then the map

$$\omega_\chi : Z(FG) \to F, \quad \sum_{g \in G} \alpha_g g \mapsto \sum_{g \in G} \alpha_g \frac{\chi(g)}{\chi(1)} + \mathrm{Rad}\,\mathscr{O}$$

is a homomorphism of algebras. There exists exactly one block B of FG with unity element e_B such that $\omega_\chi(e_B) = 1$. In this case we say that χ *belongs* to B. If $\psi \in \mathrm{Irr}(G)$ also belongs to B, then $\omega_\chi = \omega_\psi$ and $\omega_B := \omega_\chi$ is called the *central character* of B.

Definition 1.7 The set of irreducible ordinary characters belonging to the block B of G is denoted by $\mathrm{Irr}(B)$. Its cardinality is $k(B) := |\mathrm{Irr}(B)|$. For every $\chi \in \mathrm{Irr}(B)$ there is an integer $h(\chi) \ge 0$ such that $p^{h(\chi)}|G : D|_p = \chi(1)_p$ where D is a defect group of B. The number $h(\chi)$ is called the *height* of χ. We set $\mathrm{Irr}_i(B) := \{\chi \in \mathrm{Irr}(B) : h(\chi) = i\}$ and $k_i(B) := |\mathrm{Irr}_i(B)|$ for $i \ge 0$.

One can show that $k_0(B) \ge 2$ unless B has defect 0 where $k_0(B) = k(B) = l(B) = 1$ (see [216]). Therefore, the defect of B is determined by the character degrees. If B is a block of FG, the number $k(B)$ can also be expressed as $k(B) = \dim_F Z(B)$. In particular, $k(B)$ is an invariant of the algebra B. If B has defect $d \ge 0$, then $k_i(B) = 0$ for $i \ge d - 1$. Moreover, if $d \ge 3$ and $k_{d-2}(B) \ne 0$, then the defect groups of B have maximal class (see [237]).

As we have seen above, every simple $\mathscr{O}G$-module can be assigned to a uniquely determined block of G. Accordingly, the set of irreducible Brauer characters $\mathrm{IBr}(G)$ of G splits into blocks. Recall that Brauer characters are only defined on the p-regular conjugacy classes of G.

Definition 1.8 The set of irreducible Brauer characters belonging to the block B of G is denoted by $\mathrm{IBr}(B)$. Its cardinality is $l(B) := |\mathrm{IBr}(B)|$.

Here again, $l(B)$ as the number of simple B-modules is actually an invariant of the algebra structure. Also, $l(G) := |\mathrm{IBr}(G)|$ is the number of p-regular conjugacy classes in G.

The connection between ordinary characters and Brauer characters is established by (generalized) decomposition numbers.

Definition 1.9 Let $u \in G_p$, and let $\chi \in \mathrm{Irr}(G)$. Then there exist algebraic integers $d_{\chi\varphi}^u \in \mathbb{Z}[e^{2\pi i/|\langle u \rangle|}] \subseteq \mathcal{O}$ for every $\varphi \in \mathrm{IBr}(\mathrm{C}_G(u))$ such that

$$\chi(uv) = \sum_{\varphi \in \mathrm{IBr}(\mathrm{C}_G(u))} d_{\chi\varphi}^u \varphi(v) \qquad \text{for all } v \in \mathrm{C}_G(u)_{p'}.$$

These numbers are called *generalized decomposition numbers*. In case $u = 1$ we speak just of (ordinary) decomposition numbers.

Let $\mathbb{Q}_n$ be the n-th cyclotomic field over $\mathbb{Q}$. Let $\mathscr{G}$ be the Galois group of $\mathbb{Q}_{|G|}$ with fixed field $\mathbb{Q}_{|G|_{p'}}$. Restriction gives an isomorphism $\mathscr{G} \cong \mathrm{Gal}(\mathbb{Q}_{|G|_p}|\mathbb{Q}) \cong (\mathbb{Z}/|G|_p\mathbb{Z})^{\times}$, and we will often identify these groups. Then $\mathscr{G}$ acts on the irreducible characters, the generalized decomposition numbers, and on the set of p-elements of G. Here the following important relation holds

$$\gamma(d_{\chi\varphi}^u) = d_{\chi\varphi}^{u^\gamma} = d_{\gamma_{\chi\varphi}}^u$$

for $\gamma \in \mathscr{G}$. Characters χ and $^\gamma\chi \neq \chi$ are called *p-conjugate*. It can be seen that p-conjugate characters lie in the same block and have the same height. If $^\gamma\chi = \chi$ for all $\gamma \in \mathscr{G}$, then χ is called *p-rational*. In this case the numbers $d_{\chi\varphi}^u$ for all p-elements $u \in G$ and all $\varphi \in \mathrm{IBr}(\mathrm{C}_G(u))$ are (rational) integers.

1.3   Brauer's Main Theorems

In order to simplify computations one tries to replace the group G by smaller subgroups. It is crucial to understand how blocks behave under this substitution. Here the notion of Brauer correspondence gives an answer.

Definition 1.10 Let B and b be blocks of G and $H \leq G$ with central characters ω_B and ω_b respectively. If

$$\omega_B\left(\sum_{g \in G} \alpha_g g\right) = \omega_b\left(\sum_{g \in H} \alpha_g g\right)$$

for all $\sum_{g \in G} \alpha_g g \in \mathrm{Z}(FG)$, then b is a *Brauer correspondent* of B and conversely. We also write $B = b^G$.

Proposition 1.11 *Every defect group D of b (in the situation above) is contained in a defect group of b^G. If $C_G(D) \subseteq H$, then b^G is always defined. Moreover, the Brauer correspondence is transitive.*

Brauer's three main theorems relate specific sets of blocks via Brauer correspondence.

Theorem 1.12 (Brauer's First Main Theorem) *Let $P \leq G$ be a p-subgroup of G, and let $N_G(P) \leq H \leq G$. Then Brauer correspondence gives a bijection between the set of blocks of G with defect group P and the set of blocks of H with defect group P.*

Theorem 1.13 (Brauer's Second Main Theorem) *Let $u \in G_p$, and let $\chi \in \mathrm{Irr}(G)$. Assume that $\varphi \in \mathrm{IBr}(C_G(u))$ lies in a block b of $C_G(u)$. If $\chi \notin \mathrm{Irr}(b^G)$, then $d^u_{\chi\varphi} = 0$.*

Observe that b^G in Theorem 1.13 is always defined by Proposition 1.11. The Second Main Theorem allows us to arrange the generalized decomposition numbers of G in a block shape matrix

$$Q^u = \begin{pmatrix} Q^u_1 & & 0 \\ & \ddots & \\ 0 & & Q^u_m \end{pmatrix}.$$

Each Q^u_i corresponds to a block B_i of G. It is an invertible $k(B_i) \times k(B_i)$ matrix, called the *generalized decomposition matrix of B_i*. Doing the same with the ordinary decomposition numbers leads to the *(ordinary) decomposition matrix Q of a block B*. Here Q is an integral $k(B) \times l(B)$ matrix and $C := Q^T Q$ is the *Cartan matrix* of B (as an algebra). By definition, C is symmetric and positive definite. Moreover, if B has defect d, then all elementary divisors of C divide p^d, and just one of them is p^d. In particular, $p^d \leq \det C$ is a p-power.

As for ordinary character tables we have orthogonality relations of decomposition numbers.

Theorem 1.14 (Orthogonality Relations) *Let B be a block of G, and let $\mathcal{R}$ be a set of representatives of the conjugacy classes of p-elements of G. Choose $u, v \in \mathcal{R}$, blocks b_u and b_v of $C_G(u)$ resp. $C_G(v)$, and $\varphi \in \mathrm{IBr}(b_u)$ and $\psi \in \mathrm{IBr}(b_v)$. Then*

$$\sum_{\chi \in \mathrm{Irr}(B)} d^u_{\chi\varphi} \overline{d^v_{\chi\psi}} = \begin{cases} c_{\varphi\psi} & \text{if } u = v, \ b_u = b_v \text{ and } b^G_u = B \\ 0 & \text{otherwise} \end{cases}$$

where $c_{\varphi\psi}$ is the Cartan invariant of $b_u = b_v$ corresponding to $\varphi, \psi \in \mathrm{IBr}(b_u)$.

Theorem 1.15 (Brauer's Third Main Theorem) *Let $H \leq G$, and let b be a block of H with defect group D such that $C_G(D) \subseteq H$. Then b is the principal block of H if and only if b^G is the principal block of G.*

1.4 Covering and Domination

If the subgroup in the last section happens to be normal, things turn out to be easier.

Definition 1.16 Let $N \trianglelefteq G$, and let b (resp. B) be a block of N (resp. G). If $Bb \neq 0$, we say that B *covers* b.

If b is covered by B, then B has a defect group D such that $D \cap N$ is a defect group of b. If b^G is defined in the situation of Definition 1.16, then b^G covers b. The group G acts by conjugation on the set of blocks of N. The corresponding stabilizer of b is the *inertial group* $N_G(N, b)$ of b. Since blocks are ideals, we always have $N \subseteq N_G(N, b)$. If N is an arbitrary subgroup of G and b is a block of $M \trianglelefteq N_G(N)$, we define $N_G(N, b) := N_{N_G(N)}(M, b)$. If b is covered by B, then the same is true for every block in the orbit of b under G. We deduce an extended version of Brauer's First Main Theorem.

Theorem 1.17 (Extended First Main Theorem) *Let P be a p-subgroup of G. Then the Brauer correspondence induces a bijection between the blocks of G with defect group P and the $N_G(P)$-conjugacy classes of blocks b of $C_G(P)P$ with defect group P and $|N_G(P, b) : C_G(P)P| \not\equiv 0 \pmod{p}$.*

In the situation of Theorem 1.17 we define $I(B) := N_G(P, b)/C_G(P)P$ and $e(B) := |I(B)|$ for $B := b^G$. Then $I(B)$ is called *inertial quotient* and $e(B)$ is called the *inertial index* of B. Of course, these invariants do not depend on the choice of b. It known that $e(B)$ is not divisible by p. In particular, the Schur-Zassenhaus Theorem allows us to regard $I(B)$ as a subgroup of $\mathrm{Aut}(D)$.

The following important result often allows to replace G by $N_G(N, b)$.

Theorem 1.18 (Fong-Reynolds) *Let b be a block of $N \trianglelefteq G$. Then the Brauer correspondence induces a bijection α between the set of blocks of $N_G(N, b)$ covering b and the set of blocks of G covering b. Moreover, α preserves defect groups, the numbers $k(B)$ and $l(B)$, and decomposition and Cartan matrices.*

If N happens to be a defect group of B, the structure of B is well understood by a theorem of Külshammer.

Theorem 1.19 (Külshammer [163]) *Let B be a block of a finite group G with normal defect group D. Then B is Morita equivalent to a twisted group algebra*

$$\mathcal{O}_\gamma[D \rtimes I(B)]$$

where $\gamma \in O_{p'}(\mathrm{H}^2(I(B), \mathcal{O}^\times)) \cong O_{p'}(\mathrm{H}^2(I(B), \mathbb{C}^\times))$.

Recall that two rings are called *Morita equivalent* if their module categories are equivalent. Morita equivalence of blocks preserves the numbers $k(B)$, $k_i(B)$ and $l(B)$ as well as Cartan and decomposition matrices up to ordering. Recall that the *Schur multiplier* $\mathrm{H}^2(G, \mathbb{C}^\times) = \mathrm{H}_2(G, \mathbb{Z})$ is the largest group Z such that there exists a finite group L with $L/Z \cong G$ and $Z \subseteq L' \cap Z(L)$. For further properties of the

Schur multiplier we refer to Karpilovsky's book [142]. Observe that $\mathcal{O}_\gamma G \cong \mathcal{O}G$ whenever γ is trivial. For our applications we often have $\mathrm{H}^2(G, \mathbb{C}^\times) = 1$. One can replace the inconvenient twisted group algebra with the following result (see Proposition 5.15 in [222] or Proposition IV.5.37 in [19] for the statement over F).

Proposition 1.20 *Let G be a finite group, and let $1 \neq \gamma \in \mathrm{O}_{p'}(\mathrm{H}^2(G, \mathcal{O}^\times))$. Then there exists a central extension*

$$1 \to Z \to H \to G \to 1$$

such that every block of $\mathcal{O}_\gamma G$ is isomorphic to a non-principal block of H. Moreover, Z is a cyclic p'-group.

More results on twisted group algebras can be found in Conlon's paper [58].

It is also useful to go over to quotient groups.

Definition 1.21 Let B be a block of G, and let $N \trianglelefteq G$. Then the image of B under the canonical epimorphism $G \to G/N$ is a (possibly trivial) sum of blocks of G/N. Each block occurring as a summand is *dominated* by B.

In a rather special case the domination of blocks is bijective.

Theorem 1.22 *Suppose that $N \trianglelefteq G$ is a p-subgroup and $G/\mathrm{C}_G(N)$ is a p-group. Then every block B of G dominates exactly one block $\overline{B}$ of G/N. If D is a defect group of B, then D/N is a defect group of $\overline{B}$. Moreover, the Cartan matrices satisfy $C_B = |N|C_{\overline{B}}$. In particular $l(B) = l(\overline{B})$.*

In the opposite case where N is a p'-group we have at least an injective map.

Theorem 1.23 *Suppose that $N \trianglelefteq G$ is a p'-subgroup. Then every block $\overline{B}$ of G/N is dominated by exactly one block B of G. Moreover, the blocks B and $\overline{B}$ are isomorphic as algebras and have isomorphic defect groups.*

1.5 Fusion Systems

The notion of fusion systems was first formed by Puig in the eighties under the name Frobenius categories (see [224]). Later Levi, Oliver and others gave a modern approach. We refer to the books by Craven [61] and Aschbacher-Kessar-Oliver [19], as well as to a survey article by Linckelmann [184].

Definition 1.24 A (saturated) *fusion system* on a finite p-group P is a category $\mathscr{F}$ whose objects are the subgroups of P, and whose morphisms are group monomorphisms with the usual composition such that the following properties hold:

(1) For $S, T \le P$ we have

$$\mathrm{Hom}_P(S, T) := \{\varphi : S \to T : \exists y \in P : \varphi(x) = {}^y x \; \forall x \in S\} \subseteq \mathrm{Hom}_{\mathscr{F}}(S, T).$$

(2) For $\varphi \in \mathrm{Hom}_{\mathscr{F}}(S,T)$ we have $\varphi \in \mathrm{Hom}_{\mathscr{F}}(S,\varphi(S))$ and $\varphi^{-1} \in \mathrm{Hom}_{\mathscr{F}}(\varphi(S),S)$.

(3) For $S \le P$ there exists a morphism $\psi : S \to P$ such that $T := \psi(S)$ has the following properties:

(a) $\mathrm{N}_P(T)/\mathrm{C}_P(T) \in \mathrm{Syl}_p(\mathrm{Aut}_{\mathscr{F}}(T))$.

(b) Every morphism $\varphi \in \mathrm{Hom}_{\mathscr{F}}(T,P)$ can be extended to

$$\mathrm{N}_\varphi := \{y \in \mathrm{N}_P(T) : \exists z \in \mathrm{N}_P(\varphi(T)) : \varphi(^y x) = {}^z\varphi(x) \ \forall x \in T\}.$$

Part (3) in Definition 1.24 is the saturation property. Since our fusion systems are always saturated, we will omit the word "saturated" from now on. Observe that in (3) we have $T\,\mathrm{C}_P(T) \subseteq \mathrm{N}_\varphi \subseteq \mathrm{N}_P(T)$. We call subgroups $S,T \le P$ $\mathscr{F}$-conjugate if there exists an isomorphism $\varphi : S \to T$ in $\mathscr{F}$.

If G is a finite group with Sylow p-subgroup P, then we get a fusion system $\mathscr{F}_P(G)$ on P by defining $\mathrm{Hom}_{\mathscr{F}}(S,T) := \mathrm{Hom}_G(S,T)$ for $S,T \le P$. A fusion system which does not arise in this way is called *exotic*. We say that $\mathscr{F}$ is *trivial* or *nilpotent* if $\mathscr{F} = \mathscr{F}_P(P)$.

Definition 1.25 Two fusion systems $\mathscr{F}$ and $\mathscr{F}'$ on a finite p-group P are *isomorphic* if there is an automorphism $\gamma \in \mathrm{Aut}(P)$ such that

$$\mathrm{Hom}_{\mathscr{F}'}(\gamma(S),\gamma(T)) = \gamma(\mathrm{Hom}_{\mathscr{F}}(S,T)) := \{\gamma \circ \varphi \circ \gamma^{-1} : \varphi \in \mathrm{Hom}_{\mathscr{F}}(S,T)\}$$

for all subgroups $S,T \le P$.

Observe that if γ is an inner automorphism of P, then $\mathrm{Hom}_{\mathscr{F}}(\gamma(S),\gamma(T)) = \gamma(\mathrm{Hom}_{\mathscr{F}}(S,T))$ for all $S,T \le P$.

Now let B be a p-block of G with defect group D. For every subgroup $Q \le D$ there exists a Brauer correspondent b_Q of B in $\mathrm{C}_G(Q)$. The pair (Q,b_Q) is called (*B-*)*subpair*. If $Q = D$, we sometimes say *Sylow subpair* of B. These objects were developed in articles by Alperin-Broué [4] and Olsson [217]. In the latter paper, b_Q is considered as a block of $Q\,\mathrm{C}_G(Q)$ which does not make a big difference.

For two subpairs (S,b_S) and (T,b_T) we write $(S,b_S) \trianglelefteq (T,b_T)$ if $S \trianglelefteq T$ and $b_S^{T\,\mathrm{C}_G(S)} = b_T^{T\,\mathrm{C}_G(S)}$. Let $\le$ be the transitive closure of $\trianglelefteq$ (for subpairs). The group G acts on the set of subpairs in the obvious way: $^g(Q,b_Q) := (^g Q,{}^g b_Q)$. In the following we fix a Sylow B-subpair (D,b_D). Then it can be shown that there is exactly one subpair (Q,b_Q) such that $(Q,b_Q) \le (D,b_D)$ for every $Q \le D$.

Definition 1.26 The fusion system $\mathscr{F} := \mathscr{F}_D(B)$ on D is defined by

$$\mathrm{Hom}_{\mathscr{F}}(S,T) := \{\varphi : S \to T : \exists g \in G : {}^g(S,b_S) \le (T,b_T) \wedge \varphi(x) = {}^g x \ \forall x \in S\}$$

for $S,T \le D$.

If B is the principal block of G, we get $\mathscr{F}_D(B) = \mathscr{F}_D(G)$ (remember $D \in \mathrm{Syl}_p(G)$). Conversely, it is not known if every block fusion system is the fusion system of a finite group. Setting $\mathscr{F} := \mathscr{F}_D(B)$ we observe that $\operatorname{Aut}_{\mathscr{F}}(D) \cong N_G(D, b_D)/C_G(D)$ and $\operatorname{Out}_{\mathscr{F}}(D) \cong I(B)$. A fusion system $\mathscr{F}$ on a finite p-group P (or the corresponding block) is called *controlled* if $\mathscr{F} = \mathscr{F}_P(P \rtimes A)$ for a p'-subgroup $A \le \operatorname{Aut}(P)$. If P is abelian, then $\mathscr{F}$ is always controlled.

In the special case where Q is cyclic, say $Q = \langle u \rangle$, we get a $(B\text{-})subsection$ (u, b_u) where $b_u := b_Q$.

In the following we need some more concepts concerning fusion systems.

Definition 1.27 Let $\mathscr{F}$ be a fusion system on a finite p-group P, and let $Q \le P$.

- Q is called *fully $\mathscr{F}$-centralized* if $|C_P(R)| \le |C_P(Q)|$ for all $R \le P$ which are $\mathscr{F}$-isomorphic to Q.
- Q is called *fully $\mathscr{F}$-normalized* if $|N_P(R)| \le |N_P(Q)|$ for all $R \le P$ which are $\mathscr{F}$-isomorphic to Q.
- Q is called *$\mathscr{F}$-centric* if $C_P(R) = Z(R)$ for all $R \le P$ which are $\mathscr{F}$-isomorphic to Q.
- Q is called *$\mathscr{F}$-radical* if $O_p(\operatorname{Out}_{\mathscr{F}}(Q)) = 1$.

Observe that an $\mathscr{F}$-centric subgroup is also fully $\mathscr{F}$-centralized. Moreover, by Proposition I.2.5 in [19], fully $\mathscr{F}$-normalized implies fully $\mathscr{F}$-centralized. We take the opportunity to introduce two important subsystems of fusion systems.

Proposition 1.28 *Let $\mathscr{F}$ be a fusion system on a finite p-group P.*

(i) *If $Q \le P$ is fully $\mathscr{F}$-centralized, then there is a fusion system $C_{\mathscr{F}}(Q)$ on $C_P(Q)$ defined as follows: a morphism $\varphi : R \to S$ ($R, S \le C_P(Q)$) belongs to $C_{\mathscr{F}}(Q)$ if there exists a morphism $\psi : QR \to QS$ in $\mathscr{F}$ such that $\psi_{|Q} = \operatorname{id}_Q$ and $\psi_{|R} = \varphi$.*

(ii) *If $Q \le P$ is fully $\mathscr{F}$-normalized, then there is a fusion system $N_{\mathscr{F}}(Q)$ on $N_P(Q)$ defined as follows: a morphism $\varphi : R \to S$ ($R, S \le N_P(Q)$) belongs to $N_{\mathscr{F}}(Q)$ if there exists a morphism $\psi : QR \to QS$ in $\mathscr{F}$ such that $\psi(Q) = Q$ and $\psi_{|R} = \varphi$.*

A fusion system $\mathscr{F}$ on P is *constrained* if it has the form $\mathscr{F} = N_{\mathscr{F}}(Q)$ for an $\mathscr{F}$-centric subgroup $Q \le P$. It is known that every constrained fusion system is non-exotic (Theorem III.5.10 in [19]). Note that every controlled fusion system is constrained by taking $Q = P$.

If $\mathscr{F}$ is the fusion system of a block B, and $Q \le P$ is fully $\mathscr{F}$-centralized, then the block b_Q defined above has defect group $C_P(Q)$ and fusion system $C_{\mathscr{F}}(Q)$ (see Theorem IV.3.19 in [19]).

Definition 1.29 Let $\mathscr{F}$ be a fusion system on a finite p-group P. The largest subgroup $Q \le Z(P)$ such that $C_{\mathscr{F}}(Q) = \mathscr{F}$ is called the *center* $Z(\mathscr{F})$ of $\mathscr{F}$. Accordingly, we say, $\mathscr{F}$ is *centerfree* if $Z(\mathscr{F}) = 1$. The largest subgroup $Q \trianglelefteq P$ such that $N_{\mathscr{F}}(Q) = \mathscr{F}$ is denoted by $O_p(\mathscr{F})$. Obviously, $Z(\mathscr{F}) \subseteq O_p(\mathscr{F})$.

A less technical characterization of the center is given by

$$Z(\mathscr{F}) = \{x \in D : x \text{ is fixed by every morphism in } \mathscr{F}\}.$$

A fusion system $\mathscr{F}$ is constrained if and only if $C_P(O_p(\mathscr{F})) \subseteq O_p(\mathscr{F})$. The following major result is needed at several places.

Theorem 1.30 (Puig [221]) *Let B be a block of a finite group with defect group D and trivial fusion system (i.e. B is nilpotent). Then $B \cong (\mathcal{O}D)^{n \times n}$ for some $n \geq 1$. In particular B and $\mathcal{O}D$ are Morita equivalent.*

Let B be a nilpotent block with defect group D. Then it follows from Theorem 1.30 that $k_i(B) = k_i(D)$ is the number of irreducible characters of D of degree p^i for $i \geq 0$. In particular $k_0(B) = |D : D'|$ and $k(B)$ is the number of conjugacy classes of D. Moreover, $l(B) = 1$. As an example, every block B with abelian defect groups and $e(B) = 1$ is nilpotent.

Similarly to the theory of finite groups, one can define the *focal subgroup* $\mathfrak{foc}(B)$ of B (or of $\mathscr{F}$) by

$$\mathfrak{foc}(B) := \langle f(x)x^{-1} : x \in Q \leq D, \ f \in \mathrm{Aut}_{\mathscr{F}}(Q)\rangle.$$

Obviously, $D' \subseteq \mathfrak{foc}(B) \subseteq D$. It can be seen that $D/\mathfrak{foc}(B)$ acts freely on $\mathrm{Irr}_0(B)$ by the so-called $*$-construction (see [238]). As a consequence we get information on $k_0(B)$ as follows.

Proposition 1.31 (Robinson [238], Landrock [176]) *Let B be a p-block of G with defect $d > 0$. Then the following holds:*

(i) $|D : \mathfrak{foc}(B)| \mid k_0(B)$.
(ii) If $p \leq 3$, then $p \mid k_0(B)$.
(iii) If $e(B) = 1$, then $p \mid k_0(B)$.
(iv) If $p = 2$ and $d \geq 2$, then $4 \mid k_0(B)$.
(v) If $p = 2$ and $d \geq 3$, then $k_0(B) + 4k_1(B) \equiv 0 \pmod{8}$.
(vi) If $p = 2$ and $k_{d-2}(B) \neq 0$, then $k_0(B) = 4$ and $k_{d-2}(B) \leq 3$.

A recent result along these lines gives another description of nilpotent blocks.

Proposition 1.32 (Kessar-Linckelmann-Navarro [151]) *A block B of a finite group with defect group D is nilpotent if and only if $k_0(B) = |D : \mathfrak{foc}(B)|$.*

One can also define the *hyperfocal subgroup* of B (or of $\mathscr{F}$) as follows

$$\mathfrak{hyp}(B) := \langle f(x)x^{-1} : x \in Q \leq D, \ f \in O^p(\mathrm{Aut}_{\mathscr{F}}(Q))\rangle.$$

As a consequence of Alperin's Fusion Theorem (see Theorem 6.2 below), $\mathfrak{foc}(B) = D'\mathfrak{hyp}(B)$. Moreover, B is nilpotent if and only if $\mathfrak{hyp}(B) = 1$. Recently, Watanabe obtained the following very strong result which gives information for odd primes p.

Theorem 1.33 (Watanabe [283]) *Let B be a p-block of a finite group with defect group D such that $\mathfrak{hyp}(B)$ is cyclic. Then B is controlled and $I(B)$ is cyclic. Moreover, $l(B) = e(B) \mid p-1$ and $k(B) = k(D \rtimes I(B))$. The elementary divisors of the Cartan matrix of B are $|D|$ and $|C_D(I(B))|$ where $|C_D(I(B))|$ occurs with multiplicity $e(B) - 1$.*

The proof of Theorem 1.33 uses a structure result on the source algebra of B by Puig [223].

1.6 Subsections and Contributions

Now let $\mathscr{F}$ be again the fusion system of a block B. The following lemma describes the conjugation action on the subsections. I was unable to find this result in the literature. Hence, a proof is given.

Lemma 1.34 *Let $\mathscr{R}$ be a set of representatives for the $\mathscr{F}$-conjugacy classes of elements of D such that $\langle \alpha \rangle$ is fully $\mathscr{F}$-normalized for $\alpha \in \mathscr{R}$ ($\mathscr{R}$ always exists). Then*

$$\{(\alpha, b_\alpha) : \alpha \in \mathscr{R}\}$$

is a set of representatives for the G-conjugacy classes of B-subsections, where b_α has defect group $C_D(\alpha)$ and fusion system $C_{\mathscr{F}}(\langle \alpha \rangle)$.

Proof Let (α, b) be an arbitrary B-subsection. Then $(\langle \alpha \rangle, b)$ is a B-subpair which lies in some Sylow B-subpair. Since all Sylow B-subpairs are conjugate in $\mathscr{F}$, we may assume $(\langle \alpha \rangle, b) \leq (D, b_D)$. This shows $b = b_\alpha$. By the definition of $\mathscr{R}$ there exists a morphism f in $\mathscr{F}$ such that $\beta := f(\alpha) \in \mathscr{R}$. Now the definition of $\mathscr{F}$ implies that f corresponds to an element $g \in G$ such that ${}^g(\alpha, b) = (\beta, b_\beta)$.

It is also easy to see that we can always choose a representative α such that $\langle \alpha \rangle$ is fully $\mathscr{F}$-normalized.

Now suppose that (α, b_α) and (β, b_β) with $\alpha, \beta \in \mathscr{R}$ are conjugate by $g \in G$. Then (with a slight abuse of notation) we have $g \in \operatorname{Hom}_{\mathscr{F}}(\langle \alpha \rangle, \langle \beta \rangle)$. Hence, $\alpha = \beta$.

It remains to prove that b_α has defect group $C_D(\alpha)$ and fusion system $C_{\mathscr{F}}(\langle \alpha \rangle)$ for $\alpha \in \mathscr{R}$. By Proposition I.2.5 in [19], $\langle \alpha \rangle$ is also fully $\mathscr{F}$-centralized. Hence, Theorem IV.3.19 in [19] implies the claim.   □

Lemma 1.34 replaces Brauer's notion of double chains and nets. In applications it would usually be enough to assume that $\langle \alpha \rangle$ is fully $\mathscr{F}$-centralized. However, it is sometimes easier to prove that $\langle \alpha \rangle$ is fully $\mathscr{F}$-normalized. A subsection (u, b_u) is *major* if b_u also has defect group D. Thus, by Lemma 1.34 we usually assume $u \in Z(D)$ for a major subsection (u, b_u). Obviously, every subsection is major if D is abelian. However, the converse is false (cf. Chap. 15).

In order to compute invariants of blocks, the following theorem is rather important.

Theorem 1.35 (Brauer) *Let B be a block of a finite group, and let $\mathscr{R}$ be a set of representatives for the conjugacy classes of B-subsections. Then*

$$k(B) = \sum_{(u, b_u) \in \mathscr{R}} l(b_u).$$

As a consequence, we see that the difference $k(B) - l(B)$ is locally determined. Theorem 1.35 is related to the fact that the generalized decomposition matrix of B has square shape.

Now we introduce the contribution of a subsection (u, b_u); a notion introduced by Brauer [39]. Let Q_u be the part of the generalized decomposition matrix consisting of the entries $d^u_{\chi\varphi}$ where $\chi \in \mathrm{Irr}(B)$ and $\varphi \in \mathrm{IBr}(b_u)$. Assume that b_u has defect q and Cartan matrix C_u. Then the *contribution matrix* of (u, b_u) is defined as

$$M^u = (m^u_{\chi\psi})_{\chi, \psi \in \mathrm{Irr}(B)} = p^q Q_u C_u^{-1} \overline{Q_u}^{\mathrm{T}}.$$

Since $p^q C_u^{-1}$ is integral, the *contributions* $m^u_{\chi\psi}$ are algebraic integers. Hence, we may view them as elements of $\mathcal{O}$. By definition, $M^u M^u = p^q M^u$. Moreover, $\mathrm{tr}\, M^u = p^q l(b_u)$ where tr denotes the trace.

The following technical divisibility relations are quite useful. They first appeared in Broué [44] and were later generalized by Murai [193].

Proposition 1.36 *Let* (u, b_u) *be a B-subsection, and let* $\chi, \psi \in \mathrm{Irr}(B)$. *Then the following holds:*

(i) $m^u_{\chi\psi} \in \mathcal{O}^\times$ *if and only if* $h(\chi) = h(\psi) = 0$. *In particular,* $(d^u_{\chi\varphi} : \varphi \in \mathrm{IBr}(b_u)) \neq 0$ *for* $\chi \in \mathrm{Irr}_0(B)$.

(ii) *Assume that* (u, b_u) *is major. Then* $v(m^u_{\chi\psi}) \geq h(\chi)$ *were* v *is the p-adic valuation. Here equality holds if and only if* $h(\psi) = 0$. *In particular,* $(d^u_{\chi\varphi} : \varphi \in \mathrm{IBr}(b_u)) \neq 0$ *for all* $\chi \in \mathrm{Irr}(B)$.

In case $l(b_u) = 1$ this has direct consequences for the generalized decomposition numbers. Let $|\langle u \rangle| = p^k$, and let ζ be a primitive p^k-th root of unity. Let $\mathrm{IBr}(b_u) = \{\varphi_u\}$. Since $d^u_{\chi\varphi_u}$ is an algebraic integer, we can write

$$d^u_{\chi\varphi_u} = \sum_{i=0}^{\varphi(p^k)-1} a^u_i(\chi) \zeta^i \tag{1.1}$$

with $a^u_i(\chi) \in \mathbb{Z}$ (see Satz I.10.2 in [204]). Here $\varphi(p^k)$ denotes Euler's totient function.

Lemma 1.37 *Let* (u, b_u) *be a B-subsection with* $|\langle u \rangle| = p^k$ *and* $l(b_u) = 1$.

(i) *For* $\chi \in \mathrm{Irr}_0(B)$ *we have*

$$\sum_{i=0}^{\varphi(p^k)-1} a^u_i(\chi) \not\equiv 0 \pmod{p}.$$

(ii) If (u, b_u) is major and $\chi \in \mathrm{Irr}(B)$, then $p^{h(\chi)} \mid a_i^u(\chi)$ for $i = 0, \ldots, \varphi(p^k) - 1$ and

$$\sum_{i=0}^{\varphi(p^k)-1} a_i^u(\chi) \not\equiv 0 \pmod{p^{h(\chi)+1}}.$$

Proof

(i) Since $l(b_u) = 1$, we have $m_{\chi\chi}^u = d_{\chi\varphi_u}^u \overline{d_{\chi\varphi_u}^u}$. Hence, Proposition 1.36 gives $d_{\chi\varphi_u}^u \not\equiv 0 \pmod{\mathrm{Rad}\,\mathcal{O}}$. Since $\zeta \equiv 1 \pmod{\mathrm{Rad}\,\mathcal{O}}$, the claim follows from (1.1).

(ii) Let $\psi \in \mathrm{Irr}_0(B)$. Then Proposition 1.36 implies

$$h(\chi) = \nu(m_{\chi\psi}^u) = \nu(d_{\chi\varphi_u}^u) + \nu(\overline{d_{\psi\varphi_u}^u}),$$

where ν is the p-adic valuation. Thus, $h(\chi) = \nu(d_{\chi\varphi_u}^u)$ by (1.37). Now the claim is easy to see. $\qquad\square$

1.7 Centrally Controlled Blocks

In this short section we describe the notion of centrally controlled blocks which is little-known. The results are given in [168].

Theorem 1.38 (Külshammer-Okuyama [168]) *Let B be a block with fusion system $\mathscr{F}$. Let (u, b_u) be a B-subsection such that $u \in \mathrm{Z}(\mathscr{F})$. Then $k(B) \geq k(b_u)$ and $l(B) \geq l(b_u)$.*

Fusion systems controlled by centralizers also play a role in the Z^*-Theorem. In the special case where the defect group is abelian, we have the following stronger result by Watanabe (observe that the last assertion is a consequence of [168]).

Theorem 1.39 (Watanabe [280]) *Let D be abelian, and let (u, b_u) be a B-subsection such that $u \in \mathrm{Z}(\mathscr{F})$. Then $k(B) = k(b_u)$ and $l(B) = l(b_u)$. Moreover, $\mathrm{Z}(B)$ and $\mathrm{Z}(b_u)$ are isomorphic as F-algebras.*

Observe that in the situation of Theorem 1.39 we have $D = \mathrm{Z}(\mathscr{F}) \times \mathfrak{foc}(B)$.

1.8 Lower Defect Groups

The notion of lower defect groups allows us to determine the elementary divisors of the Cartan matrix of a block locally. Unfortunately, the theory is quite opaque.

We collect only the results which are necessary for the present work. We refer to [49, 81, 215, 279].

Let B be a p-block of a finite group G with defect group D and Cartan matrix C. We denote the multiplicity of an integer a as elementary divisor of C by $m(a)$. Then $m(a) = 0$ unless a is a p-power. It is well-known that $m(|D|) = 1$.

Definition 1.40 For a p-block B of G and a p-subgroup $R \leq G$ let

$$J_R(B) := \left\{ \sum_{g \in G_{p'}} \alpha_g g \in Z(B) : \alpha_g \neq 0 \Rightarrow \exists Q \in \mathrm{Syl}_p(C_G(g)), \; x \in G : xQx^{-1} \leq R \right\},$$

$$J_{<R}(B) := \sum_{Q<R} J_Q(B).$$

Then

$$m_B^{(1)}(R) := \dim_F(J_R(B)) - \dim_F(J_{<R}(B))$$

is called the 1-*multiplicity* of R as a lower defect group of B. In case $m_B^{(1)}(R) > 0$, R is called a *lower defect group* of B (this is actually a bit stronger than the usual definition).

Brauer [40] expressed $m(p^n)$ ($n \geq 0$) in terms of $m_B^{(1)}(R)$ as follows:

$$m(p^n) = \sum_{R \in \mathscr{R}} m_B^{(1)}(R) \tag{1.2}$$

where $\mathscr{R}$ is a set of representatives for the G-conjugacy classes of subgroups of G of order p^n. It is known that every lower defect group is conjugate to a subgroup of D. Since $m(|D|) = 1$, D is always a lower defect group of B. Later Eq. (1.2) was refined by Broué and Olsson by invoking the fusion system $\mathscr{F}$ of B.

Proposition 1.41 (Broué-Olsson [49]) *For $n \geq 0$ we have*

$$m(p^n) = \sum_{R \in \mathscr{R}} m_B^{(1)}(R, b_R)$$

where $\mathscr{R}$ is a set of representatives for the $\mathscr{F}$-conjugacy classes of subgroups $R \leq D$ of order p^n.

Proof This is (2S) of [49].   □

For the definition of the numbers $m_B^{(1)}(R, b_R)$ we refer to the next lemma.

Lemma 1.42 *For $R \leq D$ and $B_R := b_R^{N_G(R,b_R)}$ we have $m_B^{(1)}(R, b_R) = m_{B_R}^{(1)}(R)$. If R is fully $\mathscr{F}$-normalized, then B_R has defect group $N_D(R)$ and fusion system $N_{\mathscr{F}}(R)$.*

Proof The first claim follows from (2Q) in [49]. For the second claim we refer to Theorem IV.3.19 in [19].   □

Since we may always assume that $R \in \mathscr{R}$ is fully $\mathscr{F}$-normalized, the calculation of $m_B^{(1)}(R, b_R)$ can be done in the smaller group $N_G(R, b_R)$. Especially if the Cartan matrix of B_R is known, we may apply Proposition 1.41 with B_R instead of B. Another important reduction is given by the following lemma.

Lemma 1.43 *For $R \leq D$ we have $\sum_{Q \in \mathscr{R}} m_{B_R}^{(1)}(Q) \leq l(b_R)$ where $\mathscr{R}$ is a set of representatives for the $N_G(R, b_R)$-conjugacy classes of subgroups Q such that $R \leq Q \leq N_D(R)$.*

Proof This is implied by Theorem 5.11 in [215] and the remark following it. Notice that in Theorem 5.11 it should read $B \in \mathrm{Bl}(G)$ instead of $B \in \mathrm{Bl}(Q)$.   □

In the local situation for B_R also the next lemma is useful.

Lemma 1.44 *If R is a lower defect group of B, then $O_p(Z(G)) \subseteq R$.*

Proof See Corollary 3.7 in [215].   □

Proposition 1.45 (Watanabe [279])   *Let $u \in Q < D$. Let $\mathscr{R}$ be the set of Brauer correspondents of B in $C_G(u)$ whose defect group is larger than $|Q|$. Then*

$$m_B^{(1)}(Q) \leq \sum_{b \in \mathscr{R}} m_b^{(1)}(Q).$$

In special situations the elementary divisors of the Cartan matrix are given by the following result which is a consequence of Proposition 1.45.

Proposition 1.46 (Fujii [86])   *Let B be a p-block of a finite group with defect d and Cartan matrix C. Suppose that $l(b_u) = 1$ for every non-trivial B-subsection (u, b_u). Then $\det C = p^d$. In particular, p^d is the only non-trivial elementary divisor of C.*

Usually, it is very hard to compute $m(1)$, since this number is not locally determined. However, if the focal subgroup of B is small, one can show that $m(1) = 0$.

Proposition 1.47 (Robinson [238])   *Let B be a block of a finite group with defect group D. Then the Cartan invariants of B are divisible by $|Z(D) : Z(D) \cap \mathfrak{foc}(B)|$. In particular $m(p^n) = 0$ if $p^n < |Z(D) : Z(D) \cap \mathfrak{foc}(B)|$.*

Finally, we give a result in the opposite direction.

Proposition 1.48 (Brauer-Nesbitt [43])   *For a block B of a finite group we have $m(1) \geq 2l(B) - k(B)$.*

Chapter 2
Open Conjectures

A main theme of this work is to prove conjectures of modular representation theory in special situations. Most of these conjectures concern the relationship between local and global invariants of blocks. The first one is probably the oldest one, and will play a special role in this work.

Conjecture 2.1 (Brauer's $k(B)$-Conjecture, 1954 [35]) For a block B of a finite group with defect group D we have $k(B) \leq |D|$.

Conjecture 2.2 (Olsson, 1975 [212]) For a block B of a finite group with defect group D we have $k_0(B) \leq |D : D'|$.

One direction of the following conjecture is known to hold (see Theorem 7.14).

Conjecture 2.3 (Brauer's Height Zero Conjecture, 1956 [34]) A block B of a finite group has abelian defect groups if and only if $k(B) = k_0(B)$.

Conjecture 2.4 (Alperin-McKay Conjecture, 1975 [1]) Let B be a block of a finite group G with defect group D. Then $k_0(B) = k_0(b)$ where b is the Brauer correspondent of B in $N_G(D)$.

In a specific situation we will also consider the following refinement of the Alperin-McKay Conjecture which was proposed by Isaacs and Navarro.

Conjecture 2.5 (Galois-Alperin-McKay Conjecture, 2002 [134]) Let B and b be as in Conjecture 2.4. Then for every p-automorphism $\gamma \in \mathrm{Gal}(\mathbb{Q}_{|G|}|\mathbb{Q}_{|G|_{p'}})$ we have

$$|\{\chi \in \mathrm{Irr}_0(B) : {}^{\gamma}\chi = \chi\}| = |\{\chi \in \mathrm{Irr}_0(b) : {}^{\gamma}\chi = \chi\}|.$$

Later Navarro [199] extended the conjecture to allow not only all automorphisms of $\mathrm{Gal}(\mathbb{Q}_{|G|}|\mathbb{Q}_{|G|_{p'}})$, but also certain other automorphisms which fix $\mathrm{Irr}(B)$ as a set. As another refinement Isaacs and Navarro also proposed a congruence version of the Alperin-McKay Conjecture which takes the precise character degrees into account.

© Springer International Publishing Switzerland 2014
B. Sambale, *Blocks of Finite Groups and Their Invariants*, Lecture Notes
in Mathematics 2127, DOI 10.1007/978-3-319-12006-5_2

Since in our setting the precise degrees are usually unavailable, we will not consider this refinement.

Brauer [36] also provided a list of problems which became famous.

The following version of Alperin's Weight Conjecture [2] is particularly useful in our setting. It can be found in Sect. IV.5.7 in [19]. Here for a finite-dimensional F-algebra A, $z(A)$ denotes the number of (isomorphism classes of) simple projective A-modules. Let B be a block with defect group D and fusion system $\mathscr{F}$. Then for every $\mathscr{F}$-centric subgroup $Q \leq D$ the block b_Q has defect group $C_D(Q) \subseteq Q$ (see Sect. 1.5). Thus, b_Q dominates a block $\overline{b_Q}$ of $C_G(Q)Q/Q$ with trivial defect. Moreover, $B_Q := b_Q^{N_G(Q,b_Q)}$ dominates a block $\overline{B_Q}$ of $N_G(Q,b_Q)/Q$ which covers $\overline{b_Q}$. Hence, we are in a position to apply Theorem 7.3 below which gives us the Külshammer-Puig class γ_Q. For an explicit description of γ_Q in our special situation one can also consult Sect. IV.5.5 in [19].

Conjecture 2.6 (Alperin's Weight Conjecture (AWC), 1986 [2, 19]) Let B be a block of a finite group with defect group D and fusion system $\mathscr{F}$. Then

$$l(B) = \sum_{Q \in \mathscr{R}} z(F_{\gamma_Q} \operatorname{Out}_{\mathscr{F}}(Q))$$

where $\mathscr{R}$ is a set of representatives for the $\mathscr{F}$-conjugacy classes of $\mathscr{F}$-centric, $\mathscr{F}$-radical subgroups of D and $\gamma_Q \in \mathrm{H}^2(\operatorname{Out}_{\mathscr{F}}(Q), F^\times)$ is the Külshammer-Puig class (see Theorem 7.3).

If B is a controlled block, it can be seen that AWC reduces to $l(B) = z(F_{\gamma_D} I(B))$. If in addition $I(B)$ has trivial Schur multiplier, AWC reduces further to $l(B) = k(I(B))$. Recently, Späth [200, 263, 264] (and coauthors) has reduced the Alperin-McKay Conjecture, Brauer's Height Zero Conjecture and Alperin's Weight Conjecture to a (more involved) question about finite simple groups only.

The Ordinary Weight Conjecture, proposed by Robinson [233] and described below expresses the block invariants $k_i(B)$ locally. For this let B be a block with defect group D and fusion system $\mathscr{F}$. For an $\mathscr{F}$-centric, $\mathscr{F}$-radical subgroup $Q \leq D$ let $\mathscr{N}_Q$ be the set of chains $\sigma : 1 = R_1 < R_2 < \ldots < R_l$ of p-subgroups of $\operatorname{Out}_{\mathscr{F}}(Q)$ such that $R_i \trianglelefteq R_l$ for $i = 1, \ldots, l$. Let $|\sigma| := l$. The group $\operatorname{Out}_{\mathscr{F}}(Q)$ acts naturally on $\mathscr{N}_Q$ and on $\operatorname{Irr}(Q)$. For $\sigma \in \mathscr{N}_Q$ (resp. $\chi \in \operatorname{Irr}(Q)$) let $I(\sigma) \leq \operatorname{Out}_{\mathscr{F}}(Q)$ (resp. $I(\chi)$) be the corresponding stabilizer. Then we can restrict the Külshammer-Puig class γ_Q to $I(\sigma, \chi) := I(\sigma) \cap I(\chi)$. Define $\operatorname{Irr}^d(Q) := \{\chi \in \operatorname{Irr}(Q) : \chi(1)p^d = |Q|\}$ for $d \geq 0$. Assume that B has defect d. Then $k^i(B) := k_{d-i}(B)$ is the number of characters of *defect* $i \geq 0$.

Conjecture 2.7 (Ordinary Weight Conjecture (OWC), 1996 [19, 233]) With the notation of Conjecture 2.6 we have

$$k^i(B) = \sum_{Q \in \mathcal{R}} \sum_{\sigma \in \mathcal{N}_Q / \operatorname{Out}_{\mathscr{F}}(Q)} (-1)^{|\sigma|} \sum_{\chi \in \operatorname{Irr}^i(Q)/I(\sigma)} z(F_{\gamma_Q} I(\sigma, \chi))$$

for $i \geq 0$.

For the convenience of the reader we include two abbreviations from [19]: $\omega(Q, \sigma, \chi) := z(F_{\gamma_Q} I(\sigma, \chi))$ and

$$\mathbf{w}(Q, i) := \sum_{\sigma \in \mathcal{N}_Q / \operatorname{Out}_{\mathscr{F}}(Q)} (-1)^{|\sigma|} \sum_{\chi \in \operatorname{Irr}^i(Q)/I(\sigma)} \omega(Q, \sigma, \chi).$$

It is known that the Ordinary Weight Conjecture (for all blocks) implies Alperin's Weight Conjecture (see [235]). Also, the Ordinary Weight Conjecture is equivalent to Dade's Projective Conjecture (see [73]). We do not state the numerous versions of Dade's Conjecture here (ordinary, projective, invariant, ...).

The next conjecture on our list is of a different nature and usually harder to prove (for special cases).

Conjecture 2.8 (Donovan, 1975 [1]) For a given p-group D there are only finitely many Morita equivalence classes of p-blocks with defect group D.

In Donovan's Conjecture it is sometimes important to specify the ring (F or $\mathcal{O}$) over which the blocks are defined. Occasionally we will also mention Broué's Abelian Defect Group Conjecture which, however, will never be the objective of a proof. For this reason we go without the precise definition of Broué's Conjecture and refer to [47] instead.

Our next conjecture was proposed by Linckelmann and is also of a different nature. We will not go in the category theoretical details here.

Conjecture 2.9 (Gluing Problem, 2004 [183]) Let B be a block with defect group D and fusion system $\mathscr{F}$. Let $\overline{\mathscr{F}}$ be the orbit category of $\mathscr{F}$, and let $\overline{\mathscr{F}}^c$ be the subcategory of $\mathscr{F}$-centric subgroups. Then there exists $\gamma \in H^2(\overline{\mathscr{F}}^c, F^\times)$ such that the Külshammer-Puig classes γ_Q in Conjecture 2.6 are restrictions of γ.

In many cases it turns out that the 2-cocycle γ in the Gluing Problem is uniquely determined. However, this is not true in general by an example of Park [219].

Finally we list some more recent (and not so well-known) numerical conjectures. The first one unifies the $k(B)$-Conjecture and Olsson's Conjecture.

Conjecture 2.10 (Eaton, 2003 [72]) For a p-block B with defect group D we have

$$\sum_{i=1}^{n} k_i(B) \leq \sum_{i=0}^{n} k_i(D) p^{2i}$$

for all $n \geq 0$.

The following conjecture strengthens the Height Zero Conjecture (together with Theorem 7.14).

Conjecture 2.11 (Eaton-Moretó [76]) For a block B with non-abelian defect group D we have

$$\min\{i \geq 1 : k_i(D) > 0\} = \inf\{i \geq 1 : k_i(B) > 0\}.$$

Conjecture 2.12 (Malle-Navarro, 2006 [186]) For a block B with defect group D we have

$$k(B)/k_0(B) \leq k(D') \qquad \text{and} \qquad k(B)/l(B) \leq k(D).$$

Conjecture 2.12 is known to hold for abelian defect groups by Theorem 7.14 and Theorem V.9.17(i) in [81]. The next conjecture is explicitly stated as Conjecture 4.14.7 in [185]. It would be a consequence of the Ordinary Weight Conjecture.

Conjecture 2.13 (Robinson, 1996 [233]) If B is a p-block with non-abelian defect group D, then

$$p^{h(\chi)} < |D : Z(D)|$$

for all $\chi \in \mathrm{Irr}(B)$.

Our last conjecture only applies for $p = 2$. Here a finite group is called *rational*, if its character table is integral.

Conjecture 2.14 (Gluck, 2011 [92]) Let B be a 2-block with rational defect group of nilpotency class at most 2. Then every character in $\mathrm{Irr}(B)$ is 2-rational.

Part II
General Results and Methods

Chapter 3
Quadratic Forms

Let B be a p-block of a finite group G. Then the Cartan matrix C of B gives rise to an integral, positive definite, symmetric quadratic form $q : \mathbb{Z}^{l(B)} \to \mathbb{Z}$, $x \mapsto x^{\mathrm{T}} C x$. In this section we will briefly explore some features of q which will eventually lead to restrictions on $k(B)$. The results are taken from [245].

It is well-known that C is indecomposable as integral matrix, i.e. there is no arrangement of the indecomposable projective modules such that C splits into a direct sum of smaller matrices. However, it appears to be an open question if this is still true under more general modifications.

Question A Do there exist a Cartan matrix C of a block B and a matrix $S \in \mathrm{GL}(l(B), \mathbb{Z})$ such that $S^{\mathrm{T}} C S$ is decomposable?

The transformation $C \mapsto S^{\mathrm{T}} C S$ describes precisely Brauer's notion of basic set (see [37]). Recall that a *basic sets* is a basis for the $\mathbb{Z}$-module of generalized Brauer characters of B. For a given block it is much easier to calculate C only up to basic sets. For example, C can be obtained up to basic sets from the ordinary character table of G, i.e. the knowledge of Brauer characters is not necessary. Later we will compute C up to basic sets by means of local data.

Obviously, a change of basic sets does not affect the elementary divisors (and thus the determinant) of C. So far, we have not found an example for Question A. Nevertheless, the following example shows that the answer might be not so easy. The matrix $A = \left(\begin{smallmatrix} 1 & 1 \\ 1 & 2 \end{smallmatrix}\right)$ is indecomposable, but $\left(\begin{smallmatrix} 1 & -1 \\ 0 & 1 \end{smallmatrix}\right)^{\mathrm{T}} A \left(\begin{smallmatrix} 1 & -1 \\ 0 & 1 \end{smallmatrix}\right) = \left(\begin{smallmatrix} 1 & 0 \\ 0 & 1 \end{smallmatrix}\right)$ is not.

The motivation for Question A comes from the fact that $k(B)$ can be bounded in terms of Cartan invariants (see Theorem 4.2 below). These bounds are usually invariant under change of basic sets. The point is that the inequalities are significantly sharper for indecomposable matrices. We illustrate this fact with an example.

© Springer International Publishing Switzerland 2014
B. Sambale, *Blocks of Finite Groups and Their Invariants*, Lecture Notes
in Mathematics 2127, DOI 10.1007/978-3-319-12006-5_3

Let $l(B) = 2$ and assume that the elementary divisors of C are 2 and 16. Then C has the form

$$\begin{pmatrix} 2 & 0 \\ 0 & 16 \end{pmatrix} \text{ or } \begin{pmatrix} 6 & 2 \\ 2 & 6 \end{pmatrix}$$

up to basic sets. In the first case one can deduce $k(B) \leq 18$, while in the second case $k(B) \leq 10$ holds (see [172] or Theorem 4.2 below).

We give an answer to Question A in two special cases.

Lemma 3.1 *Let G be p-solvable and $l := l(B) \geq 2$. Then there is no matrix $S \in \mathrm{GL}(l, \mathbb{Z})$ such that $S^{\mathrm{T}} C S = \begin{pmatrix} p^d & 0 \\ 0 & C_1 \end{pmatrix}$ with $C_1 \in \mathbb{Z}^{(l-1) \times (l-1)}$. In particular C is not a diagonal matrix up to basic sets.*

Proof Assume the contrary, i.e. there is a matrix $S = (s_{ij}) \in \mathrm{GL}(l, \mathbb{Z})$ such that

$$C = (c_{ij}) = S^{\mathrm{T}} \begin{pmatrix} p^d & 0 \\ 0 & C_1 \end{pmatrix} S$$

with $C_1 \in \mathbb{Z}^{(l-1) \times (l-1)}$. Let $s_i := (s_{2i}, s_{3i}, \ldots, s_{li})$ for $i = 1, \ldots, l$. By Theorem (3H) in [83] we have

$$p^d s_{i1}^2 + s_i C_1 s_i^{\mathrm{T}} = c_{ii} \leq p^d$$

for $i = 1, \ldots, l$. Since S is invertible, there exists i such that $s_{1i} \neq 0$. We may assume $s_{11} \neq 0$. Then $s_{11} = \pm 1$ and $s_1 = (0, \ldots, 0)$, because C_1 is positive definite. Now all other columns of S are linearly independent of the first column. This gives $s_{1i} = 0$ for $i = 2, \ldots, l$. Hence, S has the form $S = \begin{pmatrix} \pm 1 & 0 \\ 0 & S_1 \end{pmatrix}$ with $S_1 \in \mathrm{GL}(l-1, \mathbb{Z})$. But then C also has the form $\begin{pmatrix} p^d & 0 \\ 0 & C_2 \end{pmatrix}$ with $C_2 \in \mathbb{Z}^{(l-1) \times (l-1)}$, a contradiction. The second claim follows at once, since p^d is always an elementary divisor of C. $\square$

The bound $c_{ij} \leq p^d$ for Cartan invariants c_{ij} used in the proof appeared as Problem 22 in Brauer's list [36]. Unfortunately it does not hold for arbitrary finite groups, since Landrock [173] gave a counterexample.

As an application, assume that the block B has abelian defect group and Cartan matrix C. Then Broué's Abelian Defect Group Conjecture would imply that C is the Cartan matrix of a block of a p-solvable group (see Theorem 1.19 and Proposition 1.20). Thus, Lemma 3.1 applies for C.

Lemma 3.2 *Let B be a p-block with defect d and Cartan matrix C. If $\det C = p^d$, then for every $S \in \mathrm{GL}(l(B), \mathbb{Z})$ the matrix $S^{\mathrm{T}} C S$ is indecomposable.*

Proof Again assume the contrary, i.e. there is a matrix $S \in \mathrm{GL}(l(B), \mathbb{Z})$ such that

$$C = S^{\mathrm{T}} \begin{pmatrix} C_1 & 0 \\ 0 & C_2 \end{pmatrix} S$$

with $C_1 \in \mathbb{Z}^{m \times m}$ and $C_2 \in \mathbb{Z}^{(l-m) \times (l-m)}$, where $l := l(B)$ and $1 \leq m < l$. In particular $l < k(B) =: k$, because $l \geq 2$. Since $\det C = p^d$, the elementary divisors of C are 1 and p^d, where p^d occurs with multiplicity one. W. l. o. g. we may assume $\det C_1 = 1$. Let $Q = (q_{ij})$ be the corresponding part of the decomposition matrix, i.e. $Q^{\mathrm{T}} Q = C_1$. By the Binet-Cauchy formula (see e.g. page 27 in [87]) we have

$$1 = \det C_1 = \sum_{\substack{V \subseteq \{1,\dots,k\}, \\ |V|=m}} \det Q_V^{\mathrm{T}} Q_V,$$

where Q_V is the $m \times m$ submatrix consisting of the entries $\{q_{ij} : i \in V, j \in \{1,\dots,m\}\}$. Since $\det Q_V^{\mathrm{T}} Q_V \geq 0$, one summand is 1 while the others are all 0. Thus we may assume, that the first m rows $q_1,\dots,q_m$ of Q are linearly independent. Now consider a row q_i for $i \in \{m+1,\dots,k\}$. Then q_i is a rational linear combination of $q_2,\dots,q_m$, because $q_2,\dots,q_m, q_i$ are linearly dependent. By the same argument, q_i is also a linear combination of $q_1,\dots,q_{j-1}, q_{j+1},\dots,q_m$ for $j = 2,\dots,m$. This forces $q_i = (0,\dots,0)$. Hence, all the rows $q_{m+1},\dots,q_k$ vanish. Now consider a column $d(u)$ of generalized decomposition numbers, where u is a nontrivial element of a defect group of B. By the orthogonality relations the scalar product of $d(u)$ and an arbitrary column of Q vanishes. This means the first m entries of $d(u)$ must be zero. Since this holds for all columns $d(u)$ with $u \neq 1$, there exists an irreducible character of B which vanishes on the p-singular elements of G. It is well-known that this is equivalent to $d = 0$. But this contradicts $l \geq 2$. $\square$

More generally, the matrix $C = (c_{ij})$ cannot have a submatrix $C_V = (c_{ij})_{i,j \in V}$ for $V \subseteq \{1,\dots,l(B)\}$ such that $\det C_V = 1$.

As an example, we have $\det C = p^d$ whenever $l(b) = 1$ for all B-subsections $(u, b) \neq (1, B)$ (Proposition 1.46). This in turn is satisfied for instance if D is abelian and $D \rtimes I(B)$ is a Frobenius group. This is true for all cyclic defect groups. In general $\det C$ can be determined locally by considering lower defect groups (see Sect. 1.8).

One often tries to choose a basic set such that C has a "nice" shape. One way to do this is given by the reduction theory of quadratic forms.

Definition 3.3 A positive definite integral quadratic form q corresponding to a symmetric matrix $(\alpha_{ij})_{1 \leq i,j \leq l}$ is called *reduced* (in the sense of Minkowski) if $\alpha_{i,i+1} \geq 0$ for $i = 1,\dots,l-1$ and for $i = 1,\dots,l$ we have

$$\alpha_{ii} \leq q(x_1,\dots,x_l)$$

whenever $\gcd(x_i,\dots,x_l) = 1$.

Apart from Minkowski's reduction there are several other approaches. For instances, the so-called LLL algorithm is much faster, but provides weaker properties.

A 2×2 matrix $C = (c_{ij})$ is reduced (i.e. its quadratic form is reduced) if and only if $0 \le 2c_{12} \le c_{11} \le c_{22}$ (see e.g. [276]). Then it is easy to see that $4c_{11}c_{22} - c_{11}^2 \le 4 \det C$. Now

$$c_{11} + c_{22} \le \frac{5}{4}c_{11} + \frac{\det C}{c_{11}} \le \frac{\det C + 5}{2} \tag{3.1}$$

follows (see proof of Theorem 1 in [245]). This will be used later. Barnes [21] has obtained similar inequalities for dimensions 3 and 4.

Every quadratic form can be reduced in the sense above. However, equivalent quadratic forms may have distinct reductions. Therefore, it is a hard problem (especially in large dimensions) to decide if two given quadratic forms are equivalent. In small dimensions lists of pairwise non-equivalent reduced quadratic forms according to their determinant appeared in book form [31, 207]. The content of these books is also available online [203].

Most of the time we will not work with reduced matrices, but usually we will choose a basic set such that $C = (c_{ij})$ has "small" entries. In particular, we may assume that $2|c_{ij}| \le \min(c_{ii}, c_{jj})$ for $i \ne j$ and $c_{11} \le \ldots \le c_{ll}$ where $l := l(B)$. Additionally, we try to minimize the number of negative entries.

The next theorem is an application of Lemma 3.2, Barnes' results [21], and a work of Külshammer and Wada [172] which we will generalize in the upcoming chapter.

Theorem 3.4 *Let B be a p-block of a finite group with defect d and Cartan matrix C. If $l(B) \le 4$ and $\det C = p^d$, then*

$$k(B) \le \frac{p^d - 1}{l(B)} + l(B).$$

Moreover, this bound is sharp.

Proof For $l := l(B) = 1$ the assertion is clear (see e.g. Corollary 5 in [216]). So let $l \ge 2$. Let $A = (a_{ij})$ be a reduced matrix in the sense above which is equivalent to C as a quadratic form. In particular we have $2|a_{ij}| \le \min\{a_{ii}, a_{jj}\}$ and $1 \le a_{11} \le a_{22} \le \ldots \le a_{ll}$. For convenience we write $\alpha := a_{11}$, $\beta := a_{22}$ and so on.

We are going to apply equation $(**)$ in [172]. In order to do so, we will bound the trace of A from above and the sum $a_{12} + a_{23} + \ldots + a_{l-1,l}$ from below.

Let $l(B) = 2$. By Lemma 3.2 we have $a_{12} \ne 0$ and $a_{12} > 0$ after a suitable change of signs (i.e. replacing A by an equivalent matrix). By Barnes [21] we have $4\alpha\beta - \alpha^2 \le 4p^d$, so that

$$\alpha + \beta \le \frac{5}{4}\alpha + \frac{p^d}{\alpha} =: f(\alpha). \tag{3.2}$$

Since $2|a_{ij}| \leq \min\{a_{ii}, a_{jj}\}$, we have $2 \leq \alpha$, and $\alpha \leq \beta$ yields $\alpha \leq 2\sqrt{p^d/3}$. The convex function $f(\alpha)$ takes its maximal value in the interval $[2, 2\sqrt{p^d/3}]$ on one of the two borders. An easy calculation shows $(p^d + 5)/2 = f(2) > f(2\sqrt{p^d/3})$ for $p^d \geq 9$. In case $p^d \leq 6$ only $\alpha = 2$ is possible. In the remaining cases we have $\alpha + \beta \leq f(2)$ for all feasible pairs (α, β) (we call a pair (α, β) feasible if it satisfies inequality (3.2)). Equation $(**)$ in [172] yields

$$k(B) \leq \alpha + \beta - a_{12} \leq f(2) - 1 = \frac{p^d - 1}{l(B)} + l(B).$$

Let $l(B) = 3$. The same discussion leads to $a_{12} + a_{23} \geq 2$ after a suitable (simultaneous) permutation of rows and columns (i.e. replacing A by $P^T A P$ with a permutation matrix P). It is not always possible to achieve $\alpha \leq \beta \leq \gamma$ additionally. But since the trace of A is symmetric in α, β and γ, we may assume $2 \leq \alpha \leq \beta \leq \gamma$ nevertheless. The inequality in [21] reads

$$4\alpha\beta\gamma - \alpha\beta^2 - \alpha^2\gamma = 2\alpha\beta\gamma + \alpha\beta(\gamma - \beta) + \alpha\gamma(\beta - \alpha) \leq 4p^d,$$

so that

$$\alpha + \beta + \gamma \leq \alpha + \beta + \frac{4p^d + \alpha\beta^2}{4\alpha\beta - \alpha^2} =: f(\alpha, \beta).$$

We describe a set which contains all feasible points. Since $2\alpha^3 \leq 2\alpha\beta\gamma + \alpha\beta(\gamma - \beta) + \alpha\gamma(\beta - \alpha) \leq 4p^d$ we get $2 \leq \alpha \leq \sqrt[3]{2p^d}$. Similarly $4\beta^2 \leq 4p^d$ and $\alpha \leq \beta \leq \sqrt{p^d}$. Thus all feasible points are contained in the convex polygon

$$\mathscr{F} := \{(\alpha, \beta) : 2 \leq \alpha \leq \sqrt[3]{2p^d}, \, \alpha \leq \beta \leq \sqrt{p^d}\}.$$

It can be shown (with the help of Maple [189]) that f is convex on $\mathscr{F}$. Hence, the maximal value of f on $\mathscr{F}$ will be attained on one of the three vertices:

$$V_1 = (2, 2),$$

$$V_2 = (2, \sqrt{p^d}),$$

$$V_3 = (\sqrt[3]{2p^d}, \sqrt[3]{2p^d}).$$

One can check that $(p^d + 14)/3 = f(V_1) \geq f(V_2)$ for $p^d \geq 10$ and $f(V_1) \geq f(V_3)$ for $p^d \geq 12$. If $p^d \leq 10$, then V_1 is the only feasible point. In the remaining case

$p^d = 11$ there is only one more feasible pair $(\alpha, \beta) = (2, 3)$. Then $\gamma = 3$ and $\alpha + \beta + \gamma \le f(V_1)$. Now $(**)$ in [172] takes the form

$$k(B) \le \alpha + \beta + \gamma - a_{12} - a_{23} \le f(V_1) - 2 = \frac{p^d - 1}{l(B)} + l(B).$$

Finally, let $l(B) = 4$. By permuting rows and columns and changing signs, we can reach (using Lemma 3.2) at least one of the two arrangements

(i) $a_{12} + a_{23} + a_{34} \ge 3$,
(ii) $a_{12} + a_{13} + a_{14} \ge 3$.

In case (i) we can use equation $(**)$ as before. Since the matrix

$$\begin{pmatrix} 2 & 1 & 1 & 1 \\ 1 & 2 & 0 & 0 \\ 1 & 0 & 2 & 0 \\ 1 & 0 & 0 & 2 \end{pmatrix}$$

is positive definite, we can use Theorem A in [172] for case (ii). Thus, for the rest of the proof we will assume that case (i) occurs. As before, we will also assume $2 \le \alpha \le \beta \le \gamma \le \delta$ and

$$4\alpha\beta\gamma\delta - \alpha^2\gamma\delta - \alpha\beta^2\delta - \alpha\beta\gamma^2 + \frac{1}{4}\alpha^2(\gamma - \beta)^2$$

$$= \alpha\beta\gamma\delta + \alpha\gamma\delta(\beta - \alpha) + \alpha\beta\delta(\gamma - \beta) + \alpha\beta\gamma(\delta - \gamma) + \frac{1}{4}\alpha^2(\gamma - \beta)^2 \le 4p^d$$

$$\tag{3.3}$$

by Barnes [21]. We search for the maximum of the function

$$f(\alpha, \beta, \gamma) := \alpha + \beta + \gamma + \frac{4p^d + \alpha\beta\gamma^2 - \frac{1}{4}\alpha^2(\gamma - \beta)^2}{4\alpha\beta\gamma - \alpha^2\gamma - \alpha\beta^2}$$

on a suitable convex polyhedron. Since $\alpha^4 \le 4p^d$ we have $2 \le \alpha \le \sqrt[4]{4p^d}$. In a similar way, we obtain the set

$$\mathscr{F} := \{(\alpha, \beta, \gamma) : 2 \le \alpha \le \sqrt[4]{4p^d}, \ \alpha \le \beta \le \sqrt[3]{2p^d}, \ \beta \le \gamma \le \sqrt{p^d}\},$$

which contains all feasible points. It can be shown that f is in fact convex on $\mathscr{F}$. The vertices of $\mathscr{F}$ are

$$V_1 := (2, 2, 2),$$

$$V_2 := (2, 2, \sqrt{p^d}),$$

$$V_3 := (2, \sqrt[3]{2p^d}, \sqrt[3]{2p^d}),$$

$$V_4 := (\sqrt[4]{4p^d}, \sqrt[4]{4p^d}, \sqrt[4]{4p^d}).$$

We fix the value $m := (p^d + 27)/4$. A calculation shows $f(V_2) \leq m$ for $p^d \geq 22$, $f(V_3) \leq m$ for $p^d \geq 20$, and $f(V_4) \leq m$ for $p^d \geq 23$. If $p^d \leq 12$, then V_1 is the only feasible point. If $p^d \leq 17$, there is only one other feasible point $(\alpha, \beta, \gamma) = (2, 2, 3)$ beside V_1. In this case $f(2, 2, 3) \leq m$ for $p^d \geq 14$. For $p^d = 13$ we have

$$\alpha + \beta + \gamma + \delta - a_{13} - a_{14} - a_{34} \leq 7 = \frac{13 - 1}{4} + 4.$$

For $p^d \leq 20$ there is one additional point $(\alpha, \beta, \gamma) = (2, 3, 3)$, which satisfies $f(2, 3, 3) \leq m$. In the remaining cases there is another additional point $(\alpha, \beta, \gamma) = (3, 3, 3)$. For this we get $f(3, 3, 3) \leq m$ if $p^d \geq 22$. Since 21 is no prime power, we can consider $f(V_1) = p^d/4 + 7$ now. If $p > 2$, then $p^d/4$ is no integer. In this case

$$\alpha + \beta + \gamma + \delta - a_{13} - a_{14} - a_{34} \leq [f(V_1)] - 3 = \frac{p^d - 1}{4} + 4,$$

where $[f(V_1)]$ is the largest integer below $f(V_1)$. Thus, let us assume $\delta = p^d/4 + 1$ (and $p = 2$). With the help of a computer one can show that up to equivalence only the possibility

$$A = \begin{pmatrix} 2 & 1 & 0 & -1 \\ 1 & 2 & 1 & 0 \\ 0 & 1 & 2 & 1 \\ -1 & 0 & 1 & 8 \end{pmatrix} \tag{3.4}$$

has the right determinant (see also the remark following the proof). By considering the corresponding decomposition matrix, one can easily deduce:

$$k(B) \leq \delta + 2 \leq \frac{p^d - 1}{l(B)} + l(B).$$

Now it remains to check, that f does not exceed m on other points of $\mathscr{F}$ (this is necessary, since $f(V_1) > m$). For that, we exclude V_1 from $\mathscr{F}$ and form a smaller polyhedron. Since only integral values for α, β, γ are allowed, we get three new vertices:

$$V_5 := (2, 2, 3),$$

$$V_6 := (2, 3, 3),$$

$$V_7 := (3, 3, 3)$$

But these points were already considered. This finishes the first part of the proof. The second part follows easily, since for blocks with cyclic defect groups equality holds.   □

In the case $l(B) = 5$ there is no inequality like (3.3). However, one can use the so called "fundamental inequality" of quadratic forms

$$\alpha\beta\gamma\delta\epsilon \le 8p^d$$

(see [21]). Of course, the complexity increases rapidly with $l(B)$.

Very recently, Theorem 3.4 has been greatly generalized in [255].

Chapter 4
The Cartan Method

4.1 An Inequality

In this section we are going to develop certain bounds on $k(B)$ for a block B of a finite group in terms of Cartan invariants. The material comes partly from [114] and partly from [252].

We begin with a result by Brandt.

Proposition 4.1 (Brandt [32]) *Let B be a block with Cartan matrix $C = (c_{ij})$. Then*

$$k(B) \le 1 - l(B) + \sum_{i=1}^{l(B)} c_{ii}.$$

The disadvantage of Proposition 4.1 is that C is usually only known up to basic sets.

The following theorem was first proved by Külshammer and Wada [172] in the special case $u = 1$. A version for $p = 2$ appeared in the author's dissertation [244]. The present form was proved in [114]. However, the proof in the latter article was incorrect (certain numbers were not algebraic integers as claimed), and we take the opportunity to give a new proof.

Theorem 4.2 *Let B be a p-block of G, and let (u, b_u) be a B-subsection. Let $C_u = (c_{ij})$ be the Cartan matrix of b_u up to basic sets. Then for every positive definite, integral quadratic form $q(x_1, \dots, x_{l(b_u)}) = \sum_{1 \le i \le j \le l(b_u)} q_{ij} x_i x_j$ we have*

$$k_0(B) \le \sum_{1 \le i \le j \le l(b_u)} q_{ij} c_{ij}.$$

© Springer International Publishing Switzerland 2014
B. Sambale, *Blocks of Finite Groups and Their Invariants*, Lecture Notes
in Mathematics 2127, DOI 10.1007/978-3-319-12006-5_4

In particular

$$k_0(B) \leq \sum_{i=1}^{l(b_u)} c_{ii} - \sum_{i=1}^{l(b_u)-1} c_{i,i+1}.$$

If (u, b_u) is major, we can replace $k_0(B)$ by $k(B)$ in these formulas.

Proof First of all, assume that C_u is the Cartan matrix of b_u (not only up to basic sets!). Let $\mathrm{IBr}(b_u) = \{\varphi_1, \ldots, \varphi_l\}$ where $l := l(b_u)$. Then we have rows $d_\chi :=$ $(d_{\chi\varphi_1}^u, \ldots, d_{\chi\varphi_l}^u)$ for $\chi \in \mathrm{Irr}(B)$. Let $Q = (\tilde{q}_{ij})_{i,j=1}^l$ with

$$\tilde{q}_{ij} := \begin{cases} q_{ij} & \text{if } i = j, \\ q_{ij}/2 & \text{if } i \neq j. \end{cases}$$

Then we have

$$\sum_{1 \leq i \leq j \leq l} q_{ij} c_{ij} = \sum_{1 \leq i,j \leq l} \tilde{q}_{ij} c_{ij} = \sum_{1 \leq i,j \leq l} \sum_{\chi \in \mathrm{Irr}(B)} \tilde{q}_{ij} d_{\chi i}^u \overline{d_{\chi j}^u}$$

$$= \sum_{\chi \in \mathrm{Irr}(B)} d_\chi Q \overline{d_\chi}^{\mathrm{T}} \geq \sum_{\chi \in \mathrm{Irr}_0(B)} d_\chi Q \overline{d_\chi}^{\mathrm{T}},$$

since Q is positive definite. Thus, it suffices to show

$$\sum_{\chi \in \mathrm{Irr}_0(B)} d_\chi Q \overline{d_\chi}^{\mathrm{T}} \geq k_0(B).$$

For this, let p^n be the order of u. Then d_{ij}^u is an integer of the p^n-th cyclotomic field $\mathbb{Q}(\zeta)$ for $\zeta := e^{2\pi i/p^n}$. It is known that $1, \zeta, \zeta^2, \ldots, \zeta^f$ with $f := p^{n-1}(p-1) - 1$ form a basis for the ring of integers of $\mathbb{Q}(\zeta)$. We fix a character $\chi \in \mathrm{Irr}_0(B)$ and set $d := d_\chi$. Then there are integral rows $a_m \in \mathbb{Z}^l$ $(m = 0, \ldots, f)$ such that $d = \sum_{m=0}^f a_m \zeta^m$. By Proposition 1.36 at least one of the rows a_m does not vanish.

Let $\mathscr{G}$ be the Galois group of $\mathbb{Q}(\zeta)$ over $\mathbb{Q}$. Then it is known that for every $\gamma \in \mathscr{G}$ there is a character $\chi' \in \mathrm{Irr}(B)$ such that $\gamma(d) = d_{\chi'}$. Thus, it suffices to show

$$\sum_{\gamma \in \mathscr{G}} \gamma(d) Q \overline{\gamma(d)}^{\mathrm{T}} = \sum_{\gamma \in \mathscr{G}} \gamma(d Q \overline{d}^{\mathrm{T}}) \geq |\mathscr{G}| = f + 1.$$

We have

$$\sum_{\gamma \in \mathscr{G}} \gamma(dQ\overline{d}^{\mathrm{T}}) = \sum_{\gamma \in \mathscr{G}} \gamma\left(\sum_{i=0}^{f} a_i Q a_i^{\mathrm{T}} + \sum_{j=1}^{f}\sum_{m=0}^{f-j} a_m Q a_{m+j}^{\mathrm{T}}(\zeta^j + \overline{\zeta}^j)\right)$$

$$= (f+1)\sum_{i=0}^{f} a_i Q a_i^{\mathrm{T}} + 2\sum_{j=1}^{f}\sum_{m=0}^{f-j} a_m Q a_{m+j}^{\mathrm{T}} \sum_{\gamma \in \mathscr{G}} \gamma(\zeta^j).$$

The p^m-th cyclotomic polynomial Φ_{p^m} has the form

$$\Phi_{p^m} = X^{p^{m-1}(p-1)} + X^{p^{m-1}(p-2)} + \ldots + X^{p^{m-1}} + 1.$$

This gives

$$\sum_{\gamma \in \mathscr{G}} \gamma(\zeta^j) = \begin{cases} -p^{n-1} & \text{if } p^{n-1} \mid j \\ 0 & \text{otherwise} \end{cases}$$

for $j \in \{1, \ldots, f\}$. It follows that

$$\sum_{\gamma \in \mathscr{G}} \gamma(dQ\overline{d}^{\mathrm{T}}) = (f+1)\sum_{i=0}^{f} a_i Q a_i^{\mathrm{T}} - 2p^{n-1}\sum_{j=1}^{p-2}\sum_{m=0}^{f-jp^{n-1}} a_m Q a_{m+p^{n-1}j}^{\mathrm{T}}$$

$$= p^{n-1}\left((p-1)\sum_{i=0}^{f} a_i Q a_i^{\mathrm{T}} - 2\sum_{j=1}^{p-2}\sum_{m=0}^{f-jp^{n-1}} a_m Q a_{m+p^{n-1}j}^{\mathrm{T}}\right).$$

$$(4.1)$$

For $p = 2$ the claim follows immediately, since then $f + 1 = 2^{n-1}$. Thus, suppose $p > 2$. Then we have

$$\{0, 1, \ldots, f - jp^{n-1}\} \cup \{(p-1-j)p^{n-1}, (p-1-j)p^{n-1}+1, \ldots, f\} = \{0, 1, \ldots, f\}$$

for all $j \in \{1, \ldots, p-2\}$. This shows that every row a_m occurs exactly $p - 2$ times in the second sum of (4.1). Hence,

$$\sum_{\gamma \in \mathscr{G}} \gamma(dQ\overline{d}^{\mathrm{T}}) =$$

$$p^{n-1}\left(\sum_{i=0}^{f} a_i Q a_i^{\mathrm{T}} + \sum_{j=1}^{p-2}\sum_{m=0}^{f-jp^{n-1}} (a_m - a_{m+jp^{n-1}})Q(a_m - a_{m+jp^{n-1}})^{\mathrm{T}}\right).$$

Now assume that a_m does not vanish for some $m \in \{0, \dots, f\}$. Then we have $a_m Q a_m^\mathrm{T} \geq 1$, since Q is positive definite. Again, a_m occurs exactly $p - 2$ times in the second sum. Let $a_m - a_{m'}$ (resp. $a_{m'} - a_m$) be such an occurrence. Then we have

$$a_{m'} Q a_{m'}^\mathrm{T} + (a_m - a_{m'}) Q (a_m - a_{m'})^\mathrm{T} \geq 1.$$

Now the first inequality of the theorem follows easily. The result does not depend on the basic set for C_u, since changing the basic set is essentially the same as taking another quadratic form q (see [172]). For the second claim we take the quadratic form corresponding to the Dynkin diagram of type A_l for q. If (u, b_u) is major, then all rows d_χ for $\chi \in \mathrm{Irr}(B)$ do not vanish (see Proposition 1.36). Hence, we can replace $k_0(B)$ by $k(B)$. $\square$

We use the opportunity to present a first application of Theorem 4.2.

Proposition 4.3 *Let (u, b_u) be a B-subsection such that b_u has defect group Q and $Q/\langle u \rangle$ is cyclic. Then*

$$k_0(B) \leq \left(\frac{|Q/\langle u \rangle| - 1}{l(b_u)} + l(b_u) \right) |\langle u \rangle| \leq |Q|.$$

Proof As usual, b_u dominates a block $\overline{b_u}$ of $\mathrm{C}_G(u)/\langle u \rangle$ with cyclic defect group $Q/\langle u \rangle$ and $l(\overline{b_u}) = l(b_u)$. By Theorem 8.6 below, the Cartan matrix b_u has the form $|\langle u \rangle|(m + \delta_{ij})_{1 \leq i,j \leq l(b_u)}$ up to equivalence where $m := (|Q/\langle u \rangle| - 1)/l(b_u)$ is the multiplicity of $\overline{b_u}$. Now the claim follows from Theorem 4.2. $\square$

Külshammer and Wada [172] have shown that there is not always a positive definite quadratic form q such that we have equality in Theorem 4.2 (for $u = 1$). However, it is not clear if there is always a quadratic form q such that

$$\sum_{1 \leq i \leq j \leq l(B)} q_{ij} c_{ij} \leq p^d \tag{4.2}$$

where d is the defect of the block B. (This would imply the $k(B)$-Conjecture in general.)

We consider an example. Let $D \cong C_2^4$, $S \in \mathrm{Syl}_3(\mathrm{Aut}(D))$, $G = D \rtimes S$ and $B = B_0(\mathcal{O}G)$. Then $k(B) = 16$, $l(B) = |S| = 9$, and the decomposition matrix Q and the Cartan matrix C of B are

$$Q = \begin{pmatrix} 1 & . & . & . & . & . & . & . & . \\ . & 1 & . & . & . & . & . & . & . \\ . & . & 1 & . & . & . & . & . & . \\ . & . & . & 1 & . & . & . & . & . \\ . & . & . & . & 1 & . & . & . & . \\ . & . & . & . & . & 1 & . & . & . \\ . & . & . & . & . & . & 1 & . & . \\ . & . & . & . & . & . & . & 1 & . \\ . & . & . & . & . & . & . & . & 1 \\ 1 & 1 & 1 & . & . & . & . & . & . \\ 1 & . & . & . & 1 & 1 & . & . & . \\ . & . & . & 1 & . & 1 & . & 1 & . \\ . & . & . & . & 1 & . & 1 & . & 1 \\ . & 1 & . & . & 1 & . & . & 1 & . \\ . & . & 1 & 1 & . & . & . & . & 1 \\ 1 & 1 & 1 & 1 & 1 & 1 & 1 & 1 & 1 \end{pmatrix}, \quad C = \begin{pmatrix} 4 & 2 & 2 & 1 & 1 & 2 & 2 & 1 & 1 \\ 2 & 4 & 2 & 1 & 2 & 1 & 1 & 2 & 1 \\ 2 & 2 & 4 & 2 & 1 & 1 & 1 & 1 & 2 \\ 1 & 1 & 2 & 4 & 1 & 2 & 1 & 2 & 2 \\ 1 & 2 & 1 & 1 & 4 & 1 & 2 & 2 & 2 \\ 2 & 1 & 1 & 2 & 1 & 4 & 2 & 2 & 1 \\ 2 & 1 & 1 & 1 & 2 & 2 & 4 & 1 & 2 \\ 1 & 2 & 1 & 2 & 2 & 2 & 1 & 4 & 1 \\ 1 & 1 & 2 & 2 & 2 & 1 & 2 & 1 & 4 \end{pmatrix}.$$

We will see that in this case there is no positive definite quadratic form q such that Inequality (4.2) is satisfied. In order to do so, we assume that q is given by the matrix $\frac{1}{2}A$ with $A = (a_{ij}) \in \mathbb{Z}^{9 \times 9}$. Since A is symmetric, we only consider the upper triangular half of A. Then the rows of Q are 1-roots of q, i.e. $rAr^{\mathrm{T}} = 2$ for every row r of Q (see Corollary B in [172]). If we take the first nine rows of Q, it follows that $a_{ii} = 2$ for $i = 1, \ldots, 9$. Now assume $|a_{12}| \geq 2$. Then

$$(1, -\operatorname{sgn} a_{12}, 0, \ldots, 0) A (1, -\operatorname{sgn} a_{12}, 0, \ldots, 0)^{\mathrm{T}} \leq 0,$$

and q is not positive definite. The same argument shows $a_{ij} \in \{-1, 0, 1\}$ for $i \neq j$. In particular there are only finitely many possibilities for q. Now the next row of Q shows

$$(a_{12}, a_{13}, a_{23}) \in \{(-1, -1, 0), (-1, 0, -1), (0, -1, -1)\}.$$

The same holds for the following triples

$$(a_{16}, a_{17}, a_{67}), \ (a_{46}, a_{48}, a_{68}), \ (a_{57}, a_{59}, a_{79}), \ (a_{25}, a_{28}, a_{58}), \ (a_{34}, a_{39}, a_{49}).$$

Finally the last row of Q shows that the remaining entries add up to 4:

$$a_{14} + a_{15} + a_{18} + a_{19} + a_{24} + a_{26} + a_{27} + a_{29} + a_{35}$$
$$+ a_{36} + a_{37} + a_{38} + a_{45} + a_{47} + a_{56} + a_{69} + a_{78} + a_{89} = 4.$$

These are too many possibilities to check by hand. So we try to find a positive definite form q with GAP [266]. To decrease the computational effort, we enumerate all positive definite 7×7 left upper submatrices of A first. There are 140,428 of them, but none can be completed to a positive definite 9×9 matrix with the given constraints.

On the positive side, one can show that any solution $C = Q_1^T Q_1$ with $Q_1 \in \mathbb{Z}^{k \times 9}$ satisfies $k \leq 16$. For the general case, we will see later that one can get a good bound on $k(B)$ by using a different approach which I like to call the "inverse Cartan method". But first we explain the Cartan method which is an application of Theorem 4.2.

4.2   An Algorithm

We explain the practical importance of Theorem 4.2. For this let B be a block with defect group D and subsection (u, b_u). After conjugation if necessary, we may assume that b_u has defect group $C_D(u)$ (see Lemma 1.34). Let C_u be the Cartan matrix of b_u. Then b_u dominates a block $\overline{b_u}$ with Cartan matrix $\frac{1}{|\langle u \rangle|} C_u$ by Theorem 1.22. Since $\overline{b_u}$ has defect group $C_D(u)/\langle u \rangle$, we can often apply induction on the defect of B in order to compute C_u. Then Theorem 4.2 gives a bound on $k_0(B)$ (or on $k(B)$).

In the following we provide an algorithm which allows us to compute even the Cartan matrix C (up to basic sets) of B in the situation above. Let $\mathscr{R}$ be a set of representatives for the conjugacy classes of B-subsections. Let $(1, B) \neq (u, b_u) \in \mathscr{R}$ such that b_u has Cartan matrix C_u. As above we may assume that C_u is known at least up to basic sets. Let Q_u be the part of the generalized decomposition matrix consisting of the numbers $d_{\chi\varphi}^u$ for $\chi \in \mathrm{Irr}(B)$ and $\varphi \in \mathrm{IBr}(b_u)$. Then the orthogonality relations imply $C_u = Q_u^T \overline{Q_u}$. Since the entries of Q_u are algebraic integers, there are only finitely many possibilities for Q_u and we can list them by computer in favorable cases. Here it is often convenient to choose a basis for the ring of algebraic integers so that we actually only need to deal with rational integers. Then one can also give a refined version of the orthogonality relations by studying the action of the Galois group of a cyclotomic field (see Sect. 5.2). More information can be gained by taking the heights of the characters and the contributions into account (see Proposition 1.36).

Suppose that we know all the possibilities for Q_u for all $(1, B) \neq (u, b_u) \in \mathscr{R}$. This means we know the generalized decomposition matrix Q except the ordinary part. Write $Q = (Q_1, Q_2)$ where Q_1 is the ordinary decomposition matrix. Strictly speaking, we only know Q_2 up to a transformation $Q_2 \to Q_2 S$ where $S \in \mathrm{GL}(k(B) - l(B), \mathbb{Z})$, since the matrices C_u are only known up to basic sets. However, this does not make much difference, since in the end we get C also only up to basic sets. It is on the other hand crucial that the numbers $k(B)$ and $l(B)$ are usually not uniquely determined by the matrices C_u.

We are now looking for integral solutions $x \in \mathbb{Z}^{k(B)}$ of the equation $Q_2^T x = 0$. By choosing a basis for the ring of algebraic integers as above we may replace Q_2

by an integral matrix $\widetilde{Q_2}$ for this purpose. Then the set of solutions of the equation above forms a free $\mathbb{Z}$-module M. We compute a basis of M by transforming $\widetilde{Q_2}$ to its Smith normal form. We write the vectors of this basis as columns of a matrix $\widetilde{Q_1}$. Since Q is invertible, the rank of Q_2 (and thus of $\widetilde{Q_2}$) is $k(B) - l(B)$. It follows that $\widetilde{Q_1}$ is a $k(B) \times l(B)$ matrix. On the other hand, also the columns of Q_1 lie in M. Hence, we find a matrix $T \in \mathbb{Z}^{l(B) \times l(B)}$ such that $Q_1 = \widetilde{Q_1}T$. It is well-known that there exists a matrix $R \in \mathbb{Z}^{l(B) \times k(B)}$ such that $RQ_1 = 1_{l(B)}$. This implies that $T \in \mathrm{GL}(l(B), \mathbb{Z})$. We conclude that the Cartan matrix $C = Q_1^{\mathrm{T}}Q_1$ of B is given by $\widetilde{Q_1}^{\mathrm{T}}\widetilde{Q_1}$ up to basic sets.

In order to reduce the number of possibilities for Q_2 we do not only replace Q_2 by Q_2S for some $S \in \mathrm{GL}(k(B) - l(B), \mathbb{Z})$, but also allow transformations of the form $Q_2 \mapsto PQ_2$ where $P \in \mathrm{GL}(k(B), \mathbb{Z})$ is orthogonal. Then $\widetilde{Q_1}$ also becomes $P\widetilde{Q_1}$ and C does not change at all. For example we can take a permutation matrix with signs for P. In other words we freely arrange the order and signs of the rows of the generalized decomposition matrix. With the matrix S above we can realize elementary column operations on Q_2. We will often apply these reductions without an explicit reference. Finally, after we have a list of possible Cartan matrices C for B, we can check if the elementary divisors are correct by computing lower defect groups (see Sect. 1.8). We can decrease the list further by reducing C as a quadratic form.

For the convenience of the reader we repeat the algorithm in a nutshell:

(1) Determine a set $\mathscr{R}$ of representatives for the conjugacy classes of B-subsections by using Lemma 1.34.
(2) Compute the Cartan matrix C_u of b_u for every $(1, B) \neq (u, b_u) \in \mathscr{R}$ by considering the dominated block $\overline{b_u}$ with defect group $\mathrm{C}_D(u)/\langle u \rangle$.
(3) Enumerate the matrices Q_u such that $Q_u^{\mathrm{T}}Q_u = C_u$ for every $(1, B) \neq (u, b_u) \in \mathscr{R}$.
(4) Form the matrix Q_2 consisting of the matrices Q_u for $u \neq 1$.
(5) Find a basis of the $\mathbb{Z}$-module $M := \{x \in \mathbb{Z}^{k(B)} : Q_2^{\mathrm{T}}x = 0\}$ and write the basis elements as columns of $\widetilde{Q_1} \in \mathbb{Z}^{k(B) \times l(B)}$.
(6) The Cartan matrix of B up to basic sets is given by $\widetilde{Q_1}^{\mathrm{T}}\widetilde{Q_1}$.
(7) Check if the elementary divisors are correct by using lower defect groups.
(8) Apply the reduction of quadratic forms.

The idea of this algorithm is not completely new. In fact, Olsson [212, Lemma 3.12] already used this approach. However, according to the author's knowledge, no one applied this algorithm systematically via computer assistance so far. We will do this in Part III. Unfortunately, the computational effort grows quickly for large defect groups. As a rule of thumb, defect groups of order at most 32 are feasible. In a recent diploma thesis [26] a defect group of order 64 was considered. Here however, many cases remained open.

4.3   The Inverse Cartan Method

In this section we present an old result by Brauer which uses the inverse of the Cartan matrix. As usual, B is a p-block of a finite group with defect d.

Theorem 4.4 (Brauer [39])   *Let (u, b_u) be a major B-subsection such that b_u has Cartan matrix $C_u = (c_{ij})$ up to basic sets. Define*

$$q(b_u) := \min\{xp^d C_u^{-1} x^T : 0 \neq x \in \mathbb{Z}^{l(b_u)}\}.$$

Then $k(B)q(b_u) \leq l(b_u)p^d$.

Since all elementary divisors of C_u divide p^d, the matrix $p^d C_u^{-1}$ is integral and positive definite. Thus, the number $q(b_u)$ is an invariant of the (equivalence class of the) quadratic form corresponding to $p^d C_u^{-1}$. At first sight it seems difficult to calculate $q(b_u)$ in praxis. Here the following lemma is quite useful.

Lemma 4.5 (Liebeck [178])   *Assume the notation in Theorem 4.4, and let $x = (x_1, \ldots, x_{l(b_u)}) \in \mathbb{Z}^{l(b_u)}$ such that $xp^d C_u^{-1} x^T \leq m \in \mathbb{N}$. Then*

$$|x_i| \leq \sqrt{\frac{c_{ii}m}{p^d}}$$

for $i = 1, \ldots, l(b_u)$.

So in order to determine $q(b_u)$ one can define m in Lemma 4.5 to be the minimal diagonal entry of $p^d C_u^{-1}$ and check (probably by computer) the defined box for smaller values. There is also a direct command in GAP to determine $q(b_u)$ (in small dimensions). One can show that there is always a basic set for b_u such that $q(b_u)$ is the first entry of $p^d C_u^{-1}$ (this follows from Definition 3.3). However, it is not clear how to construct such a basic set without the knowledge of $q(b_u)$. The combination of Theorems 4.2 and 4.4 is quite powerful as we will see in Part III.

We also add a related result by Robinson which goes in the opposite direction.

Theorem 4.6 (Robinson [231])   *Let (u, b_u) be a major B-subsection such that b_u has Cartan matrix $C_u = (c_{ij})$ up to basic sets. Let $Q(b_u)$ be the set of integers $xp^d C_u^{-1} x^T$ ($x \in \mathbb{Z}^{l(b_u)}$) which are coprime to p. Then for $q'(b_u) := \min Q(b_u)$ we have*

$$\sum_{i=0}^{\infty} k_i(B) p^{2i} \leq p^d q'(b_u).$$

One may ask if the inverse Cartan method (i.e. an application of Theorem 4.4) always gives Brauer's $k(B)$-Conjecture. However, this is not the case as one can see from the following example: Let B be a block with defect group C_2^3, Cartan matrix C and $e(B) = 21$. Then one can choose a basic set such that

$$8C^{-1} = \begin{pmatrix} 4 & 2 & 2 & 2 & 2 \\ 2 & 5 & 1 & 1 & 1 \\ 2 & 1 & 5 & 1 & 1 \\ 2 & 1 & 1 & 5 & 1 \\ 2 & 1 & 1 & 1 & 5 \end{pmatrix}.$$

Hence, $q(B) < l(B)$.

Nevertheless, we like to point out that we do not know a single Cartan matrix such that Brauer's $k(B)$-Conjecture would not follow from Theorem 4.2 or from Theorem 4.4. Since these two results are somehow related, it seems interesting to investigate the following problem: Let $C = (c_{ij}) \in \mathbb{Z}^{l \times l}$ be the Cartan matrix of a p-block with defect d. Assume that for all integral, positive definite quadratic forms $q(x_1, \ldots, x_{l(b_u)}) = \sum_{1 \leq i \leq j \leq l} q_{ij} x_i x_j$ we have

$$\sum_{1 \leq i \leq j \leq l} q_{ij} c_{ij} > p^d.$$

Then prove that $x p^d C^{-1} x^{\mathrm{T}} \geq l$ for all $0 \neq x \in \mathbb{Z}^l$. If this can be done, the $k(B)$-Conjecture would follow in full generality. A diagonal matrix shows that this argument fails for arbitrary positive definite, symmetric matrices C. This illustrates the importance of Question A.

4.4 More Inequalities

The results in this section were taken from [252]. It is obvious that Theorem 4.2 should be stronger for small values of $l(b_u)$. First we focus on major subsections. In the most elementary case we have the following special case of Theorem 4.6.

Proposition 4.7 (Robinson [231]) *Let B be a block of defect d with major subsection (u, b_u) such that $l(b_u) = 1$. Then*

$$\sum_{i=0}^{\infty} k_i(B) p^{2i} \leq p^d.$$

Moreover, in case $u = 1$ there is a result by Olsson.

Proposition 4.8 (Olsson [216]) *If $l(B) \leq 2$, then $k(B) \leq p^d$.*

However, in praxis this implication is not so useful, because usually the knowledge of $l(B)$ already implies the exact value of $k(B)$ (remember that $k(B) - l(B)$ is determined locally). In the following we generalize Olsson's result for arbitrary $u \in Z(D)$.

Theorem 4.9 *Let B be a p-block of a finite group with defect d, and let (z, b_z) be a major subsection such that $l(b_z) \leq 2$. Then one of the following holds:*

(i)

$$\sum_{i=0}^{\infty} k_i(B) p^{2i} \leq p^d.$$

(ii)

$$k(B) \leq \begin{cases} \dfrac{p+3}{2} p^{d-1} & \text{if } p > 2, \\ \dfrac{2}{3} 2^d & \text{if } p = 2. \end{cases}$$

In particular Brauer's $k(B)$-Conjecture holds for B.

Proof In case $l(b_z) = 1$, (i) holds. Hence, let $l(b_z) = 2$, and let $C_z = (c_{ij})$ be the Cartan matrix of b_z up to basic sets. We consider the number

$$q(b_z) := \min\{x p^d C_z^{-1} x^{\mathrm{T}} : 0 \neq x \in \mathbb{Z}^{l(b_z)}\} \in \mathbb{N}$$

as in Theorem 4.4. If $q(b_z) = 1$, (i) follows from Theorem 4.6. Therefore, we may assume $q(b_z) \geq 2$. Then Brauer's $k(B)$-Conjecture already holds by Theorem 4.4, but we want to obtain the stronger bound (ii). Since p^d is always an elementary divisor of C_z, we see that C_z is not a diagonal matrix. This allows us to apply Theorem 4.2. All entries of C_z are divisible by the smallest elementary divisor $\gamma := p^{-d} \det C_z$. Hence, we may consider the integral matrix $\tilde{C}_z = (\tilde{c}_{ij}) := \gamma^{-1} C_z$. After changing the basic set, we may assume that $0 < 2\tilde{c}_{12} \leq \tilde{c}_{11} \leq \tilde{c}_{22}$. Then

$$\tilde{c}_{11} + \tilde{c}_{22} \leq \frac{5}{4}\tilde{c}_{11} + \frac{\det \tilde{C}_z}{\tilde{c}_{11}} \leq \frac{p^d}{2\gamma} + \frac{5}{2}$$

by Eq. (3.1) on page 28. Now Theorem 4.2 leads to

$$k(B) \leq \gamma(\tilde{c}_{11} + \tilde{c}_{22} - \tilde{c}_{12}) \leq \frac{p^d + 3\gamma}{2}.$$

Since $\gamma \leq p^{d-1}$, we get (ii) for p odd. It remains to consider the case $p = 2$. If $\tilde{c}_{11} = 2$, we must have $\tilde{c}_{12} = 1$. Hence, under these circumstances $p > 2$, since otherwise $\det \tilde{C}_z$ is not a p-power. Now assume $\tilde{c}_{11} \geq 3$ and $p = 2$. Since

$$p^d C_z^{-1} = \frac{p^d}{\gamma} \tilde{C}_z^{-1} = \begin{pmatrix} \tilde{c}_{22} & -\tilde{c}_{12} \\ -\tilde{c}_{12} & \tilde{c}_{11} \end{pmatrix},$$

we have $q(b_z) \geq 3$. Now Theorem 4.4 implies (ii). □

It is conjectured that the matrix C_z for $l(b_z) \geq 2$ in the proof of Theorem 4.9 cannot have diagonal shape (this holds for p-solvable groups by Lemma 3.1). Hence for $l(b_z) = 2$, Theorem 4.9(ii) might always apply. Then $k(B) < p^d$ unless $p = 3$.

Olsson [216] also proved the implication

$$l(B) \leq 3 \Longrightarrow k(B) \leq p^d$$

whenever $p = 2$. We also generalize this result to arbitrary major subsections. Suppose as before that (z, b_z) is a major subsection. We denote the corresponding part of the generalized decomposition matrix by $D_z := (d_{\chi\varphi}^z : \chi \in \mathrm{Irr}(B), \varphi \in \mathrm{IBr}(b_z))$. In case $|\langle z \rangle| \leq 2$, it can be seen easily that the contribution matrix M^z is integral. Then most proofs of [216] remain true without any changes. This was more or less done in [229] (compare also with Corollary 3.5 in [231]). In the general case we have to put a bit more effort into the proof.

Theorem 4.10 *Let B be a 2-block of a finite group with defect d, and let (z, b_z) be a major B-subsection such that $l(b_z) \leq 3$. Then*

$$k(B) \leq k_0(B) + \frac{2}{3} \sum_{i=1}^{\infty} 2^i k_i(B) \leq 2^d.$$

In particular Brauer's $k(B)$-Conjecture is satisfied for B.

Proof Observe that by construction $m_{\chi\chi}^z$ is a positive real number for every $\chi \in \mathrm{Irr}(B)$, since C_z is positive definite. Let $\chi \in \mathrm{Irr}_0(B)$, and let $|\langle z \rangle| = 2^n$. In case $n \leq 1$ the proof is much easier. For this reason we assume $n \geq 2$. We write

$$m_{\chi\chi}^z = \sum_{j=0}^{2^{n-1}-1} a_j(\chi)\zeta^j$$

with $\zeta := e^{2\pi i/2^n}$ and $a_j(\chi) \in \mathbb{Z}$ for $j = 0, \ldots, 2^{n-1} - 1$. As usual, the Galois group $\mathcal{G}$ of the 2^n-th cyclotomic field acts on $\mathrm{Irr}(B)$, on the rows of D_z, and thus also on M_z in an obvious manner. Let Γ be the orbit of χ under $\mathcal{G}$. Set $m := |\Gamma|$.

Then we have

$$ma_0(\chi) = \sum_{\psi \in \Gamma} m_{\psi\psi}^z > 0.$$

Assume first that $a_0(\chi) = 1$. Since $M^z \overline{M^z}^{\mathrm{T}} = M^z M^z = 2^d M^z$, it follows that

$$m2^d = \sum_{\substack{\psi \in \Gamma, \\ \tau \in \mathrm{Irr}(B)}} |m_{\psi\tau}^z|^2.$$

Applying Galois theory gives

$$\prod_{\substack{\psi \in \Gamma, \\ \tau \in \mathrm{Irr}_i(B)}} |m_{\psi\tau}^z|^2 \in \mathbb{Q}$$

for all $i \geq 0$. By Proposition 1.36 we also know $v(m_{\psi\tau}^z) = h(\tau)$ where v is the 2-adic valuation and $\psi \in \Gamma$. Hence, also the numbers $m_{\psi\tau}^z 2^{-h(\tau)}$ are algebraic integers. This implies

$$\mathbb{Z} \ni \prod_{\substack{\psi \in \Gamma, \\ \tau \in \mathrm{Irr}_i(B)}} 2^{-2i} |m_{\psi\tau}^z|^2 \geq 1.$$

Now using the inequality of arithmetic and geometric means we obtain

$$\sum_{\substack{\psi \in \Gamma, \\ \tau \in \mathrm{Irr}_i(B)}} |m_{\psi\tau}^z|^2 \geq m2^{2i} k_i(B)$$

for all $i \geq 0$. Summing over i gives

$$m2^d = \sum_{\substack{\psi \in \Gamma, \\ \tau \in \mathrm{Irr}(B)}} |m_{\psi\tau}^z|^2 \geq m \sum_{i=0}^{\infty} 2^{2i} k_i(B)$$

which is even more than we wanted to prove.

Hence, we can assume that $a_0(\chi) \geq 2$ for all $\chi \in \mathrm{Irr}(B)$ such that $h(\chi) = 0$. It is well-known that the ring of integers of $\mathbb{Q}(\zeta) \cap \mathbb{R}$ has basis 1, $\zeta^j + \zeta^{-j} = \zeta^j - \zeta^{2^{n-1}-j}$ for $j = 1, \ldots, 2^{n-2} - 1$. In particular the numbers $a_j(\chi)$ for $j \geq 1$ come in pairs modulo 2. Since $v(m_{\chi\chi}^z) = 0$, we even have $a_0(\chi) \geq 3$. For an arbitrary character $\psi \in \mathrm{Irr}(B)$ of positive height we already know that $m_{\psi\psi}^z 2^{-h(\psi)}$ is a positive algebraic integer. Hence, $2^{h(\psi)} \mid a_j(\psi)$ for all $j \geq 0$. By Proposition 1.36 we have $v(m_{\psi\psi}^z) > h(\psi)$. Thus, we even have $2^{h(\psi)+1} \mid a_0(\psi)$. As above we also have $a_0(\psi) > 0$. This implies $\sum_{\psi \in \mathrm{Irr}_i(B)} m_{\psi\psi}^z \geq 2^{i+1} k_i(B)$ for $i \geq 1$ via Galois theory. Using $\mathrm{tr}\, M^z = 2^d l(b_z)$ it follows that

$$3 \cdot 2^d \geq \sum_{\psi \in \mathrm{Irr}(B)} m_{\psi\psi}^z \geq 3k_0(B) + \sum_{i=1}^{\infty} 2^{i+1} k_i(B).$$

This proves the claim. □

We remark that in Theorem 6(ii) in [216] it should read $l(B) \leq p^2 - 1$ (compare with Theorem 6*(ii)).

It is easy to see that the proof of Theorem 4.10 can be generalized to the following.

Proposition 4.11 *Let B be a 2-block of a finite group with defect d, and let (z, b_z) be a major B-subsection. Then for every odd number α one of the following holds:*

(i) $\sum_{i=0}^{\infty} 2^{2i} k_i(B) \leq 2^d \alpha,$

(ii) $(\alpha + 2)k_0(B) + \sum_{i=1}^{\infty} 2^{i+1} k_i(B) \leq 2^d l(b_z).$

Proof As in Theorem 4.10 let $\chi \in \mathrm{Irr}_0(B)$ and define $a_0(\chi)$ similarly. In case $a_0(\chi) \leq \alpha$ the first inequality applies. Otherwise the second inequality applies. □

Observe that Proposition 4.11 also covers (a generalization of) Theorem 8 in [216] for $p = 2$.

We now turn to arbitrary subsections. If in the situation of Theorem 4.2 the Cartan matrix is not known, one can apply the following theorem by Robinson.

Theorem 4.12 (Robinson [232]) *Let (u, b_u) be a B-subsection. If b_u has defect d, then $k_0(B) \leq p^d \sqrt{l(b_u)}$.*

We are going to improve this result for $p = 2$.

Theorem 4.13 *Let B be a 2-block of a finite group, and let (u, b_u) be a B-subsection such that b_u has defect q. Set $\alpha := \lfloor \sqrt{l(b_u)} \rfloor$ if $\lfloor \sqrt{l(b_u)} \rfloor$ is odd and $\alpha := \dfrac{l(b_u)}{\lfloor \sqrt{l(b_u)} \rfloor + 1}$ otherwise. Then $k_0(B) \leq \alpha 2^q$. In particular $k_0(B) \leq 2^q$ if $l(b_u) \leq 3$.*

Proof By Proposition 1.36 we still have $m_{\chi\psi}^u \neq 0$ as long as $h(\chi) = h(\psi) = 0$. However, in all other cases it is possible that $m_{\chi\psi}^u = 0$. So we can copy the proof of Theorem 4.10 by leaving out the characters of positive height. This gives $k_0(B) \leq \alpha 2^q$ or $k_0(B) \leq 2^q l(b_u)/(\alpha + 2)$ for every odd number α. If $\lfloor \sqrt{l(b_u)} \rfloor$ is odd, we choose $\alpha := \lfloor \sqrt{l(b_u)} \rfloor$. Otherwise we take $\alpha := \lfloor \sqrt{l(b_u)} \rfloor - 1$. The result follows. □

Finally, we generalize the "dual" inequalities in [216]. For this let $\tilde{M}^z := (\tilde{m}_{\chi\psi}^z) = 2^d 1_{k(B)} - M^z$.

Proposition 4.14 *Let B be a 2-block of a finite group with defect d, and let* (z, b_z) *be a major B-subsection. Then for every odd number* α *one of the following holds:*

(i) $\sum_{i=0}^{\infty} 2^{2i} k_i(B) \le 2^d \alpha$,

(ii) $(\alpha + 2)k_0(B) + \sum_{i=1}^{\infty} 2^{i+1} k_i(B) \le 2^d (k(B) - l(b_z))$.

In particular Brauer's $k(B)$*-Conjecture holds if* $k(B) - l(b_z) \le 3$.

Proof By Lemma V.9.3 in [81] the numbers $\tilde{m}_{\chi\chi}^z$ for $\chi \in \mathrm{Irr}(B)$ are still real, positive algebraic integers. As in Theorem 4.10 we may assume $|\langle z \rangle| = 2^n \ge 4$. Let us write

$$\tilde{m}_{\chi\chi}^z = \sum_{j=0}^{2^{n-1}-1} a_j(\chi)\zeta^j$$

with $\chi \in \mathrm{Irr}_0(B)$, $\zeta := e^{2\pi i/2^n}$ and $a_j(\chi) \in \mathbb{Z}$ for $j = 0, \ldots, 2^{n-1} - 1$. The Galois group still acts on $\tilde{M}^z$. Also the equation $\tilde{M}^z \tilde{M}^z = 2^d \tilde{M}^z$ remains true. For $\tau \in \mathrm{Irr}(B)$ we have $\nu(\tilde{m}_{\chi\tau}^z) = \nu(2^d - m_{\chi\tau}^z) = \nu(m_{\chi\tau}^z) = h(\tau)$. Hence, in case $a_0(\chi) \le \alpha$ we can carry over the arguments in Theorem 4.10.

Now assume that $a_0(\chi) > \alpha$ for all characters $\chi \in \mathrm{Irr}_0(B)$. Here the proof works also quite similar as in Theorem 4.10. In fact for a character $\psi \in \mathrm{Irr}(B)$ of positive height we have $\nu(\tilde{m}_{\psi\psi}^z) = \nu(2^d - m_{\psi\psi}^z) \ge \min(\nu(2^d), \nu(m_{\psi\psi}^z)) > h(\psi)$ by Proposition 1.36. Moreover, $\mathrm{tr}\, \tilde{M}^z = 2^d (k(B) - l(B))$. The claim follows.   □

It should be pointed out that usually $k(B) - l(B) = k(B) - l(b_1) \le k(B) - l(b_z)$ for a major subsection (z, b_z) (this holds for example if z lies in the center of the fusion system of B, see Theorem 1.38). However, this is not true in general as we will see in Theorem 13.2. Another problem is that $k(B) - l(b_z)$ for $z \ne 1$ is not locally determined (in contrast to $k(B) - l(B)$). By combining with Proposition 4.11 we can replace (ii) in the last proposition by

$$(\alpha + 2)k_0(B) + \sum_{i=1}^{\infty} 2^{i+1} k_i(B) \le 2^d \min(l(b_z), k(B) - l(b_z)).$$

Chapter 5
A Bound in Terms of Fusion Systems

In this chapter we obtain more inequalities on the invariants of a block by using local data. This time the fusion system of the block plays a role. The exposition appeared in [114].

Brauer proved Olsson's Conjecture for 2-blocks with dihedral defect groups using a Galois action on the generalized decomposition numbers (see [41]). We put his approach into an abstract framework. Let B be a p-block of a finite group G with defect group D, and let (u, b_u) be a subsection for B. Let p^k be the order of u, and let $\zeta := \zeta_{p^k}$ be a primitive p^k-th root of unity. Then there exist integral vectors $a_i^\varphi := (a_i^\varphi(\chi))_{\chi \in \mathrm{Irr}(B)} \in \mathbb{Z}^{k(B)}$ such that

$$d_{\chi\varphi}^u = \sum_{i=0}^{\varphi(p^k)-1} a_i^\varphi(\chi)\zeta^i \tag{5.1}$$

(see Sect. 1.6).

Let $\mathscr{G}$ be the Galois group of the cyclotomic field $\mathbb{Q}(\zeta)$ over $\mathbb{Q}$. Then $\mathscr{G} \cong \mathrm{Aut}(\langle u \rangle) \cong (\mathbb{Z}/p^k\mathbb{Z})^\times$ and we will often identify these groups. We will also interpret the elements of $\mathscr{G}$ as integers in $\{1, \ldots, p^k\}$ by a slight abuse of notation. Then (u^γ, b_u) for $\gamma \in \mathscr{G}$ is also a *(algebraically conjugate)* subsection and

$$\gamma(d_{\chi\varphi}^u) = d_{\chi\varphi}^{u^\gamma} = \sum_{i=0}^{\varphi(p^k)-1} a_i^\varphi(\chi)\zeta^{i\gamma}.$$

Now the situation splits naturally into characteristic 2 and odd characteristic, since the structure of the corresponding Galois groups differs significantly.

© Springer International Publishing Switzerland 2014   47
B. Sambale, *Blocks of Finite Groups and Their Invariants*, Lecture Notes
in Mathematics 2127, DOI 10.1007/978-3-319-12006-5_5

5.1   The Case $p = 2$

Let $p = 2$, and let $\mathscr{F}$ be the fusion system of B. Then by Lemma 1.34 we may assume that $\langle u \rangle$ is fully $\mathscr{F}$-normalized and $C_D(u)$ is a defect group of b_u. As before, $\langle u \rangle$ is also fully $\mathscr{F}$-centralized and

$$\mathrm{Aut}_{\mathscr{F}}(\langle u \rangle) = \mathrm{Aut}_D(\langle u \rangle) = N_D(\langle u \rangle)/C_D(u).$$

We begin with a refinement of the orthogonality relations. For a subsection (u, b_u) with $\mathrm{IBr}(b_u) = \{\varphi\}$ we set $a_i := a_i^{\varphi}$ for all i. Moreover, if $u, v \in D$ are conjugate in D, we write $u \sim_D v$.

Proposition 5.1 *Let B be a 2-block of a finite group with defect group D and fusion system $\mathscr{F}$. Let (u, b_u) be a B-subsection such that $l(b_u) = 1$ and $\langle u \rangle \neq 1$ is fully $\mathscr{F}$-normalized of order 2^k. Then*

$$(a_i, a_j) = \begin{cases} 2|N_D(\langle u \rangle) \cap C_D(u^i)/\langle u \rangle| & \text{if } u^j \sim_D u^i \sim_D u^{j+2^{k-1}}, \\ -2|N_D(\langle u \rangle) \cap C_D(u^i)/\langle u \rangle| & \text{if } u^j \nsim_D u^i \sim_D u^{j+2^{k-1}}, \\ 0 & \text{otherwise} \end{cases}$$

for $i, j \in \{0, \ldots, 2^{k-1} - 1\}$. In particular, $(a_0, a_0) = 2|N_D(\langle u \rangle)/\langle u \rangle|$.

Proof We set $d^u := (d_{\chi\varphi}^u : \chi \in \mathrm{Irr}(B))$ and $|N_D(\langle u \rangle) \cap C_D(u^i)/C_D(u)| = 2^r$. Then

$$\frac{1}{2^{k-1}} \sum_{\gamma \in \mathscr{G}} d^{u^{\gamma}} \zeta^{-i\gamma} = \frac{1}{2^{k-1}} \sum_{l=0}^{2^{k-1}-1} \sum_{\gamma \in \mathscr{G}} a_l \zeta^{(l-i)\gamma} = a_i$$

for $i = 0, \ldots, 2^{k-1} - 1$. Hence,

$$(a_i, a_j) = 2^{2(1-k)} \sum_{\gamma, \delta \in \mathscr{G}} \left(d^{u^{\gamma}}, d^{u^{\delta}} \right) \zeta^{j\delta - i\gamma}.$$

If u^{γ} and u^{δ} are conjugate under $\mathrm{Aut}_{\mathscr{F}}(\langle u \rangle)$, we have $\left(d^{u^{\gamma}}, d^{u^{\delta}} \right) = 2^d$ by Theorem 1.14. If we regard $\mathrm{Aut}_{\mathscr{F}}(\langle u \rangle)$ as a subgroup of $\mathscr{G}$, this means $\gamma\delta^{-1} \in \mathrm{Aut}_{\mathscr{F}}(\langle u \rangle)$. Therefore,

$$(a_i, a_j) = 2^{2(1-k)+d} \sum_{\gamma \in \mathscr{G}} \sum_{\delta \in \mathrm{Aut}_{\mathscr{F}}(\langle u \rangle)} \zeta^{(j\delta - i)\gamma} = 2^{2(1-k)+d} \sum_{\delta \in \mathrm{Aut}_{\mathscr{F}}(\langle u \rangle)} \sum_{\gamma \in \mathscr{G}} \zeta^{(j\delta - i)\gamma}.$$

Observe that if $|\langle u^i \rangle| \neq |\langle u^j \rangle|$, then $(a_i, a_j) = 0$. If u^i is $\mathscr{F}$-conjugate to u^j, then there is a $\delta \in \mathrm{Aut}_{\mathscr{F}}(\langle u \rangle)$ such that $j\delta - i \equiv 0 \pmod{2^k}$. In this case there are precisely 2^r such elements and the corresponding sum contributes 2^{r+k-1}. Similarly, if u^i is $\mathscr{F}$-conjugate to $u^{j+2^{k-1}}$, we get the contribution -2^{r+k-1} in the sum. All other summands vanish. This shows the result. □

Theorem 5.2 *Let B be a 2-block of a finite group G with defect group D and fusion system $\mathscr{F}$, and let (u, b_u) be a B-subsection such that $\langle u \rangle \neq 1$ is fully $\mathscr{F}$-normalized and b_u has Cartan matrix $C_u = (c_{ij})$. Let $\mathrm{IBr}(b_u) = \{\varphi_1, \ldots, \varphi_{l(b_u)}\}$ such that $\varphi_1, \ldots, \varphi_m$ are stable under $\mathrm{N}_D(\langle u \rangle)$ and $\varphi_{m+1}, \ldots, \varphi_{l(b_u)}$ are not. Then $m \geq 1$. Suppose further that u is conjugate to u^{-5^n} for some $n \in \mathbb{Z}$ in D. Then*

$$k_0(B) \leq \frac{2|\mathrm{N}_D(\langle u \rangle)/\mathrm{C}_D(u)|}{|\langle u \rangle|} \sum_{1 \leq i \leq j \leq m} q_{ij} c_{ij} \tag{5.2}$$

for every positive definite, integral quadratic form $q(x_1, \ldots, x_m) = \sum_{1 \leq i \leq j \leq m} q_{ij} x_i x_j$. In particular if $l(b_u) = 1$, we get

$$k_0(B) \leq 2|\mathrm{N}_D(\langle u \rangle)/\langle u \rangle|. \tag{5.3}$$

Proof Let $\chi \in \mathrm{Irr}_0(B)$ and $|\langle u \rangle| = 2^k$ for some $k \geq 1$. We write $d_\chi^u := (d_{\chi\varphi_1}^u, \ldots, d_{\chi\varphi_l}^u)$, where $l := l(b_u)$. Then

$$d_{\chi\varphi_i}^u \equiv \gamma(d_{\chi\varphi_i}^u) \equiv \sum_{j=0}^{2^{k-1}-1} a_j^i(\chi) \pmod{\mathrm{Rad}\,\mathcal{O}}$$

for $\gamma \in \mathscr{G}$. In particular $d_{\chi\varphi_i}^u \equiv \overline{d_{\chi\varphi_i}^u} \pmod{\mathrm{Rad}\,\mathcal{O}}$. We write $|\mathrm{C}_D(u)|C_u^{-1} = (\tilde{c}_{ij})$. Then it follows from Proposition 1.36 that

$$0 \neq m_{\chi\chi}^u \equiv \sum_{1 \leq i,j \leq l} \tilde{c}_{ij} d_{\chi\varphi_i}^u \overline{d_{\chi\varphi_j}^u} \equiv \sum_{1 \leq i \leq l} \tilde{c}_{ii} (d_{\chi\varphi_i}^u)^2$$

$$\equiv \sum_{1 \leq i \leq l} \tilde{c}_{ii} \sum_{j=0}^{2^{k-1}-1} a_j^i(\chi)^2 \equiv \sum_{1 \leq i \leq l} \tilde{c}_{ii} \sum_{j=0}^{\varphi(2^k)-1} a_j^i(\chi) \pmod{\mathrm{Rad}\,\mathcal{O}}.$$

Now every $g \in \mathrm{N}_D(\langle u \rangle)$ induces a permutation on $\mathrm{IBr}(b_u)$. Let P_g be the corresponding permutation matrix. Then g also acts on the rows $d_i^u := (d_{\chi\varphi_i}^u : \chi \in \mathrm{Irr}(B))$ for $i = 1, \ldots, l$, and it follows that $C_u P_g = P_g C_u$. Hence, we also have $C_u^{-1} P_g = P_g C_u^{-1}$ for all $g \in \mathrm{N}_D(\langle u \rangle)$. If $\{\varphi_{m_1}, \ldots, \varphi_{m_2}\}$ ($m < m_1 < m_2 \leq l$) is an orbit under $\mathrm{N}_D(\langle u \rangle)$, it follows that $d_{\chi\varphi_{m_1}}^u \equiv \ldots \equiv d_{\chi\varphi_{m_2}}^u \pmod{\mathrm{Rad}\,\mathcal{O}}$ and $\tilde{c}_{m_1 m_1} = \ldots = \tilde{c}_{m_2 m_2}$. Since the length of this orbit is even, we get

$$\sum_{1 \leq i \leq m} \tilde{c}_{ii} \sum_{j=0}^{2^{k-1}-1} a_j^i(\chi) \not\equiv 0 \pmod{2}.$$

In particular, $m \geq 1$. In case $|\langle u \rangle| = 2$ this simplifies to

$$\sum_{1 \leq i \leq m} \tilde{c}_{ii} a_0^i(\chi) \not\equiv 0 \ (\text{mod } 2).$$

We show that this holds in general. Thus, let $k \geq 2$ and $i \in \{1, \ldots, m\}$. Since (u, b_u) is conjugate to (u^{-5^n}, b_u) and φ_i is stable, we have

$$\sum_{j=0}^{2^{k-1}-1} a_j^i(\chi) \zeta^j = d_{\chi \varphi_i}^u = d_{\chi \varphi_i}^{u^{-5^n}} = \sum_{j=0}^{2^{k-1}-1} a_j^i(\chi) \zeta^{-5^n j}.$$

Moreover, for every $j \in \{0, \ldots, 2^{k-1} - 1\}$ there is some $j_1 \in \{0, \ldots, \varphi(2^k) - 1\}$ such that $\zeta^{-5^n j} = \pm \zeta^{j_1}$. In order to compare coefficients observe that

$$\zeta^j = \zeta^{-5^n j} \implies j \equiv -5^n j \ (\text{mod } 2^k) \implies 1 \equiv -5^n \ (\text{mod } 2^k / \gcd(2^k, j))$$
$$\implies j = 0.$$

Hence, the set $\{\pm \zeta^j : j = 1, \ldots, 2^{k-1} - 1\}$ splits under the action of $\langle -5^n + 2^k \mathbb{Z} \rangle$ into orbits of even length. This shows $\sum_{j=0}^{2^{k-1}-1} a_j^i(\chi) \equiv a_0^i(\chi) \ (\text{mod } 2)$. Hence,

$$\sum_{1 \leq i \leq m} \tilde{c}_{ii} a_0^i(\chi) \not\equiv 0 \ (\text{mod } 2) \tag{5.4}$$

for every $\chi \in \mathrm{Irr}_0(B)$. In particular, there is an $i \in \{1, \ldots, m\}$ such that $a_0^i(\chi) \neq 0$. This gives

$$k_0(B) \leq \sum_{1 \leq i \leq j \leq m} q_{ij}(a_0^i, a_0^j)$$

(see proof of Theorem 4.2).

Now let k again be arbitrary. Observe that $a_0^i = 2^{1-k} \sum_{\gamma \in \mathscr{G}} \gamma(d_i^u)$ for $i \in \{1, \ldots, m\}$. By the orthogonality relations for generalized decomposition numbers we have $(d_i^{u^\gamma}, d_j^{u^\delta}) = c_{ij}$ for $\gamma, \delta \in \mathscr{G}$ if u^γ and u^δ are conjugate under $\mathrm{N}_D(\langle u \rangle)$. Otherwise we have $(d_i^{u^\gamma}, d_j^{u^\delta}) = 0$. This implies

$$(a_0^i, a_0^j) = 2^{2(1-k)} \sum_{\gamma, \delta \in \mathscr{G}} (d_i^{u^\gamma}, d_j^{u^\delta}) = \frac{2|\mathrm{N}_D(\langle u \rangle)/ \mathrm{C}_D(u)|}{2^k} c_{ij},$$

and (5.2) follows. In case $l = 1$ we have $C = (|\mathrm{C}_D(u)|)$, and (5.3) is also clear. $\quad \square$

In the situation of Theorem 5.2 we have $u \in Z(C_G(u))$. Hence, all Cartan invariants c_{ij} are divisible by $|\langle u \rangle|$. This shows that the right hand side of (5.2) is always an integer. It is also known that $k_0(B)$ is divisible by 4 unless $|D| \leq 2$.

Observe that the subsection (u, b_u) in Theorem 5.2 cannot be major unless $|\langle u \rangle| \leq 2$, since then u would be contained in $Z(D)$.

If D is rational of nilpotency class (at most) 2, Gluck's Conjecture would imply $m = l(b_u)$ in Theorem 5.2. In this case it suffices to know the Cartan matrix C_u only up to basic sets. For, changing the basic set is essentially the same as taking another quadratic form q (see [172]). This must always hold in case $l(b_u) = 2$. Here we get the following simpler result.

Theorem 5.3 *Let* $p = 2$, *and let* (u, b_u) *be a B-subsection such that* $\langle u \rangle$ *is fully* $\mathscr{F}$*-normalized and u is conjugate to* u^{-5^n} *for some* $n \in \mathbb{Z}$ *in* D. *If* $l(b_u) \leq 2$, *then*

$$k_0(B) \leq 2|N_D(\langle u \rangle)/\langle u \rangle|.$$

Proof We use the notation of the proof of Theorem 5.2. We may assume that $l = 2 = m$. Here we can use (5.4) in a stronger sense. Since $|C_D(u)|$ occurs as elementary divisor of C_u exactly once, we see that the rank of $\frac{|C_D(u)|}{\det C_u} C_u$ (mod 2) is 1. Hence, $\frac{|C_D(u)|}{\det C_u} C_u$ (mod 2) has the form

$$\begin{pmatrix} 1 & 0 \\ 0 & 0 \end{pmatrix} \text{ (mod 2)}, \qquad \begin{pmatrix} 0 & 0 \\ 0 & 1 \end{pmatrix} \text{ (mod 2)}, \qquad \text{or} \qquad \begin{pmatrix} 1 & 1 \\ 1 & 1 \end{pmatrix} \text{ (mod 2)}.$$

Now it is easy to see that we may change the basic set for b_u such that $|C_D(u)|c_{11}/\det C_u$ is even and as small as possible. Then we also have to replace the rows d_1^u and d_2^u by linear combinations of each other. This gives rows $\hat{d}_i^u$ and $\hat{a}_j^i$ for $i = 1, 2$ and $j = 0, \ldots, \varphi(2^k) - 1$. Observe that the contributions do not depend on the basic set for C_u. Moreover, $\tilde{c}_{11}$ is odd and $\tilde{c}_{22}$ is even. Hence, (5.4) takes the form

$$\hat{a}_0^1(\chi) \not\equiv 0 \text{ (mod 2)}$$

for all $\chi \in \mathrm{Irr}_0(B)$. Since both φ_1 and φ_2 are stable under $N_D(\langle u \rangle)$, we have $\gamma(\hat{d}_1^u) = \hat{d}_1^u$ for all $\gamma \in \mathrm{Aut}_{\mathscr{F}}(\langle u \rangle)$. Hence,

$$k_0(B) \leq (\hat{a}_0^1, \hat{a}_0^1) = \frac{|N_D(\langle u \rangle)/ C_D(u)|c_{11}}{\varphi(2^k)}$$

as above. It remains to show that $c_{11} \leq |C_D(u)|$. The reduction theory of quadratic forms gives an equivalent matrix $C_u' = (c_{ij}')$ such that $0 \leq 2c_{12}' \leq \min(c_{11}', c_{22}')$ (see Chap. 3). In case $c_{12}' = 0$ we may assume $c_{11} \leq c_{11}' = |C_D(u)|$, since $|C_D(u)|$ is the largest elementary divisor of C_u'. Hence, let $c_{12}' > 0$. Since the entries of C_u and thus also of C_u' are divisible by $\alpha := \det C_u / |C_D(u)|$, we even have $c_{12}' \geq \alpha$. It follows that

$$3\alpha^2 \leq 3(c_{12}')^2 \leq c_{11}' c_{22}' - (c_{12}')^2 = \det C_u' \leq \frac{|C_D(u)|^2}{2}$$

and $\alpha \leq |C_D(u)|/4$. From Eq. (3.1) on page 28 we obtain

$$\max(c_{11}', c_{22}') \leq c_{11}' + c_{22}' - c_{12}' \leq c_{11}' + c_{22}' - \alpha$$

$$\leq \alpha \frac{|C_D(u)|/\alpha + 3}{2} = \frac{|C_D(u)| + 3\alpha}{2} \leq |C_D(u)|.$$

If $\alpha^{-1} c_{11}'$ or $\alpha^{-1} c_{22}'$ is even, the result follows from the minimality of c_{11}. Otherwise we replace C_u' by

$$\begin{pmatrix} 1 & -1 \\ 0 & 1 \end{pmatrix} C_u' \begin{pmatrix} 1 & 0 \\ -1 & 1 \end{pmatrix} = \begin{pmatrix} c_{11}' + c_{22}' - 2c_{12}' & c_{12}' - c_{22}' \\ c_{12}' - c_{22}' & c_{22}' \end{pmatrix}.$$

Then $c_{11} \leq c_{11}' + c_{22}' - 2c_{12}' \leq |C_D(u)|$. This finishes the proof. $\square$

If in the situation of Theorem 5.2 we have $m < l(b_u)$, we really need to know the "exact" Cartan matrix C_u which is unknown in most cases. For $p > 2$ there are not always stable characters in $\mathrm{IBr}(b_u)$ (see Proposition (2E)(ii) and the example following it in [154]).

Let us come back to our initial example. Let D be a (non-abelian) 2-group of maximal class. Then there is an element $x \in D$ such that $|D : \langle x \rangle| = 2$ and x is conjugate to x^{-5^n} for some $n \in \{0, |\langle x \rangle|/8\}$ under D. Since $\langle x \rangle \trianglelefteq D$, the subgroup $\langle x \rangle$ is fully $\mathscr{F}$-normalized, and b_x has cyclic defect group $C_D(x) = \langle x \rangle$. Since, $e(b_x) = 1$, we get $l(b_x) = 1$. Hence, Theorem 5.2 shows Olsson's Conjecture $k_0(B) \leq 4 = |D : D'|$. This was already proved in [41, 212].

5.2   The Case $p > 2$

Now we turn to the case where B is a p-block of G for an odd prime p. We fix some notation for this section. As before (u, b_u) is a B-subsection such that $|\langle u \rangle| = p^k$. Moreover, $\zeta \in \mathbb{C}$ is a primitive p^k-th root of unity. Since the situation is more complicated for odd primes, we assume further that $l(b_u) = 1$. We write $\mathrm{IBr}(b_u) = \{\varphi_u\}$. Then the generalized decomposition numbers $d_{\chi \varphi_u}^u$ for $\chi \in \mathrm{Irr}(B)$ form a column $d(u)$. Let d be the defect of b_u. Since $u \in Z(C_G(u))$, u is contained in every defect group of b_u. In particular, $k \leq d$. As in the case $p = 2$ we can write

$$d(u) = \sum_{i=0}^{\varphi(p^k)-1} a_i^u \zeta^i$$

with $a_i^u \in \mathbb{Z}^{k(B)}$ (change of notation!). We define the following matrix

$$A := \big(a_i^u(\chi) : i = 0, \ldots, \varphi(p^k) - 1, \ \chi \in \text{Irr}(B)\big) \in \mathbb{Z}^{\varphi(p^k) \times k(B)}.$$

The proof of the main theorem of this section is an application of the next proposition.

Proposition 5.4 *For every positive definite, integral quadratic form*

$$q(x_1, \ldots, x_{\varphi(p^k)}) = \sum_{1 \leq i \leq j \leq \varphi(p^k)} q_{ij} x_i x_j$$

we have

$$k_0(B) \leq \sum_{1 \leq i \leq j \leq \varphi(p^k)} q_{ij}(a_{i-1}^u, a_{j-1}^u). \tag{5.5}$$

If (u, b_u) is major, we can replace $k_0(B)$ by $\sum_{i=0}^{\infty} p^{2i} k_i(B)$ in (5.5).

Proof By Lemma 1.37(i) every column $a^u(\chi)$ of A corresponding to a character χ of height 0 does not vanish. Hence, we have

$$k_0(B) \leq \sum_{\chi \in \text{Irr}(B)} q(a^u(\chi)) = \sum_{\chi \in \text{Irr}(B)} \sum_{1 \leq i \leq j \leq \varphi(p^k)} q_{ij} a_{i-1}^u(\chi) a_{j-1}^u(\chi)$$

$$= \sum_{1 \leq i \leq j \leq \varphi(p^k)} q_{ij}(a_{i-1}^u, a_{j-1}^u).$$

If (u, b_u) is major and $\chi \in \text{Irr}(B)$, then $p^{-h(\chi)} a^u(\chi)$ is a non-vanishing integral column by Lemma 1.37(ii). In this case we have

$$\sum_{i=0}^{\infty} p^{2i} k_i(B) \leq \sum_{\chi \in \text{Irr}(B)} p^{2h(\chi)} q(p^{-h(\chi)} a^u(\chi)) = \sum_{1 \leq i \leq j \leq \varphi(p^k)} q_{ij}(a_{i-1}^u, a_{j-1}^u).$$

The second claim follows. $\square$

Notice that we have used only a weak version of Lemma 1.37 in the proof above.

In order to find a suitable quadratic form it is often very useful to replace A by UA for some integral matrix $U \in \mathrm{GL}(\varphi(p^k), \mathbb{Q})$ (observe that the argument in the proof of Proposition 5.4 remains correct).

However, we need a more explicit expression of the scalar products (a_i^u, a_j^u). For this reason we introduce an auxiliary lemma about inverses of Vandermonde matrices. Let $\mathscr{G} = \{\sigma_1, \ldots, \sigma_{\varphi(p^k)}\}$. For an integer $i \in \mathbb{Z}$ there is $i' \in \{1, \ldots, p^{k-1}\}$ such that $-i \equiv i' \pmod{p^{k-1}}$. We will use this notation for the rest of the section.

Lemma 5.5 *The inverse of the Vandermonde matrix* $V := \left(\sigma_i(\zeta)^{j-1}\right)_{i,j=1}^{\varphi(p^k)}$ *is given by*

$$V^{-1} = p^{-k}\left(\sigma_j(t_{i-1})\right)_{i,j=1}^{\varphi(p^k)},$$

where $t_i = \zeta^{-i} - \zeta^{i'}$.

Proof For $i, j \in \{0, \ldots, \varphi(p^k) - 1\}$ we have

$$\sum_{l=1}^{\varphi(p^k)} \sigma_l(t_i)\sigma_l(\zeta)^j = \sum_{l=1}^{\varphi(p^k)} \sigma_l(\zeta^{j-i} - \zeta^{j+i'}).$$

Assume first that $i = j$. Then $\zeta^{j-i} = 1$ and $j + i' = i + i'$ is divisible by p^{k-1} but not by p^k. Hence, $\zeta^{j+i'}$ is a primitive p-th root of unity. Since the second coefficient of the p-th cyclotomic polynomial $\Phi_p(X) = X^{p-1} + X^{p-2} + \ldots + X + 1$ is 1, we get $\sum_{l=1}^{\varphi(p^k)} \sigma_l(\zeta^{j+i'}) = -p^{k-1}$. This shows that

$$\sum_{l=1}^{\varphi(p^k)} \sigma_l(1 - \zeta^{i+i'}) = \varphi(p^k) + p^{k-1} = p^k.$$

Now let $i \neq j$. Then $j - i \not\equiv 0 \pmod{p^k}$ and $j + i' \not\equiv 0 \pmod{p^k}$. Moreover, $j - i \equiv j + i' \pmod{p^{k-1}}$, since $i + i' \equiv 0 \pmod{p^{k-1}}$. Assume first that $j - i \not\equiv 0 \pmod{p^{k-1}}$. Then ζ^{j-i} is a primitive p^s-th root of unity for some $s \geq 2$. Since the second coefficient of the p^s-th cyclotomic polynomial $\Phi_{p^s}(X) = X^{(p-1)p^{s-1}} + X^{(p-2)p^{s-1}} + \ldots + X^{p^{s-1}} + 1$ (see Lemma I.10.1 in [204]) is 0, we have $\sum_{l=1}^{\varphi(p^k)} \sigma_l(\zeta^{j-i}) = 0$. The same holds for $j + i'$. Finally let $j - i \equiv 0 \pmod{p^{k-1}}$. Then we have (as in the first part of the proof)

$$\sum_{l=1}^{\varphi(p^k)} \sigma_l(\zeta^{j-i} - \zeta^{j+i'}) = -p^{k-1} + p^{k-1} = 0.$$

This proves the claim. $\square$

Now let $\mathcal{A} := \mathrm{Aut}_{\mathcal{F}}(\langle u \rangle) \leq \mathcal{G}$. The next proposition shows that the scalar products (a_i^u, a_j^u) only depend on p, $k - d$ and $\mathcal{A}$.

Proposition 5.6 *We have*

$$p^{k-d}(a_i^u, a_j^u) = |\{\tau \in \mathcal{A} : p^k \mid i - j\tau\}| - |\{\tau \in \mathcal{A} : p^k \mid i + j'\tau\}|$$

$$+ |\{\tau \in \mathcal{A} : p^k \mid i' - j'\tau\}| - |\{\tau \in \mathcal{A} : p^k \mid i' + j\tau\}|. \tag{5.6}$$

Proof Let $W := \left(d_{\chi\varphi_u}^{\sigma_i(u)} : i = 1, \ldots, \varphi(p^k), \chi \in \mathrm{Irr}(B)\right)$ be a part of the generalized decomposition matrix. If V is the Vandermonde matrix in Lemma 5.5, we have $VA = W$ and $A = V^{-1}W$. This shows

$$\left((a_{i-1}^u, a_{j-1}^u)\right)_{i,j=1}^{\varphi(p^k)} = AA^{\mathrm{T}} = V^{-1}WW^{\mathrm{T}}V^{-\mathrm{T}} = V^{-1}W\overline{W}^{\mathrm{T}}\overline{V}^{-\mathrm{T}}.$$

Now let $S := (s_{ij})_{i,j=1}^{\varphi(p^k)}$, where

$$s_{ij} := \begin{cases} 1 & \text{if } \sigma_i\sigma_j^{-1} \in \mathcal{A}, \\ 0 & \text{otherwise.} \end{cases}$$

Then the orthogonality relations (see proof of Theorem 5.2) imply $W\overline{W}^{\mathrm{T}} = p^d S$. It follows that

$$p^{2k-d}(a_i^u, a_j^u) = \sum_{l=1}^{\varphi(p^k)} \sigma_l(t_i) \sum_{m=1}^{\varphi(p^k)} s_{lm}\sigma_m(\overline{t_j}) = \sum_{l=1}^{\varphi(p^k)} \sum_{\tau \in \mathcal{A}} \sigma_l(t_i \tau(\overline{t_j}))$$

$$= \sum_{\tau \in \mathcal{A}} \sum_{l=1}^{\varphi(p^k)} \sigma_l((\zeta^{-i} - \zeta^{i'})\tau(\zeta^j - \zeta^{-j'}))$$

$$= \sum_{\tau \in \mathcal{A}} \sum_{l=1}^{\varphi(p^k)} \sigma_l(\zeta^{j\tau-i} + \zeta^{i'-j'\tau} - \zeta^{-i-j'\tau} - \zeta^{i'+j\tau}). \tag{5.7}$$

As in the proof of Lemma 5.5 we have

$$\sum_{l=1}^{\varphi(p^k)} \sigma_l(\zeta^{j\tau-i}) = \begin{cases} \varphi(p^k) & \text{if } p^k \mid j\tau - i, \\ 0 & \text{if } p^{k-1} \nmid j\tau - i, \\ -p^{k-1} & \text{otherwise.} \end{cases}$$

This can be combined to

$$\sum_{\tau \in \mathscr{A}} \sum_{l=1}^{\varphi(p^k)} \sigma_l(\zeta^{j\tau-i}) = p^k |\{\tau \in \mathscr{A} : p^k \mid j\tau - i\}| - p^{k-1} |\{\tau \in \mathscr{A} : p^{k-1} \mid j\tau - i\}|.$$

We get similar expressions for the other numbers $i' - j'\tau$, $-i - j'\tau$ and $i' + j\tau$. Since $i + i' \equiv j + j' \equiv 0 \pmod{p^{k-1}}$, we have $j\tau - i \equiv i' - j'\tau \equiv -i - j'\tau \equiv i' + j\tau \pmod{p^{k-1}}$. Thus, the terms of the form $p^{k-1} |\{\ldots\}|$ in (5.7) cancel out each other. This proves the proposition.   □

Since the group $\mathrm{Aut}(\langle u \rangle)$ is cyclic, $\mathscr{A}$ is uniquely determined by its order. We introduce a notation.

Definition 5.7 Let $\mathscr{A}$ be as in Proposition 5.6. Then we define $\Gamma(d, k, |\mathscr{A}|)$ as the minimum of the expressions

$$\sum_{1 \le i \le j \le \varphi(p^k)} q_{ij}(a_{i-1}^u, a_{j-1}^u)$$

where q ranges over all positive definite, integral quadratic forms. By Proposition 5.4 we have $k_0(B) \le \Gamma(d, k, |\mathscr{A}|)$, and $\sum_{i=0}^{\infty} p^{2i} k_i(B) \le \Gamma(d, k, |\mathscr{A}|)$ if (u, b_u) is major.

We will calculate $\Gamma(d, k, |\mathscr{A}|)$ by induction on k. First we collect some easy facts.

Lemma 5.8 Let $\mathscr{H} \le (\mathbb{Z}/p^k\mathbb{Z})^\times$ where we regard $\mathscr{H}$ as a subset of $\{1, \ldots, p^k\}$. Then $|\{\sigma \in \mathscr{H} : \sigma \equiv 1 \pmod{p^j}\}| = \gcd(|\mathscr{H}|, p^{k-j})$ for $1 \le j \le k$.

Proof The canonical epimorphism $(\mathbb{Z}/p^k\mathbb{Z})^\times \to (\mathbb{Z}/p^j\mathbb{Z})^\times$ has kernel $\mathscr{K}$ of order p^{k-j}. Hence, $|\{\sigma \in \mathscr{H} : \sigma \equiv 1 \pmod{p^j}\}| = |\mathscr{H} \cap \mathscr{K}| = \gcd(|\mathscr{H}|, p^{k-j})$, since the p-subgroups of the cyclic group $(\mathbb{Z}/p^k\mathbb{Z})^\times$ are totally ordered by inclusion.   □

Lemma 5.9 *We have*

$$(a_0^u, a_0^u) = (|\mathscr{A}| + |\mathscr{A}|_p) p^{d-k}$$

and

$$\frac{p^{k-d}}{\gcd(|\mathscr{A}|_p, j)} (a_i^u, a_j^u) \in \{0, \pm 1, \pm 2\}$$

for $i + j > 0$. If $a_i^u \ne 0$ for some $i \ge 1$, then $(a_i^u, a_i^u) = 2p^{d-k} \gcd(|\mathscr{A}|_p, i)$. Moreover, $(a_i^u, a_j^u) = 0$ whenever $\gcd(i, p^{k-1}) \ne \gcd(j, p^{k-1})$.

Proof For $i = j = 0$ we have $i + j'\tau = p^{k-1}\tau \not\equiv 0 \pmod{p^k}$ and $i' + j\tau = p^{k-1} \not\equiv 0 \pmod{p^k}$ for all $\tau \in \mathscr{A}$. Moreover, by Lemma 5.8 there are precisely $|\mathscr{A}|_p$ elements $\tau \in \mathscr{A}$ such that $i' - j'\tau = p^{k-1}(1 - \tau) \equiv 0 \pmod{p^k}$. The first claim follows from Proposition 5.6.

Now let $i + j > 0$ and $\tau \in \mathscr{A}$ such that $i \equiv j\tau \pmod{p^k}$. Then we have $j \neq 0$. Assume that also $\tau_1 \in \mathscr{A}$ satisfies $i \equiv j\tau_1 \pmod{p^k}$. Then $j(\tau - \tau_1) \equiv 0 \pmod{p^k}$ and $\tau^{-1}\tau_1 \equiv 1 \pmod{p^k / \gcd(p^k, j)}$. Thus, Lemma 5.8 implies

$$|\{\tau \in \mathscr{A} : p^k \mid i - j\tau\}| \in \{0, \gcd(|\mathscr{A}|_p, j)\}.$$

The same argument also works for the other summands in (5.6), since $\gcd(|\mathscr{A}|_p, j) = \gcd(|\mathscr{A}|_p, j')$. This gives

$$p^{k-d}(a_i^u, a_j^u) \in \{0, \pm \gcd(|\mathscr{A}|_p, j), \pm 2\gcd(|\mathscr{A}|_p, j)\}$$

whenever $i + j > 0$.

Suppose $i \geq 1$ and $i \equiv i\tau \pmod{p^k}$ for some $\tau \in \mathscr{A}$. Then $\tau \equiv 1 \pmod{p}$ and thus $i \equiv i\tau - (i + i')(\tau - 1) \equiv -i'\tau + i + i' \pmod{p^k}$. Hence, $i' \equiv i'\tau \pmod{p^k}$. This shows $|\{\tau \in \mathscr{A} : p^k \mid i - i\tau\}| = |\{\tau \in \mathscr{A} : p^k \mid i' - i'\tau\}|$. Moreover, we have $|\{\tau \in \mathscr{A} : p^k \mid i + i'\tau\}| = |\{\tau \in \mathscr{A} : p^k \mid i\tau^{-1} + i'\}| = |\{\tau \in \mathscr{A} : p^k \mid i' + i\tau\}|$. This shows $a_i^u = 0$ or $(a_i^u, a_i^u) = 2p^d \gcd(|\mathscr{A}|_p, i)/p^k$.

Finally suppose that $\gcd(i, p^{k-1}) \neq \gcd(j, p^{k-1})$. Then $i \not\equiv j\tau \pmod{p^{k-1}}$ and thus $p^k \nmid i - j\tau$ for all $\tau \in \mathscr{A}$. The same holds for the other terms in (5.6), since $i + i' \equiv j + j' \equiv 0 \pmod{p^{k-1}}$. The last claim follows. $\quad\square$

Proposition 5.10 *We have*

$$\Gamma(d, 1, |\mathscr{A}|) = \big(|\mathscr{A}| + (p - 1)/|\mathscr{A}|\big)p^{d-1}.$$

Proof Since $|\mathscr{A}| \mid p - 1$, we have $|\mathscr{A}|_p = 1$. Hence, $(a_0^u, a_0^u) = (|\mathscr{A}| + 1)p^{d-1}$ and $(a_i^u, a_j^u) \in \{0, \pm p^{d-1}, \pm 2p^{d-1}\}$ for $i + j > 0$ by Lemma 5.9. First we determine the indices i such that $a_i^u = 0$. For this we use Proposition 5.6. Observe that we always have $i' = 1$. In particular for all i, j we have $p \mid i' - j'\tau$ for $\tau = 1$. It follows that $a_i^u = 0$ if and only if $-i \equiv \tau \pmod{p}$ for some $\tau \in \mathscr{A}$. We write this condition in the form $-i \in \mathscr{A}$. This gives exactly $|\mathscr{A}| - 1$ vanishing rows and columns. Thus, all the scalar products (a_i^u, a_j^u) with $-i \in \mathscr{A}$ or $-j \in \mathscr{A}$ vanish. Hence, assume that $-i \notin \mathscr{A}$ and $-j \notin \mathscr{A}$. Then $(a_i^u, a_j^u) \in \{p^{d-1}, 2p^{d-1}\}$ for $i + j > 0$. In case $(a_i^u, a_j^u) = 2p^{d-1}$ we have $a_i^u = a_j^u$. This happens exactly when $j \neq 0$ and $ij^{-1} \in \mathscr{A}$. Since $-i \notin \mathscr{A}$, the coset $i\mathscr{A}$ in $\mathscr{G}$ does not contain -1. Hence, there are precisely $|\mathscr{A}|$ choices for j such that $ij^{-1} \in \mathscr{A}$.

Hence, we have shown that the rows a_i^u for $i = 1, \ldots, p - 2$ split into $|\mathscr{A}| - 1$ zero rows and $(p - 1)/|\mathscr{A}| - 1$ groups consisting of $|\mathscr{A}|$ equal rows each. If we replace the matrix A by UA for a suitable matrix $U \in \mathrm{GL}(p - 1, \mathbb{Z})$, we get a new matrix with exactly $(p - 1)/|\mathscr{A}|$ non-vanishing rows (this is essentially the same as taking another (positive definite) quadratic form in (5.5), see [172]). After leaving out the zero rows we get a $(p-1)/|\mathscr{A}| \times (p-1)/|\mathscr{A}|$ matrix

$$AA^{\mathrm{T}} = p^{d-1} \begin{pmatrix} |\mathscr{A}|+1 & 1 & \cdots & 1 \\ 1 & 2 & \ddots & \vdots \\ \vdots & & \ddots & 1 \\ 1 & \cdots & 1 & 2 \end{pmatrix}.$$

Now we can apply the quadratic form q corresponding to the Dynkin diagram $A_{(p-1)/|\mathscr{A}|}$ in Eq. (5.5). This gives

$$\Gamma(d, 1, |\mathscr{A}|) \leq \big(|\mathscr{A}| + (p-1)/|\mathscr{A}|\big) p^{d-1}.$$

On the other hand, $p^{1-d} AA^{\mathrm{T}}$ is the square of the matrix

$$\begin{pmatrix} 1 & \cdots & 1 & \\ 1 & & 1 & \\ \vdots & & & \ddots \\ 1 & & & & 1 \end{pmatrix}$$

which has exactly $|\mathscr{A}| + (p-1)/|\mathscr{A}|$ columns. This shows that $\Gamma(d, 1, |\mathscr{A}|)$ cannot be smaller.   □

The next proposition gives an induction step.

Proposition 5.11 *If $|\mathscr{A}|_p \neq 1$, then*

$$\Gamma(d, k, |\mathscr{A}|) = \Gamma(d, k-1, |\mathscr{A}|/p).$$

Proof Since $|\mathscr{A}|_p \neq 1$, we have $k \geq 2$. Let $i \in \{1, \ldots, \varphi(p^k) - 1\}$ such that $\gcd(i, p) = 1$. We will see that $(a_i^u, a_i^u) = 0$ and thus $a_i^u = 0$. By Lemma 5.9 and Eq. (5.6) it suffices to show that there is some $\tau \in \mathscr{A}$ such that $p^k \mid i' + i\tau$. We can write this in the form $-i^{-1}i' \in \mathscr{A}$, since i represents an element of $(\mathbb{Z}/p^k\mathbb{Z})^\times$. Now let $-i' = i + \alpha p^{k-1}$ for some $\alpha \in \mathbb{Z}$. Then $-i^{-1}i' = 1 + i^{-1}\alpha p^{k-1}$ is an element of order p in $\mathscr{G}$. Since $\mathscr{G}$ has only one subgroup of order p, it follows that $-i^{-1}i' \in \mathscr{A}$.

Hence, in order to apply Proposition 5.4 it remains to consider the indices which are divisible by p. Let $\overline{\mathscr{A}}$ be the image of the canonical map $(\mathbb{Z}/p^k\mathbb{Z})^\times \to (\mathbb{Z}/p^{k-1}\mathbb{Z})^\times$ under $\mathscr{A}$. Then $|\overline{\mathscr{A}}| = |\mathscr{A}|/p$ (cf. Lemma 5.8). If i and j are divisible by p, we have

$$|\{\tau \in \mathscr{A} : p^k \mid i + j\tau\}| = p \cdot |\{\tau \in \overline{\mathscr{A}} : p^{k-1} \mid (i/p) + (j/p)\tau\}|.$$

A similar equality holds for the other summands in (5.6). Here observe that $(i/p)' = i'/p$, where the dash on the left refers to the case p^{k-1}. Thus, the remaining matrix is just the matrix in case p^{k-1}. Hence, $\Gamma(d, k, |\mathscr{A}|) = \Gamma(d, k-1, |\overline{\mathscr{A}}|) = \Gamma(d, k-1, |\mathscr{A}|/p)$. □

Now we are in a position to prove the main theorem of this section.

Theorem 5.12 *Let B be a p-block of a finite group where p is an odd prime, and let (u, b_u) be a B-subsection such that $l(b_u) = 1$ and b_u has defect d. Moreover, let $\mathscr{F}$ be the fusion system of B, and let $|\mathrm{Aut}_\mathscr{F}(\langle u \rangle)| = p^s r$ where $p \nmid r$ and $s \geq 0$. Then we have*

$$k_0(B) \leq \frac{|\langle u \rangle| + p^s(r^2 - 1)}{|\langle u \rangle| \cdot r} p^d. \tag{5.8}$$

If (in addition) (u, b_u) is major, we can replace $k_0(B)$ by $\sum_{i=0}^\infty p^{2i} k_i(B)$ in (5.8).

Proof As before let $|\langle u \rangle| = p^k$. We will prove by induction on k that

$$\Gamma(d, k, p^s r) = \frac{p^k + p^s(r^2 - 1)}{p^k r} p^d.$$

By Proposition 5.10 we may assume $k \geq 2$. By Proposition 5.11 we can also assume that $s = 0$. As before we consider the matrix A. Like in the proof of Proposition 5.11 it is easy to see that the indices divisible by p form a block of the matrix AA^T which contributes $\Gamma(d, k-1, r)/p$ to $\Gamma(d, k, r)$. It remains to deal with the matrix $\tilde{A} := (a_i^u : \gcd(i, p) = 1)$. By Lemma 5.9 the entries of $p^{k-d} \tilde{A} \tilde{A}^\mathrm{T}$ lie in $\{0, \pm 1, \pm 2\}$. Moreover, if $\gcd(i, p) = 1$ we have $(a_i^u, a_i^u) = 2p^{d-k}$ (see proof of Proposition 5.11).

With the notation of the proof of Proposition 5.6 we have $VA = W$. In particular $\mathrm{rk}\, AA^\mathrm{T} = \mathrm{rk}\, A = \mathrm{rk}\, W = |\mathscr{G} : \mathscr{A}|$. If we set $A_1 := (a_i^u : p \mid i)$, it also follows that $\mathrm{rk}\, A_1 A_1^\mathrm{T} = \mathrm{rk}\, A_1 = \varphi(p^{k-1})/r$. Since the rows of $\tilde{A}$ are orthogonal to the rows of A_1 (see Lemma 5.9), we see that $\mathrm{rk}\, \tilde{A} = (\varphi(p^k) - \varphi(p^{k-1}))/r = p^{k-2}(p-1)^2/r$.

Now we will find $p^{k-2}(p-1)^2/r$ linearly independent rows of $\tilde{A}$. For this observe that $\mathscr{A}$ acts on $\Omega := \{i : 1 \leq i \leq p^{k-1}, \gcd(i, p) = 1\}$ by ${}^\tau i := \tau \cdot i$ $(\mathrm{mod}\ p^{k-1})$ for $\tau \in \mathscr{A}$. Since $p \nmid r$, every orbit has length r (see Lemma 5.8). We choose a set of representatives Δ for these orbits. Then $|\Delta| = p^{k-2}(p-1)/r$. Finally for $i \in \Delta$ we set $\Delta_i := \{i + jp^{k-1} : j = 0, \ldots, p-2\}$. We claim that the rows a_i^u with $i \in \bigcup_{j \in \Delta} \Delta_j$ are linearly independent. We do this in two steps.

Step 1: $(a_i^u, a_j^u) = 0$ for $i, j \in \Delta$, $i \neq j$.

We will show that all summands in (5.6) vanish. First assume that $i \equiv j\tau$ $(\mathrm{mod}\ p^k)$ for some $\tau \in \mathscr{A}$. Then of course we also have $i \equiv j\tau$ $(\mathrm{mod}\ p^{k-1})$ which contradicts the choice of Δ. Exactly the same argument works for the other summands. For the next step we fix some $i \in \Delta$.

Step 2: a_j^u for $j \in \Delta_i$ are linearly independent.

It suffices to show that the matrix $A' := p^{k-d}(a_l^u, a_m^u)_{l,m\in\Delta_i}$ is invertible. We already know that the diagonal entries of A' equal 2. Now write $m = l + jp^{k-1}$ for some $j \neq 0$. We consider the summands in (5.6). Assume that there is some $\tau \in \mathscr{A}$ such that $l \equiv m\tau \equiv (l + jp^{k-1})\tau \pmod{p^k}$. Then we have $\tau \equiv 1 \pmod{p^{k-1}}$ which implies $\tau = 1$. However, this contradicts $j \neq 0$. On the other hand we have $l' \equiv m'\tau \equiv l'\tau \pmod{p^k}$ for $\tau = 1 \in \mathscr{A}$. Now assume $-l \equiv m'\tau \pmod{p^k}$. Then the argument above implies $\tau = 1$ and $l + l' \equiv 0 \pmod{p^k}$ which is false. Similarly the last summand in (5.6) equals 0. Thus, we have shown that $A' = (1 + \delta_{lm})_{l,m\in\Delta_i}$ is invertible.

Therefore we have constructed a basis for the row space of $\tilde{A}$. Hence, there exists an integral matrix $U \in \mathrm{GL}(p^{k-2}(p-1)^2, \mathbb{Q})$ such that the only non-zero rows of $U\tilde{A}$ are a_i^u for $i \in \bigcup_{j\in\Delta} \Delta_j$. Then we can leave out the zero rows and obtain a matrix (still denoted by $\tilde{A}$) of dimension $p^{k-2}(p-1)^2/r$. Moreover, the two steps above show that $p^{k-d}\tilde{A}\tilde{A}^{\mathrm{T}}$ consists of $p^{k-2}(p-1)/r$ blocks of the form $(1 + \delta_{ij})_{1\le i,j\le p-1}$. Thus, an application of the quadratic form q corresponding to the Dynkin diagram $A_{p^{k-2}(p-1)^2/r}$ in Eq. (5.5) gives

$$\Gamma(d,k,r) \le \frac{\Gamma(d,k-1,r)}{p} + \frac{p^{k-1}(p-1)}{p^k r}p^d = \frac{p^k + r^2 - 1}{p^k r}p^d.$$

The minimality of $\Gamma(d,k,r)$ is not so clear as in the proof of Proposition 5.10, since here we do not know if $\det U \in \{\pm 1\}$. However, it suffices to give an example where $k_0(B) = \Gamma(d,k,r)$. By Proposition 5.6 we already know that $\Gamma(d,k,r) = p^{d-k}\Gamma(k,k,r)$. Hence, we may assume $d = k$. Let $G = \langle u \rangle \rtimes C_r$ and B be the principal block of G. Then it is easy to see that the hypothesis of the theorem is satisfied. Moreover,

$$k_0(B) = k(B) = \frac{|D| - 1}{r} + r = \Gamma(d,k,r).$$

Hence, the proof is complete.   □

We add some remarks. It is easy to see that the right hand side of (5.8) is always an integer. Moreover, if $\mathscr{A} = \mathscr{G}$ (i.e. $s = k - 1$ and $r = p - 1$) or $\mathscr{A}$ is a p-group (i.e. $r = 1$), we get the same bound as in Theorem 4.12 and Proposition 4.7. In all other cases Theorem 5.12 really improves Theorem 4.12 and Proposition 4.7. For $k \ge 2$ the case $s = 0$ and $r = p - 1$ gives the best bound for $k_0(B)$. If k tends to infinity, $\Gamma(d,k,p-1)$ goes to $p^d/(p-1)$.

Regarding Olsson's Conjecture, we have to say (in contrast to the case $p = 2$) that Olsson's Conjecture does not follow from Theorem 5.12 if it does not already follow from Theorem 4.12, since the right hand side of (5.8) is always larger than p^{d-1}.

In the proof we already saw that Inequality (5.8) is sharp for blocks with cyclic defect groups. Perhaps it is possible that this can provide a more elementary proof of Dade's Theorem 8.6. For this it would be sufficient to bound $l(B)$ from below, since the difference $k(B) - l(B)$ is locally determined.

As an application of Theorem 5.12 we give a concrete example. Let B be an 11-block with defect group $D \cong C_{11} \times C_{11}$ and inertial index $e(B) = 5$ (for smaller primes results by Usami and Puig give more complete information, e.g. [227, 270]). Assume that $\mathrm{Aut}_{\mathscr{F}}(D)$ acts diagonally (and thus fixed point freely) on both factors C_{11}. Then we have $l(b_u) = 1$ for all non-trivial subsections (u, b_u). Then Theorem 5.12 gives $k(B) \leq 77$ while Theorem 4.2 only implies $k(B) \leq 121$. Also Theorem 1.39 is useless here. However, for the principal block B of $G = D \rtimes \mathrm{Aut}_{\mathscr{F}}(D)$ we have $k(B) = 29$.

As it was pointed out earlier, for odd primes p and $l(b_u) > 1$ there is not always a stable character in $\mathrm{IBr}(b_u)$ under $N_G(\langle u \rangle, b_u)$, even for $l(b_u) = 2$. However, the situation is better if we consider the principal block.

Proposition 5.13 *Let B be the principal p-block of G for an odd prime p, and let (u, b_u) be a B-subsection such that $l(b_u) = 2$, and b_u has defect d and Cartan matrix $C_u = (c_{ij})$. Then we may choose a basic set for C_u such that $p^d c_{11} / \det C_u$ is divisible by p. Moreover, let $\mathscr{F}$ be the fusion system of B and $|\mathrm{Aut}_{\mathscr{F}}(\langle u \rangle)| = p^s r$, where $p \nmid r$ and $s \geq 0$. Then we have*

$$k_0(B) \leq \frac{|\langle u \rangle| + p^s(r^2 - 1)}{|\langle u \rangle| \cdot r} c_{11}.$$

Proof By Brauer's Third Main Theorem, b_u is the principal block of $C_G(u)$ and so $\mathrm{IBr}(b_u)$ contains the trivial Brauer character. Hence, both characters of $\mathrm{IBr}(b_u)$ are stable under $N_G(\langle u \rangle)$. As in the proof of Theorem 5.3, $\frac{p^d}{\det C_u} C_u \pmod{p}$ has rank 1. Hence, we can choose a basic set for C_u such that $p^d c_{11} / \det C_u$ and $p^d c_{12} / \det C_u$ are divisible by p. As in the proof of Theorem 5.3, the rows d_i^u and a_j^i become $\hat{d}_i^u$ and $\hat{a}_j^i$ for $i = 1, 2$ and $j = 0, \ldots, \varphi(|\langle u \rangle|) - 1$. Write $p^d C_u^{-1} = (\tilde{c}_{ij})$. For $\chi \in \mathrm{Irr}_0(B)$ we have

$$0 \not\equiv m_{\chi\chi}^u \equiv \tilde{c}_{11} \left(\hat{d}_{\chi\varphi_1}^u \right)^2 \pmod{\mathrm{Rad}\, \mathcal{O}}.$$

In particular, $\hat{a}_j^1(\chi) \neq 0$ for some $j \in \{0, \ldots, \varphi(p^k) - 1\}$. Now since

$$(\hat{d}_1^u, \gamma(\hat{d}_1^u)) = \begin{cases} c_{11} & \text{if } \gamma \in \mathscr{A}, \\ 0 & \text{if } \gamma \in \mathscr{G} \setminus \mathscr{A}, \end{cases}$$

the proof works as in case $l(b_u) = 1$. $\square$

Chapter 6
Essential Subgroups and Alperin's Fusion Theorem

In this chapter we provide a version of Alperin's Fusion Theorem which is one of the main tools for studying fusion systems. Later we will investigate the structure of essential subgroups. The material for $p = 2$ comes from [257]. The results on essential subgroups for odd primes are unpublished so far. We remark further some results for $p = 2$ also appeared in [211]. However, the proofs there are quite different.

Let $\mathcal{F}$ be a fusion system on a finite p-group P. We begin with the definition of an $\mathcal{F}$-essential subgroup.

Definition 6.1 A subgroup $Q < P$ is called $\mathcal{F}$-*essential* if the following properties hold:

(i) Q is fully $\mathcal{F}$-normalized.
(ii) Q is $\mathcal{F}$-centric.
(iii) $\mathrm{Out}_{\mathcal{F}}(Q) := \mathrm{Aut}_{\mathcal{F}}(Q)/\mathrm{Inn}(Q)$ contains a *strongly p-embedded* subgroup H, i.e. $p \mid |H| < |\mathrm{Out}_{\mathcal{F}}(Q)|$ and $p \nmid |H \cap {}^x H|$ for all $x \in \mathrm{Out}_{\mathcal{F}}(Q) \setminus H$.

Notice that in [184] the first property is not required. Also Property (iii) implies that Q is $\mathcal{F}$-radical. Let $\mathcal{E}$ be a set of representatives for the $\mathrm{Aut}_{\mathcal{F}}(P)$-conjugacy classes of $\mathcal{F}$-essential subgroups of P. Then the number $|\mathcal{E}|$ is sometimes called the *essential rank* of the fusion system. The following theorem says basically that $\mathcal{F}$ is controlled by $\mathcal{E}$ and P.

Theorem 6.2 (Alperin's Fusion Theorem) *Let $\mathcal{F}$ be a fusion system on a finite p-group P. Then every isomorphism in $\mathcal{F}$ is a composition of finitely many isomorphisms of the form $\varphi : S \to T$ such that $S, T \le Q \in \mathcal{E} \cup \{P\}$ and there exists $\psi \in \mathrm{Aut}_{\mathcal{F}}(Q)$ with $\psi_{|S} = \varphi$. Moreover, if $Q \neq P$, we may assume that ψ is a p-element.*

Proof This is a slightly stronger version as Theorem I.3.5 in [19]. First, we show that it suffices to take the set $\mathcal{E}$ instead of all $\mathcal{F}$-essential subgroups. For this let Q be $\mathcal{F}$-essential and $\alpha(Q) \in \mathcal{E}$ for some $\alpha \in \mathrm{Aut}_{\mathcal{F}}(P)$. Moreover, let $S, T \le Q$,

© Springer International Publishing Switzerland 2014

B. Sambale, *Blocks of Finite Groups and Their Invariants*, Lecture Notes in Mathematics 2127, DOI 10.1007/978-3-319-12006-5_6

$\psi \in \mathrm{Aut}_{\mathcal{F}}(Q)$ and $\psi_{|S} = \varphi : S \to T$. Then $\alpha\psi\alpha^{-1} \in \mathrm{Aut}_{\mathcal{F}}(\alpha(Q))$. Hence, $\varphi = \alpha^{-1} \circ (\alpha\psi\alpha^{-1})_{|\alpha(S)} \circ \alpha_{|S}$ is a composition of isomorphisms which have the desired form.

In order the prove the last claim, it remains to show that $\varphi \in \mathrm{Aut}_{\mathcal{F}}(Q)$ for $Q \in \mathcal{E}$ can be written in the stated form. By induction on $|P : Q|$, we may assume that the claim holds for all $\mathcal{F}$-automorphisms of $\mathrm{N}_P(Q)$. Let

$$A := \langle f \in \mathrm{Aut}_{\mathcal{F}}(Q) \; p\text{-element}\rangle = O^{p'}(\mathrm{Aut}_{\mathcal{F}}(Q)) \trianglelefteq \mathrm{Aut}_{\mathcal{F}}(Q).$$

Since $\mathrm{Aut}_P(Q)$ is a Sylow p-subgroup of $\mathrm{Aut}_{\mathcal{F}}(Q)$ (see for example Proposition I.2.5 in [19]), the Frattini argument implies $\mathrm{Aut}_{\mathcal{F}}(Q) = A \, \mathrm{N}_{\mathrm{Aut}_{\mathcal{F}}(Q)}(\mathrm{Aut}_P(Q))$. Hence, we can write $\varphi = \alpha\beta$ such that $\alpha \in A$ and $\beta \in \mathrm{N}_{\mathrm{Aut}_{\mathcal{F}}(Q)}(\mathrm{Aut}_P(Q))$. With the notation of Definition 1.24 we have $\mathrm{N}_\beta = \mathrm{N}_P(Q)$. Then β can be extended to a morphism β' on $\mathrm{N}_P(Q)$. By induction, β' is a composition of isomorphisms of the stated form and so is $\beta = \beta'_{|Q}$ and β^{-1}. Thus, after replacing φ by $\varphi \circ \beta^{-1}$, we may assume $\varphi \in A$. Then it is obvious that φ is a composition of isomorphisms as desired.   □

We deduce some necessary conditions for a subgroup $Q \leq P$ to be $\mathcal{F}$-essential. Since Q is $\mathcal{F}$-centric, we have $\mathrm{C}_P(Q) \subseteq Q$. In particular if Q is abelian, it must be a maximal abelian subgroup. Conversely every maximal (normal) abelian subgroup $R \leq P$ satisfies $\mathrm{C}_P(R) = R$.

Since $\mathrm{Out}_{\mathcal{F}}(Q)$ contains a strongly p-embedded subgroup, $\mathrm{Out}_{\mathcal{F}}(Q)$ is not a p-group and not a p'-group. Moreover, $\mathrm{N}_P(Q)/Q$ is isomorphic to a Sylow p-subgroup of $\mathrm{Out}_{\mathcal{F}}(Q)$. We also have $O_p(\mathrm{Aut}_{\mathcal{F}}(Q)) = \mathrm{Inn}(Q)$. Consider the canonical homomorphism

$$\Psi : \mathrm{Aut}_{\mathcal{F}}(Q) \to \mathrm{Aut}(Q/\Phi(Q)).$$

It is well-known that $\mathrm{Ker}\,\Psi$ is a p-group. On the other hand, $\mathrm{Inn}(Q)$ acts trivially on the abelian group $Q/\Phi(Q)$. This gives $\mathrm{Ker}\,\Psi = \mathrm{Inn}(Q)$ and $\mathrm{Out}_{\mathcal{F}}(Q) \leq \mathrm{Aut}(Q/\Phi(Q))$. In particular $\mathrm{N}_P(Q)/Q$ acts faithfully on $Q/\Phi(Q)$. Hence, $[\langle x \rangle, Q] \not\subseteq \Phi(Q)$ for all $x \in \mathrm{N}_P(Q) \setminus Q$.

Recall that the *rank* of a p-group P is the minimal number of generators of P, i.e. $\log_p |P/\Phi(P)|$. In contrast the *p-rank* is the maximal rank of an abelian subgroup.

Proposition 6.3 *Let $\mathcal{F}$ be a fusion system on a finite p-group P. If $Q \leq P$ is $\mathcal{F}$-essential of rank r, then $\mathrm{Out}_{\mathcal{F}}(Q) \leq \mathrm{GL}(r, p)$ and $|\mathrm{N}_P(Q)/Q| \leq p^{r(r-1)/2}$. Moreover, $\mathrm{N}_P(Q)/Q$ has nilpotency class at most $r - 1$ and exponent at most $p^{\lceil \log_p(r) \rceil}$. In particular $|\mathrm{N}_P(Q)/Q| = p$ if $r = 2$.*

Proof A Sylow p-subgroup of $\mathrm{GL}(r, p)$ is given by the group U of upper triangular matrices with ones on the main diagonal. We may assume $\mathrm{N}_P(Q)/Q \leq U$. Then U

has order $p^{r(r-1)/2}$ and nilpotency class $r-1$ (see Sect. III.16 in [128]). Let $x \in U$ and $m := \lceil \log_p(r) \rceil$. Then we have

$$x^{p^m} - 1 = (x-1)^{p^m} = 0.$$

This shows that U has exponent at most (precisely) p^m.   □

If p is odd or Q is abelian, a similar argument shows that $\mathrm{Out}_{\mathscr{F}}(Q)$ is isomorphic to a quotient of $\mathrm{Aut}(\Omega(Q))$ (see Theorems 5.2.4 and 5.3.10 in [94]). In this case we have $\Omega(Q) \not\subseteq Z(N_P(Q))$. In the general case one can replace $\Omega(Q)$ by a so-called "critical" subgroup (see Theorem 5.3.11 in [94]).

In the following we will improve Proposition 6.3 by taking a closer look at the strongly p-embedded subgroups. The case $p = 2$ in the next theorem is a result by Bender [22] and the odd case can be found in [18, 96].

Theorem 6.4 *A finite group H contains a strongly p-embedded subgroup if and only if one of the following holds:*

(1) $O_p(H) = 1$ *and the Sylow p-subgroups of H have p-rank 1, i.e. they are non-trivial cyclic or quaternion (where $p = 2$).*
(2) $O^{p'}(H/O_{p'}(H))$ *is isomorphic to one of the following (almost) simple groups (see Theorem 7.7 for notation):*

   (a) $\mathrm{PSL}(2, p^n)$,
   (b) $\mathrm{PSU}(3, p^n)$,
   (c) $\mathrm{Sz}(2^{2n+1})$ *for $p = 2$ and $n \geq 1$,*
   (d) ${}^2G_2(3^{2n-1})$ *for $p = 3$ and $n \geq 1$,*
   (e) A_{2p} *for $p \geq 5$,*
   (f) $\mathrm{PSL}_3(4)$, M_{11} *for $p = 3$,*
   (g) $\mathrm{Aut}(\mathrm{Sz}(32))$, ${}^2F_4(2)'$, McL, Fi_{22} *for $p = 5$,*
   (h) J_4 *for $p = 11$.*

Let H be as in Theorem 6.4 for $p = 2$, and let S be a Sylow 2-subgroup of H. Then S is a Suzuki 2-group, i.e. S has an automorphism which permutes the involutions of S transitively (see page 201 in [174] for $H = \mathrm{PSU}(3, 2^n)$). This will be used later. The next lemmas are important to bound the order of $N_P(Q)/Q$, where Q is $\mathscr{F}$-essential.

Lemma 6.5 *If $\mathrm{PSL}(2, p^n)$ is isomorphic to a section of $\mathrm{GL}(r, p)$, then $n \leq r/2$.*

Proof The group $\mathrm{PSL}(2, p^n)$ has exponent $p(p^{2n} - 1)/\gcd(2, p-1)^2$ (see 8.6.9 in [159] for example), while $\mathrm{GL}(r, p)$ has exponent $p^{\lceil \log_p(r) \rceil} \mathrm{lcm}\{p^i - 1 : i = 1, \ldots, r\}$ (see [155]). Hence,

$$p^{2n} - 1 \mid \gcd(2, p-1)^2 \mathrm{lcm}\{p^i - 1 : i = 1, \ldots, r\}. \tag{6.1}$$

Assume $2n > r$. Since $\mathrm{PSL}(2, p^n)$ is non-abelian, we certainly have $r > 1$ and $n > 1$. Therefore, Zsigmondy's Theorem (see for example Theorem 3 in [239]) implies $p = 2$ and $n = 3$. Then, however, the left hand side of (6.1) is divisible by 9 while the right hand side is not. □

Lemma 6.6 *If* $\mathrm{PSU}(3, p^n)$ *is simple and isomorphic to a section of* $\mathrm{GL}(r, p)$, *then* $3n \leq r/2$.

Proof Since $|\mathrm{PSU}(3, p^n)|$ is divisible by $p^{3n} + 1$, we obtain

$$p^{3n} + 1 \mid |\mathrm{GL}(r, p)| = \prod_{i=1}^{r} p^i - 1.$$

It follows that

$$p^{6n} - 1 = (p^{3n} - 1)(p^{3n} + 1) \mid (p^{3n} - 1) \prod_{i=1}^{r} p^i - 1.$$

Assume $6n > r$. As in Lemma 6.5, Zsigmondy's Theorem shows $p = 2$ and $n = 1$. But then $\mathrm{PSU}(3, p^n)$ is not simple. □

Lemma 6.7 *If* $\mathrm{Sz}(2^{2n-1})$ *is isomorphic to a section of* $\mathrm{GL}(r, 2)$, *then* $4n - 2 \leq r/2$.

Proof The order of $\mathrm{Sz}(2^{2n-1})$ is divisible by $2^{4n-2} + 1$. Hence,

$$2^{8n-4} - 1 = (2^{4n-2} - 1)(2^{4n-2} + 1) \mid (2^{4n-2} - 1) \prod_{i=1}^{r} 2^i - 1.$$

Assume $8n - 4 > r$. Then Zsigmondy's Theorem gives a contradiction. □

Lemma 6.8 *If* $^2G_2(3^{2n+1})$ *is isomorphic to a section of* $\mathrm{GL}(r, 3)$, *then* $3(2n + 1) \leq r/2$.

Proof Since $|^2G_2(3^{2n+1})|$ is divisible by $3^{6n+3} + 1$, we get

$$3^{12n+6} - 1 = (3^{6n+3} - 1)(3^{6n+3} + 1) \mid (3^{6n+3} - 1) \prod_{i=1}^{r} 3^i - 1.$$

Suppose $6(2n + 1) > r$. Then Zsigmondy's Theorem gives a contradiction. □

Theorem 6.9 *Let* $\mathscr{F}$ *be a fusion system on a p-group* P. *If* $Q \leq P$ *is* $\mathscr{F}$-*essential of rank* r, *then one of the following holds for* $N := \mathrm{N}_P(Q)/Q$:

  (i)   *N is cyclic of order at most* $p^{\lceil \log_p(r) \rceil}$.
  (ii) *N is elementary abelian of order at most* $p^{\lfloor r/2 \rfloor}$.
  (iii) $p = 2$ *and N is quaternion of order at most* $p^{\lceil \log_p(r) \rceil + 1}$.
  (iv) $p = 2$, $\Omega(N) = \mathrm{Z}(N) = \Phi(N) = N'$ *and* $|N| = |\Omega(N)|^2 \leq p^{\lfloor r/2 \rfloor}$.
  (v) $p = 2$, $\Omega(N) = \mathrm{Z}(N) = \Phi(N) = N'$ *and* $|N| = |\Omega(N)|^3 \leq p^{\lfloor r/2 \rfloor}$.

(vi) $p > 2$, N has order $p^{3n} \leq p^{\lfloor r/2 \rfloor}$, exponent p, p-rank $2n$ and $Z(N) = N' = \Phi(N) \cong C_p^n$ for some $n \geq 1$.

(vii) $p = 3$, $N \cong p_+^{1+2}$ and $r \geq 6$.

(viii) $p = 3$, $|N| = p^{6n+3} \leq p^{\lfloor r/2 \rfloor}$, $|Z(N)| = p^{2n+1}$, $\Omega(N) = \Phi(N) = N' = Z_2(N) \cong C_p^{4n+2}$ for some $n \geq 1$.

Proof By definition, $\mathrm{Out}_{\mathscr{F}}(Q)$ contains a strongly p-embedded subgroup and N is a Sylow p-subgroup of $\mathrm{Out}_{\mathscr{F}}(Q)$. By Theorem 6.4, N is cyclic, quaternion or a Sylow p-subgroup of an almost simple group S. In the first two cases the order of N is bounded by Proposition 6.3. In the remaining case, we need to discuss the various possibilities for S. Since Q has rank r, we may assume $\mathrm{Out}_{\mathscr{F}}(Q) \leq \mathrm{GL}(r, p)$.

First suppose that $S \cong \mathrm{PSL}(2, p^n)$. Then we get $n \leq r/2$ by Lemma 6.5. In particular, N is elementary abelian of rank at most $r/2$. In case $S \cong \mathrm{PSU}(3, p^n)$ we obtain $3n \leq r/2$ by Lemma 6.6. Thus, N has order $p^{3n} \leq p^{\lfloor r/2 \rfloor}$. If $p = 2$, then we are in case (6.9) by Higman [115]. In case $p > 2$ it is easy to see that N has exponent p and $Z(N) = N' = \Phi(N) \cong C_p^n$. The p-rank of S (and thus N) can be found in Table 3.3.1 on page 108 in [97].

Next, let $S \cong \mathrm{Sz}(2^{2n+1})$ and $p = 2$. Then the order of N is bounded by Lemma 6.7, and [115] implies that we are in case (iv). If $p = 3$ and $S \cong {}^2G_2(3^{2n+1})$, then Lemma 6.8 shows that N has order $3^{6n+3} \leq 3^{\lfloor r/2 \rfloor}$. Since S has a faithful, seven-dimensional representation over a field of characteristic 3, we get $\exp N \leq 9$. For $n = 0$ one can compute $\exp N = 9$. Hence, the same must be true for all n. Moreover, the 3-rank of N can be found in Table 3.3.1 on page 108 in [97]. Other properties can be derived from [144].

Now let $S \cong A_{2p}$ for some $p \geq 5$. Then of course N is elementary abelian of order p^2. In order to prove $r \geq 4$, it suffices to show that A_{2p} is not involved in $\mathrm{GL}(3, p)$. Observe that $\gcd(2p - 1, (p^3 - 1)(p^2 - 1)(p - 1)) \mid 21$. This leaves the possibility $p = 11$. But here $13 \nmid |\mathrm{GL}(3, 11)|$.

The remaining cases for S are of exceptional nature. In particular $p \leq 11$. It is easy to see that $|N| \leq p^3$ and $\exp N = p$ in all instances. Hence, N occurs in one of the cases already covered. However, it remains to verify the bound on r. But this follows just by comparing the orders of these groups. □

It may happen that $\lceil \log_p(r) \rceil > \lfloor r/2 \rfloor$, however we have the following addition to Theorem 6.9.

Proposition 6.10 (Lemma 1.7 in [211]) *In the situation of Theorem 6.9 we have* $|N| \leq p^{\lfloor r/2 \rfloor}$

Since essential subgroups cannot have rank 1 (otherwise the automorphism group would be abelian), we take a closer look at the essential subgroups of rank 2. The next proposition generalizes Lemma 4.1 in [241].

Proposition 6.11 *Let $\mathscr{F}$ be a fusion system on a p-group P. If $Q \leq P$ is $\mathscr{F}$-essential of rank at most 2, then $\mathrm{SL}(2, p) \leq \mathrm{Out}_{\mathscr{F}}(Q) \leq \mathrm{GL}(2, p)$ and one of the following holds*

(i) $|Q| \leq p^3$ and P has maximal class. In case $p = 2$ we have $P \cong \{D_{2^n}, SD_{2^n}, Q_{2^n}\}$ for some $n \geq 3$.

(ii) $p \in \{2, 3\}$, $Q \cong C_{p^r} \times C_{p^r}$ and $|P : Q| = p$ for some $r \geq 2$. Moreover, P is non-metacyclic. In case $p = 2$, we have $P \cong C_{2^r} \wr C_2$. In case $p = 3$, P has maximal class.

(iii) $Q/\mathrm{K}_3(Q)\Phi(Q')$ is minimal non-abelian of type (r, r), i.e.

$$Q/\mathrm{K}_3(Q)\Phi(Q') = \langle x, y \mid x^{p^r} = y^{p^r} = [x, y]^p = [x, x, y] = [y, x, y] = 1 \rangle$$

for some $r \geq 1$. In case $p = 2$, we have $r \geq 2$. In particular Q is non-metacyclic (for all primes p).

Proof As usual we may regard $\mathrm{Out}_{\mathscr{F}}(Q)$ as a subgroup of $\mathrm{GL}(2, p)$. Since $\mathrm{Out}_{\mathscr{F}}(Q)$ contains at least two Sylow p-subgroups, we get $\mathrm{SL}(2, p) \leq \mathrm{Out}_{\mathscr{F}}(Q)$ from 8.6.7 in [159]. In particular $Q \not\cong C_p \times C_{p^2}$ (see also Proposition 3.3 in [265]). Hence, in case $|Q| \leq p^3$ Propositions 1.8 and 10.17 in [23] imply that P has maximal class. The additional statement for $p = 2$ is well-known. Suppose next that Q is abelian of order at least p^4. Again we must have $Q \cong C_{p^r} \times C_{p^r}$ for some $r \geq 2$. By Proposition 6.3 it holds that $|N_P(Q) : Q| = p$. Choose $g \in N_P(Q) \setminus Q$. Then g (as an element of $\mathrm{Out}_{\mathscr{F}}(Q)$) acts non-trivially on $\Omega(Q)$. It follows that Q is the only abelian maximal subgroup of $N_P(Q)$. Hence, Q is characteristic in $N_P(Q)$ and $N_P(Q) = P$. Now let $p = 2$ and $Q = \langle x, y \rangle$. Then we may assume that $^g x = y$ and $^g y = x$. We can write $g^2 = (xy)^i$ for some $i \in \mathbb{Z}$, because g centralizes g^2. Then an easy calculation shows that gx^{-i} has order 2. Hence, $P \cong C_{2^r} \wr C_2$. Now let $p \geq 3$. Since g acts non-trivially on $\Omega(Q)$, we conclude that P has p-rank 2. It follows from Proposition 3.13 in [67] that $p = 3$. It is known that fusion systems on metacyclic 3-groups are always controlled (see [265]). Hence, P is non-metacyclic. Blackburn classified all those groups (see Theorem 11.5 below). We need to exclude the groups $C(p, n)$ and $G(P, n, \epsilon)$. This follows from Lemmas A.6 and A.8 in [67].

Finally, assume that Q is non-abelian (and p is arbitrary again). Since $\mathrm{Out}_{\mathscr{F}}(Q)$ acts faithfully on Q/Q', we get $Q/Q' \cong C_{p^r} \times C_{p^r}$ for some $r \geq 1$. Moreover, by Hilfssatz III.1.11 in [128] we know that $Q'/\mathrm{K}_3(Q)$ is cyclic and thus $|Q'/\mathrm{K}_3(Q)\Phi(Q')| = p$. Therefore, the group $\overline{Q} := Q/\mathrm{K}_3(Q)\Phi(Q')$ is minimal non-abelian by Lemma 12.1. Now the structure of $\overline{Q}$ follows from [242] (see Theorem 12.2). For $p = 2$ and $r = 1$, Taussky's Theorem (see Satz III.11.9 in [128]) implies that Q is metacyclic. Then however, we have $Q \cong Q_8$ by Lemma 1 in [190]. Thus, we end up in case (i). It is easy to see that $Q/\mathrm{K}_3(Q)\Phi(Q')$ (and therefore Q) is not metacyclic.   □

In the situation of Proposition 6.11, all maximal subgroups of Q are isomorphic. Hence, we are in a position to apply [111, 188]. In particular, there are only finitely many such groups Q for a given coclass.

We note a corollary of Theorem 6.9.

Proposition 6.12 *Let $\mathscr{F}$ be a fusion system on a p-group P. If $Q \leq P$ is $\mathscr{F}$-essential of rank at most 3, then $|N_P(Q) : Q| = p$. Moreover, if $p = 2$, then $\mathrm{Out}_{\mathscr{F}}(Q) \cong S_3$.*

Proof The first claim follows from Proposition 6.10. Now let $p = 2$. Then we may assume that Q has rank 3 and $\mathrm{Out}_{\mathscr{F}}(Q) \leq \mathrm{GL}(3,2)$. By Theorem 6.9, the Sylow 2-subgroups of $\mathrm{Out}_{\mathscr{F}}(Q)$ are cyclic. In particular, $\mathrm{Out}_{\mathscr{F}}(Q)$ is 2-nilpotent. If $7 \nmid |\mathrm{Out}_{\mathscr{F}}(Q)|$, then $\mathrm{Out}_{\mathscr{F}}(Q)$ lies in the normalizer of a Sylow 3-subgroup of $\mathrm{GL}(3,2)$, and the claim follows. Otherwise $\mathrm{Out}_{\mathscr{F}}(Q)$ lies in the normalizer N of a Sylow 7-subgroup of $\mathrm{GL}(3,2)$. However, $|N| = 21$. Contradiction.   □

For $p = 2$ it is worthwhile to note the rank 4 and rank 5 cases.

Lemma 6.13 *Let $\mathscr{F}$ be a fusion system on a finite 2-group P, and let $Q \leq P$ is an $\mathscr{F}$-essential subgroup.*

(i) *If Q has rank 4, then $\mathrm{Out}_{\mathscr{F}}(Q)$ is isomorphic to one of the following groups: S_3, D_{10}, $S_3 \times C_3$, $C_3^2 \rtimes C_2$ (where C_2 acts as inversion), $C_5 \rtimes C_4$, $D_{10} \times C_3$, $C_3^2 \rtimes C_4$, A_5, $C_{15} \times C_4$ (with trivial center), $\mathrm{GL}(2,4)$.*
(ii) *If Q has rank 5, then $\mathrm{Out}_{\mathscr{F}}(Q)$ is isomorphic to one of the following groups: S_3, D_{10}, $S_3 \times C_3$, $C_3^2 \rtimes C_2$ (where C_2 acts as inversion), $C_5 \rtimes C_4$, $D_{10} \times C_3$, $C_3^2 \rtimes C_4$, $S_3 \times C_7$, A_5, $C_{15} \times C_4$ (with trivial center), $S_3 \times (C_7 \rtimes C_3)$, $\mathrm{GL}(2,4)$.*

Proof If Q has rank 4, then $\mathrm{Out}_{\mathscr{F}}(Q) \leq \mathrm{GL}(4,2) \cong A_8$. Here the claim can be showed by computer. Now assume that Q has rank 5. Then it is too costly to run through all subgroups of $\mathrm{GL}(5,2)$. Let $H := \mathrm{Out}_{\mathscr{F}}(Q)$ and $S \in \mathrm{Syl}_2(H)$. By Proposition 6.10 we have $|S| \leq 4$. If S is cyclic, then H is solvable. Hence, H lies in a local subgroup of $\mathrm{GL}(5,2)$ and we can enumerate them with GAP. Now suppose that $S \cong C_2^2$. Then by Theorem 6.4, $N := O^{2'}(H / O_{2'}(H))$ is a simple group. By a theorem of Gorenstein and Walter [98], N is isomorphic to $\mathrm{PSL}(2,q)$ where $q \equiv \pm 3 \pmod 8$. In particular $|N| = \frac{1}{2}(q-1)q(q+1)$. Since $H \leq \mathrm{GL}(5,2)$, this forces $q = 5$ and $N \cong A_5$. Since $\mathrm{Out}(N) \cong C_2$, we conclude that $H / O_{2'}(H) \cong A_5$. By Feit-Thompson $O_{2'}(H)$ is solvable. Hence, in case $O_{2'}(H) \neq 1$, H lies in a local subgroup of $\mathrm{GL}(5,2)$. Since these cases were already handled, we end up with $H \cong A_5$.   □

We also have information in special cases.

Lemma 6.14 *Let $\mathscr{F}$ be a fusion system on a finite 2-group P. If $Q \in \{C_2^3, D_8 * D_8\}$ is an $\mathscr{F}$-essential subgroup of P, then every subgroup of P has rank at most 4.*

Proof This follows from the Lemmas 99.3 and 99.7 in [25].   □

In the situation of Proposition 6.12 it is hard to say something about $\mathrm{Out}_{\mathscr{F}}(Q)$ for $p > 2$. The existence of a strongly p-embedded subgroup is equivalent to the fact that $\mathrm{Out}_{\mathscr{F}}(Q)$ contains a non-normal Sylow p-subgroup of order p. For $p = 3$ we have the following result.

Lemma 6.15 *Let $\mathscr{F}$ be a fusion system on a finite 3-group P, and let $Q \leq P$ is an $\mathscr{F}$-essential subgroup.*

(i) *If Q has rank 3, then $\mathrm{Out}_{\mathscr{F}}(Q)$ is isomorphic to one of the following groups: A_4, S_4, $\mathrm{SL}(2,3)$, $A_4 \times C_2$, $C_{13} \rtimes C_3$, $\mathrm{GL}(2,3)$, $S_4 \times C_2$, $\mathrm{SL}(2,3) \times C_2$, $(C_{13} \rtimes C_3) \times C_2$, $\mathrm{GL}(2,3) \times C_2$.*

(ii) *If Q has rank 4 and $\mathrm{N}_P(Q)/Q$ is not cyclic, then A_6 is involved in $\mathrm{Out}_{\mathscr{F}}(Q)$. In particular, $5 \mid |\mathrm{Aut}(Q)|$.*

Proof The first part follows by a computer enumeration over all subgroups of $\mathrm{GL}(3,3)$. Now assume that Q has rank 4. Then $|\mathrm{N}_P(Q)/Q| \leq 9$ by Proposition 6.10. If $\mathrm{N}_P(Q)/Q$ is not cyclic, $\mathrm{Out}_{\mathscr{F}}(Q)$ must involved a simple group S given by Theorem 6.4. Considering the order of $\mathrm{GL}(4,3)$ gives $S \cong A_6$. $\square$

On the other hand, the search for non-nilpotent fusion systems for odd primes is simplified by the following result.

Theorem 6.16 (Glesser [91]) *Let P be a p-group for $p > 2$. If there is a non-nilpotent fusion system on P, then there exists a non-nilpotent constrained fusion system on P. In particular, there exists a finite group G such that $P \in \mathrm{Syl}_p(G)$ and G is not p-nilpotent.*

Therefore, the following algorithm helps to find non-nilpotent fusion systems on P:

(1) Check if $\mathrm{Aut}(P)$ is a p-group (otherwise there are non-nilpotent controlled fusion systems).
(2) Make a list $\mathscr{L}$ of all candidates of essential subgroups up to P-conjugation.
(3) For each $Q \in \mathscr{L}$, check if there is a subgroup $N \trianglelefteq P$ such that $\mathrm{C}_P(N) \subseteq N \subseteq Q$, $P/\mathrm{Z}(N) \leq \mathrm{Aut}(N)$ and $\mathrm{Aut}(N)$ has no normal Sylow p-subgroup.
(4) For each N above construct finite groups $\mathrm{Z}(N).A$ where $A \leq \mathrm{Aut}(N)$ and check if they have the desired fusion system.

Chapter 7
Reduction to Quasisimple Groups and the Classification

7.1 Fong Reductions

An application of Theorem 1.18 gives us the so-called First Fong Reduction. The statement about Morita equivalence can be found in Proposition 3.8 in [48], and the claim about the fusion system comes from Proposition IV.6.3 in [19].

Theorem 7.1 (First Fong Reduction) *Let B be a block of a finite group G. Then B is Morita equivalent to a* quasiprimitive *block $\tilde{B}$ of a finite group $\tilde{G}$, i.e. for every normal subgroup $\tilde{N}$ of $\tilde{G}$, $\tilde{B}$ covers just one block of $\tilde{N}$. Moreover, B and $\tilde{B}$ have the same defect group and the same fusion system.*

The Second Fong Reduction can be stated as follows.

Theorem 7.2 (Second Fong Reduction) *Let $N \trianglelefteq G$, and let B be a p-block of G which covers a stable block of N with trivial defect. Then there is a central extension*

$$1 \to Z \to H \to G/N \to 1$$

such that B is Morita equivalent to a block of H with the same defect group and the same fusion system. Moreover, Z is a cyclic p'-group.

7.2 Extensions of Nilpotent Blocks

Theorem 7.3 (Külshammer-Puig [169]) *Let B be a p-block of G with defect group D. Suppose that B covers a nilpotent block b of $N \trianglelefteq G$. Then B is Morita equivalent to a block of a twisted group algebra $\mathcal{O}_\gamma L$ where L is an extension of $D \cap N$ with $N_G(N, b)/N$.*

© Springer International Publishing Switzerland 2014
B. Sambale, *Blocks of Finite Groups and Their Invariants*, Lecture Notes in Mathematics 2127, DOI 10.1007/978-3-319-12006-5_7

The 2-cocycle γ appearing in Theorem 7.3 is sometimes called the *Külshammer-Puig class*. By combining the result with Proposition 1.20 we see that B is Morita equivalent to a block $\tilde{B}$ of a group $\tilde{L}$ with defect group D and $D \cap N \trianglelefteq \tilde{L}$. In particular, if b also has defect group D, we are in a position to use Theorem 1.19.

Later, Puig [225] obtained results in the opposite direction, i.e. results about blocks *covered* by nilpotent blocks.

Proposition 7.4 (Puig) *Let B be a nilpotent block of a finite group G, and suppose that B covers a block b of $N \trianglelefteq G$ with defect group D. Then b is Morita equivalent to the unique block of $\mathrm{N}_N(D)$ with Brauer correspondent b.*

7.3 Components

A finite group H is called *quasisimple* if $H/\mathrm{Z}(H)$ is simple and $H' = H$. A subgroup $U \le G$ is *subnormal* in G if there exists a series $U \trianglelefteq U_1 \trianglelefteq \ldots \trianglelefteq U_m = G$. A subgroup $C \le G$ is a *component* of G if C is quasisimple and subnormal in G. The *layer* $\mathrm{E}(G)$ of a finite group G is the subgroup generated by all components of G. It is known that $\mathrm{E}(G)$ is a central product of components. Hence, the following lemma is relevant.

Lemma 7.5 *Let $G = G_1 * G_2$ be a central product of finite groups G_1 and G_2, and let B be a block of G. For $i = 1, 2$, let B_i be the (unique) block of $G_i \trianglelefteq G$ covered by B. Then the following holds*

(i) *If B_i has defect group D_i for $i = 1, 2$, then $D_1 D_2$ is a defect group of B.*
(ii) *If $G_1 \cap G_2$ is a p'-group, then $B \cong B_1 \otimes B_2$.*
(iii) *B is nilpotent if and only if both B_1 and B_2 are.*

Proof The first two parts follow from Proposition 1.5 in [71]. We quote the proof of the third part from [75]: We may write $G = E/Z$ where $E = G_1 \times G_2$ and $Z \le \mathrm{Z}(E)$. Let B_E be the unique block of E dominating B, so $\mathrm{O}_{p'}(Z)$ is in the kernel of B_E, and B_E has defect group D_E such that $D_E Z/Z$ is a defect group for B. By An and Eaton [13, 2.6], B_E is nilpotent if and only if B is. Note that B_E is a product of blocks of G_1 and G_2 which are nilpotent if and only if B_1 and B_2 are. Hence, it suffices to consider the case $G = G_1 \times G_2$. However, the result follows easily in this case since the normalizer and centralizer of a subgroup Q of $G_1 \times G_2$ are $\mathrm{N}_{G_1}(\pi_1(Q)) \times \mathrm{N}_{G_2}(\pi_2(Q))$ and $\mathrm{C}_{G_1}(\pi_1(Q)) \times \mathrm{C}_{G_2}(\pi_2(Q))$, where $\pi_i(Q)$ is the image of the projection onto G_i (we leave the details to the reader). □

As usual we denote the Fitting subgroup of G by $\mathrm{F}(G)$ and the generalized Fitting subgroup by $\mathrm{F}^*(G) := \mathrm{E}(G)\mathrm{F}(G)$. It is known that $[\mathrm{E}(G), \mathrm{F}(G)] = 1$ and $\mathrm{C}_G(\mathrm{F}^*(G)) \subseteq \mathrm{F}^*(G)$.

Lemma 7.6 *Let Q be a quasisimple group. Then $\mathrm{Aut}(Q) \le \mathrm{Aut}(Q/\mathrm{Z}(Q))$.*

Proof Let $S := Q/Z(Q)$ (a simple group). Consider the canonical map $f :$ $\mathrm{Aut}(Q) \to \mathrm{Aut}(S)$. Let $\alpha \in \mathrm{Ker}\, f$. Then $\alpha(g)g^{-1} \in Z(Q)$ for all $g \in Q$. Hence, we get a map $\beta : Q \to Z(Q)$, $g \mapsto \alpha(g)g^{-1}$. Moreover, it is easy to see that β is a homomorphism. Since Q is perfect, we get $\beta = 1$ and thus $\alpha = \mathrm{id}_Q$. This shows $\mathrm{Aut}(Q) \le \mathrm{Aut}(S)$. □

In the following we sketch the reduction to quasisimple groups. For this let B be a block of a finite group G. By Theorem 7.1 we may assume that B is quasiprimitive. Moreover, the Second Fong Reduction allows to assume that $O_{p'}(G)$ is central and cyclic. In some cases we can use Theorem 7.3 to show that $O_p(G) = 1$. Then it follows that $\mathrm{F}(G) = Z(G) = O_{p'}(G)$. We consider the unique block b of $\mathrm{E}(G)$ covered by B. Write $\mathrm{E}(G) = L_1 * \ldots * L_n$ where $L_1, \ldots, L_n$ are the components of G. For $i = 1, \ldots, n$, b covers a block b_i of L_i with defect group D_i. Then by Lemma 7.5, $D_1 \times \ldots \times D_n$ is a defect group of b and thus contained in D. Again in favorable cases (for instance if the p-rank of D is small) we obtain $n = 1$. This means that $\mathrm{E}(G)$ is quasisimple. Then $Z(G) \subseteq C_G(\mathrm{E}(G)) = C_G(\mathrm{E}(G)Z(G)) = C_G(\mathrm{F}^*(G)) \subseteq \mathrm{F}(G) = Z(G)$ and $G/Z(G) = G/C_G(\mathrm{E}(G)) \le \mathrm{Aut}(\mathrm{E}(G))$. Moreover, by Lemma 7.6, $\mathrm{Aut}(\mathrm{E}(G)) \le \mathrm{Aut}(\mathrm{E}(G)/Z(\mathrm{E}(G)))$.

So G is a central extension of a subgroup of the automorphism group of a simple group S and a cyclic p'-group. Using Schreier's Conjecture (which follows from the classification of the finite simple groups) we deduce that S is the only non-abelian composition factor of G.

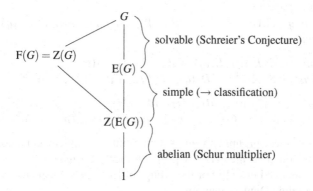

The following arguments often help to remove composition factors lying "above" S (so that G is in fact quasisimple). Suppose that we have a normal subgroup $N \trianglelefteq G$ of prime index q. Then B covers a unique block B_N of N with defect group $N \cap D$. The situation splits naturally into two cases. In the first case, B is the only block covering B_N (see for example [105]). Using the action of G on N, the set $\mathrm{Irr}(B_N)$ consists of α orbits of length q and β stable characters. Clifford theory yields $k(B_N) = \alpha q + \beta$ and $k(B) = \alpha + \beta q$. This can often be used the derive a contradiction. Similar considerations can be applied to $\mathrm{IBr}(B_N)$ and $\mathrm{IBr}(B)$. In the second case, B_N is covered by several blocks of G. Then [196, Corollary 5.5.6]

shows that $q \neq p$. In particular, B_N also has defect group D. Moreover, $G = N \, C_G(D)$, and all blocks of G covering B_N have defect group D. If there is a non-stable character $\psi \in \mathrm{Irr}(B_N)$, then $\psi^G \in \mathrm{Irr}(B_N^G) = \mathrm{Irr}(B)$ without loss of generality by Nagao and Tsushima [196, Lemma 5.3.1]. For another non-stable character $\tilde{\psi} \in \mathrm{Irr}(B_N)$ we also have $\tilde{\psi}^G \in \mathrm{Irr}(B)$. In particular, $\mathrm{Irr}(B_N)$ also contains a stable character which extends in q many ways to an irreducible character of G. One of these extensions must lie in $\mathrm{Irr}(B)$. Again counting arguments apply.

As a final remark, we note that in order to prove Donovan's Conjecture it suffices to assume $O^{p'}(G) = G$ (see [166]).

7.4   The Classification of the Finite Simple Groups

After we have reduced the situation to quasisimple groups, it is time to apply the classification of the finite simple groups which we state as follows.

Theorem 7.7 (CFSG)   *Every finite simple group belongs to one of the following families:*

(1) cyclic groups C_p of prime order.
(2) alternating groups A_n for $n \geq 5$.
(3) groups of Lie type which split further into the following classes:

    (a) classical groups
        $\mathrm{PSL}(n, q)$, $\mathrm{PSU}(n, q)$, $\mathrm{P}\Omega_{2n+1}(q)$, $\mathrm{P}\Omega_{2n}^+(q)$, $\mathrm{P}\Omega_{2n}^-(q)$, $\mathrm{PSp}_{2n}(q)$.
    (b) exceptional groups

        • *untwisted:* $E_6(q)$, $E_7(q)$, $E_8(q)$, $F_4(q)$, $G_2(q)$.
        • *twisted:* $\mathrm{Sz}(2^{2n+1})$, ${}^3D_4(q)$, ${}^2E_6(q)$, ${}^2F_4(2^{2n+1})'$, ${}^2G_2(3^{2n+1})$.

(4) 26 sporadic groups: M_{11}, M_{12}, M_{22}, M_{23}, M_{24}, J_1, J_2, J_3, J_4, HS, He, McL, Suz, Ly, Ru, $O'N$, Co_1, Co_2, Co_3, Fi_{22}, Fi_{23}, Fi_{24}', HN, Th, BM, M.

The orders of these groups as well as their Schur multipliers and automorphism groups can be found in the ATLAS [59]. Further information (for example the p-ranks) are contained in [97]. For the groups of Lie type we sometimes also refer to the corresponding Dynkin diagram.

In the following we will list known results concerning the representation theory of simple groups. The groups of prime order are certainly uninteresting. For the alternating groups it is often useful to study the corresponding symmetric groups S_n first. We refer to [139, 218]. For a non-negative integer n we denote the number of partitions of n by $p(n)$. Here we set $p(0) := 1$ and $S_0 := 1$ (the symmetric group on an empty set).

Theorem 7.8   *Let B be a p-block of S_n. Then there exists a non-negative integer ω called the* weight *of B with the following properties:*

(i) The defect group D of B is isomorphic to a Sylow p-subgroup of $S_{p\omega}$.

(ii)   *The fusion system of B is* $\mathscr{F}_D(S_{p\omega})$.
(iii) $k(B) = \sum p(\omega_1)\dots p(\omega_p)$ *where the sum is taken over all non-negative tuples*
   $(\omega_1,\dots,\omega_p)$ *such that* $\sum \omega_i = \omega$.
(iv) $l(B) = \sum p(\omega_1)\dots p(\omega_{p-1})$ *where the sum is taken over all non-negative tuples* $(\omega_1,\dots,\omega_{p-1})$ *such that* $\sum \omega_i = \omega$.

Olsson [213] showed that in the situation of Theorem 7.8 also the numbers $k_i(B)$ can be expressed in terms of ω. However, the formulas are too complicated to state here. In fact, Enguehard [77] constructed a perfect isometry between B and $B_0(\mathscr{O}S_{p\omega})$. For the complicated definition of a perfect isometry we refer to [48].

The Sylow subgroups of the symmetric groups are given by the following basic result.

Proposition 7.9 *Define* $P_i := C_p \wr \dots \wr C_p$ *(i factors) for $i \geq 1$. Let $n = \sum_{i=0}^{\infty} a_i p^i$ be the p-adic expansion of n (i.e. $0 \leq a_i \leq p-1$). Then a Sylow p-subgroup of S_n is isomorphic to*

$$\prod_{i=1}^{\infty} P_i^{a_i}.$$

Information about the Morita equivalence class of a block of a symmetric group can be obtained from its *core*. In particular, Donovan's Conjecture is true for blocks of symmetric groups (see Scopes [262]). The essential rank of block fusion systems of symmetric groups is determined in [9]. Moreover, the elementary divisors of the Cartan matrix of a block of a symmetric group were computed in Theorem 4.5 in [27].

Now we turn to alternating groups.

Theorem 7.10 *Let B be a p-block of A_n with defect group D. Then one of the following holds:*

(i) $p \neq 2$ *and B is covered by two blocks $\hat{B}$ and $\hat{B}'$ of S_n with defect group D. Moreover, B and $\hat{B}$ are isomorphic as algebras and $\mathscr{F}_D(B) = \mathscr{F}_D(\hat{B})$.*
(ii) B *is covered by a unique block $\hat{B}$ of S_n with defect group $\hat{D}$ and weight ω. Then $D = \hat{D} \cap A_n$ and $\mathscr{F}_D(B) = \mathscr{F}_D(A_{p\omega})$.*

Proof The distinction into the two cases is well-known. The isomorphism in the first case can be found for example in [46, Théorème 0.1]. It remains to prove the claim about the fusion system of B. Here we use Jacobsen [138]. Let $\hat{B}$ be a block of S_n with weight ω which covers B. Assume that $\mathscr{F}_D(B) = \mathscr{F}_D(\hat{B}) = \mathscr{F}_D(S_{p\omega})$. Then by the proof of Theorem 28 in [138] we have $N := n - p\omega \geq 2$ and B has a unique Brauer correspondent b in $C_{A_n}(D)$. By the proof of Lemma 27 in [138], b corresponds to a block of defect 0 in A_N (i.e. a character $\psi \in \mathrm{Irr}(A_N)$). Let $\sigma \in S_N \setminus A_N$. Since b is unique, ψ is fixed by σ. Hence, ψ extends to two irreducible characters of S_N. Translating this to S_n means that B is covered by two blocks of S_n. It is easy to see that also the converse holds.   □

Again precise formulas for $k(B)$, $l(B)$ and $k_i(B)$ can be given in terms of sophisticated combinatorial objects. Also perfect isometries are known to exist by Brunat and Gramain [50]. Donovan's Conjecture is also known to hold for the blocks of alternating groups by Hiss [116]. Later, Kessar [145, 146] extended these results to covering groups.

The block theory of simple groups of Lie type is vastly more complicated. But at least in the defining characteristic we have the following strong theorem.

Theorem 7.11 (Humphreys [125], An-Dietrich [10]) *Every p-block B of a simple group G of Lie type in characteristic p has maximal or trivial defect. In the former case, $\mathscr{F}_P(B) = \mathscr{F}_P(G)$ for $P \in \mathrm{Syl}_p(G)$.*

In the general case it is often possible to go over to a general linear group. Here we use a paper by Fong and Srinivasan [85].

Theorem 7.12 *Let B be a p-block of $G := \mathrm{GL}(n,q)$ with defect group D and $p \nmid q$. Then there exists a semisimple element $s \in G$ such that D is a Sylow p-subgroup of $\mathrm{C}_G(s)$. Moreover, $\mathrm{C}_G(s)$ is a direct product of groups of the form $\mathrm{GL}(m,q^k)$.*

Hence, in order to understand the structure of defect groups of blocks of linear groups, we may study Sylow subgroups of $\mathrm{GL}(n,q)$.

Proposition 7.13 (Weir [285], Leedham-Green-Plesken [177]) *Let P be a Sylow p-subgroup of $\mathrm{GL}(n,q)$. Then one of the following holds*

- (i) $p \mid q$ and P is conjugate in $\mathrm{GL}(n,q)$ to the group of upper triangular matrices with ones on the main diagonal.
- (ii) $2 \neq p \nmid q$ and $P \cong C_{p^r} \wr Q$. Here $Q \in \mathrm{Syl}_p(S_{\lfloor n/e \rfloor})$ and $p^r m = q^e - 1$ where $p \nmid m$ and e is the order of q modulo p.
- (iii) $p = 2$, $4 \mid q - 1$ and $P \cong C_{2^r} \wr Q$. Here $Q \in \mathrm{Syl}_2(S_n)$ and $2^r m = q - 1$ where $2 \nmid m$.
- (iv) $p = 2 \nmid q$, $4 \nmid q - 1$, $2 \mid n$ and $P \cong SD_{2^{r+1}} \wr Q$. Here $Q \in \mathrm{Syl}_2(S_{n/2})$ and $2^r m = q^2 - 1$ where $2 \nmid m$.
- (v) $p = 2 \nmid q$, $4 \nmid q - 1$, $2 \nmid n$ and $P \cong C_2 \times (SD_{2^{r+1}} \wr Q)$. Here $Q \in \mathrm{Syl}_2(S_{(n-1)/2})$ and $2^r m = q^2 - 1$ where $2 \nmid m$.

Similar results hold for other classical groups. Hiss and Kessar [117, 118] and Waldmüller [277] have obtained partial results on Donovan's Conjecture for classical groups.

Finally, we turn to sporadic groups. Here for $p = 2$ the defect groups of blocks are listed in Landrock's paper [175] (see also [208]). Results on the essential rank of sporadic groups are contained in [8]. The possible Brauer trees of blocks of sporadic groups with cyclic defect groups are determined in [119, 194]. Moreover, many of the sporadic groups can be handled with GAP [266]. Information on blocks with specific properties can be found in articles by An and Eaton [7, 12–14].

In addition to Donovan's Conjecture mentioned above, several of the other conjectures from Chap. 2 have been checked for some of the finite simple groups. We do not give references here.

We collect some applications of the classification.

Theorem 7.14 (Kessar-Malle [152]) *Every block B of a finite group with abelian defect group satisfies $k(B) = k_0(B)$.*

Theorem 7.15 (Navarro-Tiep [201]) *Brauer's Height Zero Conjecture holds for 2-blocks of maximal defect.*

The proof of the next theorem relies on constructions of perfect isometries.

Theorem 7.16 (Fong, Harris, Sawabe, Usami, Watanabe [84, 258, 259, 274]) *Let B be a principal p-block with abelian defect group.*

  (i) *If $p = 2$, then $l(B) = k(I(B))$.*
 (ii) *If $e(B)$ is a prime or the square of a prime, then $l(B) = e(B)$.*
(iii) *If $p \neq 3$ and $I(B)$ is an elementary abelian 2-group or D_8, then $l(B) = k(I(B))$.*

In particular, in all three cases Alperin's Weight Conjecture holds.

Compare also with Table 1 in the introduction.

The final result of this section gives information about so-called *TI defect blocks*.

Theorem 7.17 (An-Eaton [11]) *Let B be a p-block of a finite group G with defect group D such that $D \cap gDg^{-1} = O_p(G)$ for all $g \in G \setminus N_G(D)$. Then Alperin's Weight Conjecture and the Alperin-McKay Conjecture hold for B.*

7.5   Blocks of p-Solvable Groups

For sake of completeness we state a few results concerning the opposite situation where G is a p-solvable group, i.e. the non-abelian composition factors of G are p'-groups. In this case the block theory of G is well-understood. One of the main theorems comes from Külshammer [161]. We enhance this old result by invoking fusion systems.

Theorem 7.18 *Let B be a p-block of a p-solvable group G with defect group D and fusion system $\mathscr{F}$. Then B is Morita equivalent to a twisted group algebra $\mathcal{O}_\gamma H$ where H is an extension of $P := O_p(\mathscr{F})$ with $\mathrm{Out}_{\mathscr{F}}(P)$. Moreover, $D \in \mathrm{Syl}_p(H)$.*

Proof By the Fong reductions, we may assume that B is quasiprimitive and $O_{p'}(G)$ is central. Then we are in a position to apply Proposition J in [161] which also works over $\mathcal{O}$ instead of F (as B. Külshammer informed me). In particular, $D \in \mathrm{Syl}_p(G)$. We need to show that the normal subgroup P in Theorem A in [161] coincides with $O_p(\mathscr{F})$. By Proposition J in [161] we have

$$P = D \cap O_{p'p}(G) = D \cap (O_p(G) \times O_{p'}(G)) = O_p(G) \subseteq O_p(\mathscr{F}).$$

On the other hand, it follows from Parts (iii) and (v) in Theorem A in [161] that $O_p(\text{Out}_{\mathscr{F}}(P)) = 1$, i.e. P is $\mathscr{F}$-radical. Moreover, the Hall-Higman Lemma implies $C_D(P) \subseteq P$. Hence, P is also $\mathscr{F}$-centric. By Theorem 5.39 in [61] we obtain $O_p(\mathscr{F}) \subseteq P$. $\square$

In the situation of Theorem 7.18, we have $C_D(O_p(\mathscr{F})) \subseteq O_p(\mathscr{F})$ (see also [60]). This gives the following consequence.

Corollary 7.19 *The fusion system of a p-block of a p-solvable group is constrained.*

As an immediate corollary of Theorem 7.18 we obtain Donovan's Conjecture restricted to blocks of p-solvable groups. Most of the other conjectures introduced in Chap. 2 are also satisfied for p-solvable groups. We list some references:

- Brauer's $k(B)$-Conjecture for p-solvable groups reduces to what is known as the $k(GV)$-Problem (see Nagao [195]). This problem was settled recently by work of several authors (see [260]).
- Brauer's Height Zero Conjecture for p-solvable groups was verified by Gluck and Wolf [93].
- The Alperin-McKay Conjecture for p-solvable groups was proved by Okuyama and Wajima [210].
- Isaacs and Navarro [133] obtained the Galois-Alperin-McKay Conjecture for p-solvable groups.
- Külshammer [164] showed that the Alperin-McKay Conjecture (for a specific block) would imply Olsson's Conjecture (for the same block). Hence, also Olsson's Conjecture for p-solvable groups is true.
- A proof of Alperin's Weight Conjecture for p-solvable groups (and its mysterious history) appeared in [20].
- Eaton [73] has shown that the Ordinary Weight Conjecture is equivalent to Dade's Projective Conjecture (also if one restricts to p-solvable groups). The latter conjecture holds for p-solvable groups by work of Robinson [234]. Thus, the OWC is also correct for p-solvable groups.
- In particular, also Robinson's Conjecture is satisfied for p-solvable groups. There is an even stronger bound on the heights of characters given in [192].
- It was mentioned in Linckelmann [183] that the Gluing Problem for p-solvable groups has a unique solution.
- Concerning the Eaton-Moretó Conjecture for p-solvable groups it is at least known that

$$\min\{i \geq 1 : k_i(D) > 0\} \leq \inf\{i \geq 1 : k_i(B) > 0\}$$

(see [76]).
- Gluck's Conjecture is at least true for solvable groups by Gluck [92].

Part III
Applications

Chapter 8
Metacyclic Defect Groups

As a first application of the methods in Part II we investigate blocks with metacyclic defect groups. We remark that all metacyclic p-groups are classified (see e.g. [179]), but we will not make use of this fact.

8.1 The Case $p = 2$

The following theorem collects the knowledge about 2-blocks with metacyclic defect groups.

Theorem 8.1 *Let B be a 2-block of a finite group with metacyclic defect group D. Then one of the following holds:*

(1) B *is nilpotent. Then $k_i(B)$ is the number of ordinary characters of D of degree 2^i. In particular $k(B)$ is the number of conjugacy classes of D and $k_0(B) = |D : D'|$. Moreover, $l(B) = 1$.*
(2) D *is a dihedral group of order $2^n \geq 8$. Then $k(B) = 2^{n-2} + 3$, $k_0(B) = 4$ and $k_1(B) = 2^{n-2} - 1$. According to two different fusion systems, $l(B)$ is 2 or 3. The Cartan matrix of B is*

$$\begin{pmatrix} 2^{n-2} + 1 & 2 \\ 2 & 4 \end{pmatrix} \qquad or \qquad \begin{pmatrix} 2^{n-2} + 1 & 1 & 1 \\ 1 & 2 & . \\ 1 & . & 2 \end{pmatrix}$$

up to basic sets. Moreover, the characters of height 0 in B are 2-rational and $\mathrm{Irr}_1(B)$ splits in $n - 2$ families of 2-conjugate characters of lengths $1, 2, 4, \ldots, 2^{n-3}$ respectively.

© Springer International Publishing Switzerland 2014
B. Sambale, *Blocks of Finite Groups and Their Invariants*, Lecture Notes in Mathematics 2127, DOI 10.1007/978-3-319-12006-5_8

(3) *D is a quaternion group of order 8. Then $k(B) = 7$, $k_0(B) = 4$ and $k_1(B) = l(B) = 3$. The Cartan matrix of B is*

$$2 \begin{pmatrix} 2 & 1 & 1 \\ 1 & 2 & . \\ 1 & . & 2 \end{pmatrix}$$

up to basic sets. Moreover, there is one pair of 2-conjugate characters of height 1 and all other characters in B are 2-rational.

(4) *D is a quaternion group of order $2^n \geq 16$. Then $k_0(B) = 4$ and $k_1(B) = 2^{n-2} - 1$. Moreover, $\mathrm{Irr}_1(B)$ splits in $n - 2$ families of 2-conjugate characters of lengths $1, 2, 4, \ldots, 2^{n-3}$ respectively, and all other characters in B are 2-rational. According to two different fusion systems, one of the following holds*

(a) *$k(B) = 2^{n-2} + 4$, $k_{n-2}(B) = 1$ and $l(B) = 2$. The Cartan matrix of B is*

$$2 \begin{pmatrix} 2^{n-3} + 1 & 2 \\ 2 & 4 \end{pmatrix}$$

up to basic sets.

(b) *$k(B) = 2^{n-2} + 5$, $k_{n-2}(B) = 2$ and $l(B) = 3$. The Cartan matrix of B is*

$$2 \begin{pmatrix} 2^{n-3} + 1 & 1 & 1 \\ 1 & 2 & . \\ 1 & . & 2 \end{pmatrix}$$

up to basic sets.

(5) *D is a semidihedral group of order $2^n \geq 16$. Then $k_0(B) = 4$ and $k_1(B) = 2^{n-2} - 1$. Moreover, $\mathrm{Irr}_1(B)$ splits in $n - 2$ families of 2-conjugate characters of lengths $1, 2, 4, \ldots, 2^{n-3}$ respectively, and all other characters in B are 2-rational. According to three different fusion systems, one of the following holds*

(a) *$k(B) = 2^{n-2} + 3$ and $l(B) = 2$. The Cartan matrix of B is*

$$2 \begin{pmatrix} 2^{n-3} + 1 & 2 \\ 2 & 4 \end{pmatrix}$$

up to basic sets.

(b) *$k(B) = 2^{n-2} + 4$, $k_{n-2}(B) = 1$ and $l(B) = 2$. The Cartan matrix of B is*

$$\begin{pmatrix} 2^{n-2} + 1 & 2 \\ 2 & 4 \end{pmatrix}$$

up to basic sets.

(c) $k(B) = 2^{n-2} + 4$, $k_{n-2}(B) = 1$ and $l(B) = 3$. The Cartan matrix of B is

$$\begin{pmatrix} 2^{n-2} + 1 & 1 & 1 \\ 1 & 3 & -1 \\ 1 & -1 & 3 \end{pmatrix}$$

up to basic sets.

(6) $D \cong C_{2^n} \times C_{2^n}$ is homocyclic. Then $k(B) = k_0(B) = (|D| + 8)/3$ and $l(B) = 3$. The Cartan matrix of B is

$$\frac{1}{3} \begin{pmatrix} 2^{2n} + 2 & 2^{2n} - 1 & 2^{2n} - 1 \\ 2^{2n} - 1 & 2^{2n} + 2 & 2^{2n} - 1 \\ 2^{2n} - 1 & 2^{2n} - 1 & 2^{2n} + 2 \end{pmatrix}$$

up to basic sets.

Proof Let $\mathscr{F}$ be the fusion system of B. It was shown in the author's dissertation [244] (see also [248]) that $\mathscr{F}$ is nilpotent unless homocyclic or a 2-group of maximal class. Independently, this was also obtained by Robinson [237] and Craven-Glesser [63]. In fact, Propositions 6.11 and 10.2 below imply the claim. Moreover, we will generalize this result in Theorem 10.17.

Now let D be a 2-group of maximal class. Then Brauer [41] and Olsson [212] computed $k(B)$, $k_i(B)$ and $l(B)$. This will be generalized in Chap. 9. They also obtained the distribution into 2-rational and 2-conjugate characters. The statement about Cartan matrices can be extracted from Erdmann [80] and Cabanes-Picaronny [52].

Suppose next that D is homocyclic. Then results by Usami [270] show that B is perfectly isometric to its Brauer correspondent in $N_G(D)$. Observe that Usami assumes $p \neq 2$ in her paper. However, in a later paper together with Puig [227, Introduction] they claim without proof that the case $p = 2$ can be handled similarly. An explicit proof for the special case above was given in the author's dissertation [244]. Hence, the invariants and the Cartan matrix can be obtained from Theorem 1.19. □

If D is dihedral (including C_2^2), semidihedral or quaternion, then B has tame representation type. We deduce the conjectures.

Corollary 8.2 *Every 2-block B with metacyclic defect group satisfies the following conjectures:*

- *Alperin's Weight Conjecture*
- *Brauer's $k(B)$-Conjecture*
- *Brauer's Height-Zero Conjecture*
- *Olsson's Conjecture*
- *Galois-Alperin-McKay Conjecture*
- *Ordinary Weight Conjecture*

- *Gluck's Conjecture*
- *Eaton's Conjecture*
- *Eaton-Moretó Conjecture*
- *Malle-Navarro Conjecture*
- *Robinson's Conjecture*

Moreover, the Gluing Problem for B has a unique solution.

Proof Most conjectures follow straight from Theorem 8.1. The Galois-Alperin-McKay Conjecture asserts (for $p = 2$) that every Galois automorphism γ has the same number of fixed points in $\mathrm{Irr}_0(B)$ and in $\mathrm{Irr}_0(b)$ where b is the Brauer correspondent of B in $\mathrm{N}_G(D)$. This has been checked for nilpotent blocks in [134]. If D has maximal class, then the action of γ on $\mathrm{Irr}_0(B)$ is trivial by Theorem 8.1. Thus, we may assume that D is homocyclic. By Brauer's Permutation Lemma, the number of fixed points in $\mathrm{Irr}(B)$ under the action of the cyclic group $\langle \gamma \rangle$ is the same as the number of fixed columns of the generalized decomposition matrix. For $1 \neq u \in D$ we have $l(b_u) = 1$ and u is not conjugate to any of its proper powers under $I(B)$. Therefore, the number of fixed points of γ is locally determined, and the Galois-Alperin-McKay Conjecture follows.

For Alperin's Weight Conjecture we may refer to [147]. The Ordinary Weight Conjecture for tame cases was shown in [235] (we may also refer to Theorems 9.10, 9.30 and 9.39 from the next chapter). Even Dade's Invariant Conjecture holds here by a result of Uno [268]. In the homocyclic case, the OWC reduces to

$$k(B) = \sum_{\chi \in \mathrm{Irr}(D)/I(B)} |I(\chi)|$$

which is true. Now we settle Gluck's Conjecture. This is easy to see in the nilpotent case, since we have $l(b_u) = 1$ for every B-subsection (u, b_u) here. In the tame cases, D has nilpotency class 2 and thus, order 8. Here Gluck's Conjecture holds by Theorem B in [92]. Also the abelian case was handled in [92, Corollary 3.2].

It remains to consider the Gluing Problem. If B is a controlled block, the Gluing Problem has a unique solution by Example 5.3 in [183]. This solves the nilpotent case and the homocyclic case. For the tames cases the claim follows from [219]. □

Now we discuss Donovan's Conjecture. The nilpotent case follows by Puig's Theorem 1.30 at once. For dihedral and semidihedral defect groups, Holm [121,122] proved Donovan's Conjecture by using Erdmann's work [80] (at least over the field F; preliminary work was done by Donovan [69] and Linckelmann [181]). He also gave an argument which shows Donovan's Conjecture for quaternion defect groups provided $l(B) = 3$ (see also [150]). A version for $\mathcal{O}$ and Q_8 can be found in [123]. Unfortunately, the case (4a) in Theorem 8.1 appears to be open. However, the Morita equivalences are determined up to certain scalars.

In the smallest homocyclic case C_2^2, the blocks also have tame representation type. Here Donovan's Conjecture over F follows from [79] and over $\mathcal{O}$ by Linckelmann [182] (for a stronger result see [62]). For the general case of a homocyclic defect group we have results by Usami [270] about the existence of perfect isometries. Brauer [38] has shown that a group with a homocyclic Sylow 2-subgroup of order at least 16 is solvable.

In a recent paper [74] we addressed Donovan's Conjecture for these defect groups by making use of the classification of the finite simple groups. We provide the details without the long and complicated proof.

Theorem 8.3 *Let B be a block of a finite group G with defect group $D \cong C_{2^m} \times C_{2^m}$ for some $m \geq 2$. Then B is Morita equivalent to its Brauer correspondent in $N_G(D)$.*

Corollary 8.4 *Let B be a 2-block of a finite group with abelian defect group D of rank at most 2. Then one of the following holds:*

(i) B is nilpotent and thus Morita equivalent to $\mathcal{O}D$.
(ii) B is Morita equivalent to $\mathcal{O}[D \rtimes C_3]$.
(iii) $D \cong C_2^2$ and B is Morita equivalent to $B_0(\mathcal{O}A_5)$.

In particular, Donovan's Conjecture holds for D and Broué's Abelian Defect Group Conjecture holds for B.

Apart from these results, the work [74] also contains the following strong theorem.

Theorem 8.5 *Donovan's Conjecture holds 2-blocks with elementary abelian defect groups.*

8.2 The Case $p > 2$

For sake of completeness we start with the cyclic case. Brauer [33] obtained the invariants of blocks with defect 1. This was extended by Dade [65] to cyclic defect groups. Later, also Broué's Abelian Defect Group Conjecture was established for cyclic defect groups (see [180, 228, 240]).

Theorem 8.6 (Dade [65]) *Let B be a p-block of a finite group with cyclic defect group D. Then*

$$l(B) = e(B) \mid p - 1, \qquad k(B) = k_0(B) = \frac{|D| - 1}{e(B)} + e(B).$$

The Cartan matrix of B is given by $(m + \delta_{ij})_{1 \leq i,j \leq e(B)}$ up to basic sets where $m := (|D| - 1)/e(B)$ is the multiplicity of B.

In the next interesting case the defect group is elementary abelian of order 9. Here the block invariants are not determined completely (see [154]). Nevertheless, Donovan's Conjecture is known to hold for all principal 3-blocks with abelian defect groups (see [156]). Also, Broué's Conjecture is true for principal blocks with defect group C_3^2 (see [158]). It is also easy to see that Alperin's Weight Conjecture for the defect groups $C_{3^n} \times C_{3^m}$ where $n \neq m$ follows from the Usami-Puig results [226, 270]. This was explicitly carried out in [288].

Despite these obstacles in the abelian case, Brauer's $k(B)$-Conjecture and Olsson's Conjecture were proved for all blocks with metacyclic defect groups by Gao [88] and Yang [287]. In this section we will add some more conjectures to the list. One important ingredient is the following result by Stancu.

Theorem 8.7 (Stancu [265]) *Let $p > 2$, and let B be a p-block of a finite group with metacyclic defect group. Then B is controlled.*

Now we are in a position to prove the main theorem of this section.

Theorem 8.8 *Let B be a p-block of a finite group with a metacyclic, non-abelian defect group D for an odd prime p Then one of the following holds*

(1) B is nilpotent.s
(2) D has the following form

$$D = \langle x, y \mid x^{p^m} = y^{p^n} = 1, \ yxy^{-1} = x^{1+p^l} \rangle \cong C_{p^m} \rtimes C_{p^n} \qquad (8.1)$$

with $0 < l < m$ and $m - l \leq n$. Moreover, $l(B) = e(B) \mid p - 1$ and

$$k(B) = \left(\frac{p^l + p^{l-1} - p^{2l-m-1} - 1}{e(B)} + e(B) \right) p^n.$$

In particular, Alperin's Weight Conjecture holds for B.

Proof If D is a non-split extension of two cyclic groups, then a result by Dietz [68] says that B is nilpotent. Hence, we may assume that D is split, and B is non-nilpotent. Then it is easy to see that D has a presentation as in Eq. (8.1). Assume that the map $x \rightarrow x^{\alpha_1}$ defines an automorphism of $\langle x \rangle$ of order $p - 1$. Then by Theorem 2.5 in [88] the map α with $\alpha(x) = x^{\alpha_1}$ and $\alpha(y) = y$ defines an automorphism of D of order $p - 1$, and we may assume that $I(B) \leq \langle \alpha \rangle$. In particular, $e(B) \mid p - 1$. Moreover, it is easy to see that $\mathrm{foc}(B) = \langle x \rangle$ is cyclic. Hence, Theorem 1.33 implies $l(B) = e(B)$ and $k(B) = k(D \rtimes I(B))$. Since $D \rtimes I(B)$ has only one block (namely the principal block), we obtain $k(B)$ from [89] (this can also be obtained via [113]).   □

It remains to consider the numbers $k_i(B)$. Apart from [89] we mention a result by Watanabe [124], which states that two principal blocks with a common metacyclic, non-abelian defect group and the same inertial index are perfectly isometric.

We will make use of the parameters m, n, l introduced in Theorem 8.8. The following result is taken from [254] (using [251]).

Proposition 8.9 *Let B be a p-block of a finite group with a non-abelian split metacyclic defect group D for an odd prime p. Then*

$$2p^n \leq k_0(B) \leq \left(\frac{p^l - 1}{e(B)} + e(B) \right) p^n \leq p^{n+l} = |D : D'|,$$

$$\sum_{i=0}^{\infty} p^{2i} k_i(B) \leq \left(\frac{p^l - 1}{e(B)} + e(B) \right) p^{n+m-l} \leq p^{n+m} = |D|,$$

$$p^n \mid k_0(B), \quad p^{n-m+l} \mid k_i(B) \quad \text{for } i \geq 1,$$

$$k_i(B) = 0 \quad \text{for} \quad i > \min\left\{ 2(m - l), \frac{m + n - 1}{2} \right\}.$$

Proof We continue the notation from the proof of Theorem 8.8. By Proposition 1.31 we have $p^n \mid |D : \mathfrak{foc}(B)| \mid k_0(B)$. In particular $p^n \leq k_0(B)$. In case $k_0(B) = p^n$ it follows from Proposition 1.32 that B is nilpotent. However then we would have $k_0(B) = |D : D'| = p^{n+l} > p^n$. Therefore $2p^n \leq k_0(B)$. Proposition 1.31 also implies $p^{n-m+l} \mid |Z(D) : Z(D) \cap \mathfrak{foc}(B)| \mid k_i(B)$ for $i \geq 1$.

Now consider the subsection (y, b_y). Since B is controlled, b_y has defect group $C_D(y)$ and fusion system $C_{\mathscr{F}}(\langle y \rangle)$ where $\mathscr{F}$ is the fusion system of B. It follows that $e(b_y) = e(B)$. As usual, b_y dominates a block of $C_G(y)/\langle y \rangle$ with cyclic defect group $C_D(y)/\langle y \rangle \cong \langle x^{p^{m-l}} \rangle$ of order p^l. Hence, Proposition 4.3 implies the first inequality. For the second we consider $u := x^{p^{m-l}} \in Z(D)$. Since $I(B)$ acts non-trivially on $\langle u \rangle$, we observe that b_u is nilpotent and $l(b_u) = 1$. Moreover, $|\mathrm{Aut}_{\mathscr{F}}(\langle u \rangle)| = e(B)$. Thus, Theorem 5.12 shows the second claim. Since $k_0(B) > 0$, it follows at once that $k_i(B) = 0$ for $i > (n + m - 1)/2$. On the other hand Corollary V.9.10 in [81] implies $k_i(B) = 0$ for $i > 2(m - l)$. □

Theorem 8.10 *Let B be a p-block of a finite group with a non-abelian split metacyclic defect group D for an odd prime p. Then the lengths of the families of p-conjugate characters in $\mathrm{Irr}(B)$ are the same as in $\mathrm{Irr}(D \rtimes I(B))$.*

Proof It suffices to show that the distribution of $\mathrm{Irr}(B)$ into 3-rational and 3-conjugate characters only depends on D and $e(B)$. By Brauer's Permutation Lemma (Lemma IV.6.10 in [81]) we may study the action of the Galois group $\mathscr{G}$ (see Sect. 1.2) on the columns of the generalized decomposition matrix. For the elements $u \in D$ such that $l(b_u) = 1$ there is no difficulty in determine the action of $\mathscr{G}$ on the powers of u. Now assume that $l(b_u) > 1$. Let $\mathscr{F}$ be the fusion system of B. Then there is a non-trivial p'-element $\gamma \in \mathrm{Aut}_{\mathscr{F}}(D)$ which centralizes u. Since $\mathrm{Aut}_{\mathscr{F}}(D) \cong \mathrm{Inn}(D) \rtimes I(B)$, γ is D-conjugate to a non-trivial power of α (see proof of Theorem 8.8). Since $C_D(\alpha) = \langle y \rangle$, it follows that u is D-conjugate to a power of y. Thus, we may assume $u \in \langle y \rangle$. Suppose that u is $\mathscr{F}$-conjugate to u^i for some $i \in \mathbb{Z}$. Then u and u^i are even conjugate in D. Therefore $u\langle x \rangle$ and $u^i\langle x \rangle$ are conjugate in the abelian group $D/\langle x \rangle$. Hence, $u^i = u$. Thus, we have seen that the powers of u are pairwise non-conjugate in $\mathscr{F}$. As in the proof of Proposition 8.9

we obtain $l(b_u) = e(B)$. Let $|\langle u \rangle| = p^k$. Then the action of $\mathscr{G}$ gives $e(B)$ p-rational characters (corresponding to u^0) and $e(B)$ orbits of length $(p-1)p^i$ for each $i = 0, \ldots, k - 1$. $\square$

In the situation of Theorem 8.10 it is of course possible to determine the lengths of the families in terms of p, n, m, l and $e(B)$. Since this is quite tedious, we will only do so in special cases (see next section).

Corollary 8.11 *Let B be a block of a finite group with metacyclic defect group. Then Brauer's $k(B)$-Conjecture, Brauer's Height Zero Conjecture and Olsson's Conjecture are satisfied for B.*

Proof Assume first that B has abelian defect group D. Then Brauer's $k(B)$-Conjecture follows from Theorem 4.2 (cf. Proposition 4.3 or Theorem VII.10.13 in [81]). Then of course also Olsson's Conjecture is true. For the Height Zero Conjecture we refer to Theorem 7.14. Now let D be non-abelian. Then we may assume that D is split by Theorem 8.8. We need to show that $k_0(B) < k(B)$. By Proposition 8.9 it suffices to show

$$\left(\frac{p^l - 1}{e(B)} + e(B) \right) p^n < \left(\frac{p^l + p^{l-1} - p^{2l-m-1} - 1}{e(B)} + e(B) \right) p^n.$$

This reduces to $l < m$, one of our hypotheses. $\square$

Apart from a special case covered in [261], it seems that there are no results about B in the literature for p-solvable groups. We take the opportunity to give such a result which also holds in a more general situation.

Theorem 8.12 *Let B be a controlled block of a p-solvable group. If $I(B)$ is cyclic, then B is Morita equivalent to the group algebra $\mathcal{O}[D \rtimes I(B)]$ where D is the defect group of B.*

Proof This follows from Theorem 7.18. $\square$

Let us consider the opposite situation where G is quasisimple. Then the main theorem of [7] tells us that B cannot have non-abelian metacyclic defect groups. Thus, in order to settle the general case it would be sufficient to reduce the situation to quasisimple groups. As a concrete example we note that the principal 3-block of $^2G_2(3)$ has defect group $C_9 \rtimes C_3$ (see Example 4.3 in [120]).

In the next section we restrict the parameter l in order to compute $k_i(B)$.

8.2.1   Metacyclic, Minimal Non-abelian Defect Groups

In this section we assume that $l = m - 1$, i.e.

$$D = \langle x, y \mid x^{p^m} = y^{p^n} = 1, \; yxy^{-1} = x^{1+p^{m-1}} \rangle \cong C_{p^m} \rtimes C_{p^n}$$

where $m \geq 2$ and $n \geq 1$. These are precisely the metacyclic, minimal non-abelian defect groups (see Chap. 12 below). By Lemma 12.1 these are also the metacyclic p-groups such that $|D'| = p$. The material is an improved version of [254].

Theorem 8.13 *Let B be a p-block of a finite group with metacyclic, minimal non-abelian defect group D for an odd prime p. Then*

$$k_0(B) = \left(\frac{p^{m-1}-1}{e(B)} + e(B)\right)p^n \qquad k_1(B) = \frac{p-1}{e(B)}p^{n+m-3}$$

$$k(B) = \left(\frac{p^{m-1} + p^{m-2} - p^{m-3} - 1}{e(B)} + e(B)\right)p^n \quad l(B) = e(B).$$

In particular the following conjectures are satisfied for B (in addition to those listed in previous results):

- *Alperin-McKay Conjecture*
- *Ordinary Weight Conjecture*
- *Eaton's Conjecture*
- *Eaton-Moretó Conjecture*
- *Robinson's Conjecture*
- *Malle-Navarro Conjecture*

Proof By Proposition 8.9 we have

$$p^n \mid k_0(B) \leq \left(\frac{p^{m-1}-1}{e(B)} + e(B)\right)p^n.$$

Thus, by way of contradiction we may assume that

$$k_0(B) \leq \left(\frac{p^{m-1}-1}{e(B)} + e(B) - 1\right)p^n.$$

Then Theorem 8.8 implies the following contradiction

$$\left(\frac{p^m-1}{e(B)} + p^2 + e(B) - 1\right)p^n$$

$$= \left(\frac{p^{m-1}-1}{e(B)} + e(B) - 1\right)p^n + \left(\frac{p^{m-2} - p^{m-3}}{e(B)} + 1\right)p^{n+2}$$

$$\leq \sum_{i=0}^{\infty} p^{2i} k_i(B) \leq \left(\frac{p^m - p}{e(B)} + pe(B)\right)p^n < \left(\frac{p^m-1}{e(B)} + p^2\right)p^n.$$

This gives $k_0(B)$. By way of contradiction suppose that $k_i(B) > 0$ for some $i \geq 2$. By Proposition 8.9, $k_1(B)$ and $k_i(B)$ are divisible by p^{n-1}. This shows

$$\left(\frac{p^{m-1}-1}{e(B)} + e(B)\right)p^n + \left(\frac{p^{m-1}-p^{m-2}}{e(B)} - 1\right)p^{n+1} + p^{n+3} \leq \sum_{i=0}^{\infty} p^{2i} k_i(B)$$

$$\leq \left(\frac{p^{m-1}-1}{e(B)} + e(B)\right)p^{n+1}.$$

Hence, we derive the following contradiction

$$p^{n+3} - p^{n+1} \leq \left(\frac{1-p}{e(B)} + e(B)(p-1)\right)p^n \leq p^{n+2}.$$

This shows $k_1(B) = k(B) - k_0(B)$.

Since the Brauer correspondent of B in $N_G(D)$ has the same fusion system, the Alperin-McKay Conjecture follows. Now we prove the Ordinary Weight Conjecture. Let $Q \leq D$ be an $\mathscr{F}$-radical subgroup. If $I(B)$ does not restrict to Q, we derive the contradiction $\mathrm{Out}_{\mathscr{F}}(Q) \cong N_D(Q)/Q \neq 1$. Hence, $I(B)$ acts on Q and thus also on $N_D(Q)$. In particular, $N_D(Q)/Q \trianglelefteq \mathrm{Out}_{\mathscr{F}}(Q)$. This shows $N_D(Q) = Q$ and $Q = D$. Since $I(B)$ is cyclic, we conclude that $\mathrm{H}^2(\mathrm{Out}_{\mathscr{F}}(Q), F^\times) = 1$. Thus all 2-cocycles appearing in the OWC are trivial. Therefore the conjecture asserts that $k_i(B)$ only depends on $\mathscr{F}$ and thus on $e(B)$. Since the conjecture is known to hold for the principal block of the solvable group $G = D \rtimes I(B)$, the claim follows. Eaton's Conjecture is equivalent to Brauer's $k(B)$-Conjecture and Olsson's Conjecture. Both are known to hold by Corollary 8.11. Also the Eaton-Moretó Conjecture and Robinson's Conjecture are trivially satisfied for B. The Malle-Navarro Conjecture asserts that $k(B)/k_0(B) \leq k(D') = p$ and $k(B)/l(B) \leq k(D)$. The first inequality is easy to see. For the second inequality we observe that every conjugacy class of D has at most p elements, since $|D : Z(D)| = p^2$. Hence,

$$k(D) = |Z(D)| + \frac{|D| - |Z(D)|}{p} = p^{n+m-1} + p^{n+m-2} - p^{n+m-3}.$$

We deduce

$$\frac{k(B)}{l(B)} \leq k(B) = \left(\frac{p^{m-1} + p^{m-2} - p^{m-3} - 1}{e(B)} + e(B)\right)p^n$$

$$\leq (p^{m-1} + p^{m-2} - p^{m-3})p^n = k(D).$$

$\square$

Theorem 8.13 already gives new information for the non-abelian defect group of order p^3 and exponent p^2 (completing results by Hendren [107]). We will denote this group by p_-^{1+2}.

Corollary 8.14 *Let B be a p-block of a finite group with defect group p_-^{1+2} for an odd prime p. Then the Galois-Alperin-McKay Conjecture holds for B.*

Proof Let D be a defect group of B, and let $\mathscr{G}$ be the relevant Galois group. Let $\gamma \in \mathscr{G}$ be a p-element. Then it suffices to show that γ acts trivially on $\mathrm{Irr}_0(B)$. By Lemma IV.6.10 in [81] it is enough to prove that γ acts trivially on the $\mathscr{F}$-conjugacy classes of subsections of B via $^{\gamma}(u, b_u) := (u^{\overline{\gamma}}, b_u)$ where $u \in D$ and $\overline{\gamma} \in \mathbb{Z}$. Since γ is a p-element, this action is certainly trivial unless $|\langle u \rangle| = p^2$. Here however, the action of γ on $\langle u \rangle$ is just the D-conjugation. The result follows. $\square$

In the situation of Corollary 8.14 one can say a bit more: By Theorem 8.10, it is easy to see that $\mathrm{Irr}(B)$ splits into the following families of p-conjugate characters:

- $(p - 1)/e(B) + e(B)$ orbits of length $p - 1$,
- two orbits of length $(p - 1)/e(B)$,
- $e(B)$ p-rational characters.

Without loss of generality, let $e(B) > 1$. Since $k_1(B) < p - 1$, all orbits of length $p - 1$ of p-conjugate characters must lie in $\mathrm{Irr}_0(B)$. In case $e(B) = p - 1$ the remaining $(p - 1)/e(B) + e(B)$ characters in $\mathrm{Irr}_0(B)$ must be p-rational. Now let $e(B) < \sqrt{p - 1}$. Then it is easy to see that $\mathrm{Irr}_0(B)$ contains just one orbit of length $(p - 1)/e(B)$ of p-conjugate characters. Unfortunately, it is not clear if this also holds for $e(B) \geq \sqrt{p - 1}$.

For the prime $p = 3$ we have $e(B) = p - 1$ unless B is nilpotent. This allows us to obtain more information for one particular family (where $m = 2$ and $l = 1$).

Theorem 8.15 *Let B be a non-nilpotent block of a finite group with (non-abelian) defect group $C_9 \rtimes C_{3^n}$ for some $n \geq 1$. Then $\mathrm{Irr}_0(B)$ splits into three 3-rational characters and three families of 3-conjugate characters of size $2 \cdot 3^i$ for $i = 0, \dots, n - 1$. The irreducible characters of height 1 split into one 3-rational character and one family of 3-conjugate characters of size $2 \cdot 3^i$ for $i = 0, \dots, n - 2$. In particular the Galois-Alperin-McKay Conjecture holds for B. The Cartan matrix of B is given by*

$$3^n \begin{pmatrix} 2 & 1 \\ 1 & 5 \end{pmatrix}$$

up to basic sets. Moreover, the Gluing Problem for B has a unique solution.

Proof By Theorem 8.13 we have $k_0(B) = 3^{n+1}$, $k_0(B) = 3^{n-1}$ and $l(B) = 2$. The elements y^i, $x^3 y^{3j}$ and xy^i for $i = 0, \dots, 3^n - 1$ and $j = 0, \dots, 3^{n-1} - 1$ form a set of representatives for the $\mathscr{F}$-conjugacy classes. We have $l(b_u) = 2$ for all $u \in \langle y \rangle$. This gives two 3-rational characters and two orbits (of 3-conjugate characters) of length $2 \cdot 3^i$ for $i = 0, \dots, n - 1$. For all other elements u we have $l(b_u) = 1$. Since x^3 and x^{-3} are $\mathscr{F}$-conjugate, we have another 3-rational character. On the other hand, for $u = x^3 y^{3j}$ with $j = 1, \dots, 3^{n-1} - 1$ the powers of u are pairwise non-conjugate in $\mathscr{F}$. This yields one orbit of length $2 \cdot 3^i$ for $i = 0, \dots, n-2$. The element

x corresponds to a 3-rational character, while the elements $xy^{3^{n-1}}$ and $xy^{-3^{n-1}}$ form one orbit of length 2. Finally the elements xy^j for $j \in \{1, \dots, 3^n-1\} \backslash \{3^{n-1}, 2\cdot 3^{n-1}\}$ form one orbit of length $2\cdot 3^i$ for $i = 1, \dots, n-1$. The three families of length $2\cdot 3^{n-1}$ must certainly lie in $\mathrm{Irr}_0(B)$.

Now we consider the action of $\mathrm{Irr}(D/\mathfrak{foc}(B))$. The characters in $\mathrm{Irr}_0(B)$ form three orbits under the action of $\mathrm{Irr}(D/\mathfrak{foc}(B))$, while the characters in $\mathrm{Irr}_1(B)$ form just one orbit. Observe that $\mathrm{Z}(D)\mathfrak{foc}(B)/\mathfrak{foc}(B) = \langle y^3\mathfrak{foc}(B)\rangle$ is cyclic. Let $\lambda \in \mathrm{Irr}(\mathrm{Z}(D)\mathfrak{foc}(B)/\mathfrak{foc}(B))$ and $u := y^3$. Then $(\lambda * \chi)(u) = \lambda(u)\chi(u)$ (see [238]) and $d^u_{\lambda * \chi, \varphi} = \lambda(u)d^u_{\chi\varphi}$ for $\chi \in \mathrm{Irr}_1(B)$ and $\varphi \in \mathrm{IBr}(b_u)$. This yields orbits (of 3-conjugate characters of height 1) of lengths $1, 2, 2\cdot 3, \dots, 2\cdot 3^{n-2}$. Since no proper sum of these numbers results in $2 \cdot 3^i$ for some i, we see that these orbits do not merge further. This shows that $\mathrm{Irr}_1(B)$ consists of one 3-rational character and one family of 3-conjugate characters of length $2\cdot 3^i$ for $i = 0, \dots, n-2$. The distribution of $\mathrm{Irr}_0(B)$ follows from the arguments above. As a byproduct, it is interesting to note that every irreducible character of B can be obtained from a 3-rational character via the $*$-construction.

By Theorem 1.33, the elementary divisors of the Cartan matrix C of B are 3^n and 3^{n+2}. Hence, $\tilde{C} := 3^{-n}C = \left(\begin{smallmatrix} a & b \\ b & c \end{smallmatrix}\right)$ is an integral matrix with elementary divisors 1 and 9. We may assume that $\tilde{C}$ is reduced as binary quadratic form by changing the basic set if necessary. This means $0 \le 2b \le a \le c$. We derive $3a^2/4 \le ac - b^2 = \det\tilde{C} = 9$ and $a \in \{1, 2, 3\}$. This gives only the following two possibilities for $\tilde{C}$:

$$\begin{pmatrix} 2 & 1 \\ 1 & 5 \end{pmatrix}, \qquad\qquad \begin{pmatrix} 1 & 0 \\ 0 & 9 \end{pmatrix}.$$

It remains to exclude the second matrix (which contradicts Question A on page 25). So assume by way of contradiction that this matrix occurs for $\tilde{C}$. Two irreducible characters of B in the same orbit under $\mathrm{Irr}(D/\mathfrak{foc}(B))$ have the same ordinary decomposition numbers. Hence, the decomposition matrix of B consists of 3^n rows of the form (α_1, α_2), 3^n rows (β_1, β_2), 3^n rows (γ_1, γ_2), and 3^{n-1} rows of the form (δ_1, δ_2) (for the characters of height 1). Consequently,

$$3^n = 3^n(\alpha_1^2 + \beta_1^2 + \gamma_1^2) + 3^{n-1}\delta_1^2,$$

$$3^{n+2} = 3^n(\alpha_2^2 + \beta_2^2 + \gamma_2^2) + 3^{n-1}\delta_2^2,$$

$$0 = 3^n(\alpha_1\alpha_2 + \beta_1\beta_2 + \gamma_1\gamma_2) + 3^{n-1}\delta_1\delta_2.$$

Since $3 \mid \delta_i$, we deduce $\delta_1 = 0$. Since no row of the decomposition matrix vanishes, $\delta_2 = \pm 3$. Without loss of generality, $\alpha_1 = \pm 1$ and $\beta_1 = \gamma_1 = 0$. Then the third equation implies $\alpha_2 = 0$. Thus, $6 = \beta_2^2 + \gamma_2^2$. A contradiction.

Finally we investigate the Gluing Problem for B. For this we use the notation of [219]. Up to conjugation there are four $\mathscr{F}$-centric subgroups $Q_1 := \langle x^3, y\rangle$, $Q_2 := \langle x, y^3\rangle$, $Q_3 := \langle xy, x^3\rangle$ and D. This gives seven chains of $\mathscr{F}$-centric subgroups. It can be shown that $\mathrm{Aut}_{\mathscr{F}}(Q_1) \cong S_3$, $\mathrm{Aut}_{\mathscr{F}}(Q_2) \cong C_6$, $\mathrm{Aut}_{\mathscr{F}}(Q_3) \cong C_3$ and

$\mathrm{Aut}_{\mathscr{F}}(D) \cong C_3 \times S_3$. It follows that $\mathrm{H}^2(\mathrm{Aut}_{\mathscr{F}}(\sigma), F^\times) = 0$ for all chains σ of $\mathscr{F}$-centric subgroups of D. Consequently, $\mathrm{H}^0([S(\mathscr{F}^c)], \mathscr{A}_{\mathscr{F}}^2) = 0$. Hence, by Theorem 1.1 in [219] the Gluing Problem has at least one solution. (Obviously, this should hold in a more general context.)

Now we determine $\mathrm{H}^1([S(\mathscr{F}^c)], \mathscr{A}_{\mathscr{F}}^1)$. For a finite group A it is known that $\mathrm{H}^1(A, F^\times) = \mathrm{Hom}(A, F^\times) = \mathrm{Hom}(A/A'O^{p'}(A), F^\times)$. Using this we observe that $\mathrm{H}^1(\mathrm{Aut}_{\mathscr{F}}(\sigma), F^\times) \cong C_2$ for all chains except $\sigma = Q_3$ and $\sigma = (Q_3 < D)$ in which case we have $\mathrm{H}^1(\mathrm{Aut}_{\mathscr{F}}(\sigma), F^\times) = 0$. Since $[S(\mathscr{F}^c)]$ is partially ordered by taking subchains, one can view $[S(\mathscr{F}^c)]$ as a category where the morphisms are given by the pairs of ordered chains. In particular $[S(\mathscr{F}^c)]$ has exactly 13 morphisms. With the notation of [284] the functor $\mathscr{A}_{\mathscr{F}}^1$ is a *representation* of $[S(\mathscr{F}^c)]$ over $\mathbb{Z}$. Hence, we can view $\mathscr{A}_{\mathscr{F}}^1$ as a module $\mathscr{M}$ over the incidence algebra of $[S(\mathscr{F}^c)]$. More precisely, we have

$$\mathscr{M} := \bigoplus_{a \in \mathrm{Ob}[S(\mathscr{F}^c)]} \mathscr{A}_{\mathscr{F}}^1(a) \cong C_2^5.$$

At this point we can apply Lemma 6.2(2) in [284]. For this let $d : \mathrm{Hom}[S(\mathscr{F}^c)] \to \mathscr{M}$ a derivation. Then by definition we have $d(\beta) = 0$ for $\beta \in \{(Q_3, Q_3), (Q_3, Q_3 < D), (D, Q_3 < D), (Q_3 < D, Q_3 < D)\}$. For all identity morphisms $\beta \in \mathrm{Hom}([S(\mathscr{F}^c)])$ we have $d(\beta) = d(\beta\beta) = \mathscr{A}_{\mathscr{F}}^1(\beta)d(\beta) + d(\beta) = 2d(\beta) = 0$. Since $\beta\gamma$ for $\beta, \gamma \in \mathrm{Hom}([S(\mathscr{F}^c)])$ is only defined if β or γ is an identity, we see that there are no further restrictions on d. On the four morphisms $(Q_1, Q_1 < D)$, $(D, Q_1 < D)$, $(Q_2, Q_2 < D)$ and $(D, Q_2 < D)$ the value of d is arbitrary. It remains to show that d is an inner derivation. For this observe that the map $\mathscr{A}_{\mathscr{F}}^1(\beta)$ is bijective if β is one of the four morphisms above. Now we construct a set $u = \{u_a \in \mathscr{A}_{\mathscr{F}}^1(a) : a \in \mathrm{Ob}[S(\mathscr{F}^c)]\}$ such that d is the inner derivation induced by u. Here we can set $u_{Q_1 < D} = 0$. Then the equation $d((Q_1, Q_1 < D)) = \mathscr{A}_{\mathscr{F}}^1((Q_1, Q_1 < D))(u_{Q_1})$ determines u_{Q_1}. Similarly, $d((D, Q_1 < D)) = \mathscr{A}_{\mathscr{F}}^1(u_D)$ determines u_D. Then $d((D, Q_2 < D)) = \mathscr{A}_{\mathscr{F}}^1(u_D) - u_{Q_2 < D}$ gives $u_{Q_2 < D}$ and finally $d((Q_2, Q_2 < D)) = \mathscr{A}_{\mathscr{F}}^1(u_{Q_2}) - u_{Q_2 < D}$ determines u_{Q_2}. Hence, Lemma 6.2(2) in [284] shows $\mathrm{H}^1([S(\mathscr{F}^c)], \mathscr{A}_{\mathscr{F}}^1) = 0$. So the Gluing Problem has only one solution by Theorem 1.1 in [219]. $\square$

Proposition 8.16 *Let B be a 3-block of a finite group with defect group D. Assume that $D/\langle z \rangle$ is metacyclic but not homocyclic for some $z \in \mathrm{Z}(D)$. Then Brauer's $k(B)$-Conjecture holds for B.*

Proof If $D/\langle z \rangle$ is metacyclic and non-abelian, then $l(b_z) \le 2$ by Theorem 8.8. Hence the claim follows from Theorem 4.9. Now assume that $D/\langle z \rangle \cong C_{3^m} \times C_{3^n}$ where $m \ne n$. Let $\overline{b_z}$ be the block with defect group $D/\langle z \rangle$ dominated by b_z. Then $e(b_z) \le 4$. Now the claim follows from Lemmas 14.4 and 14.5 below. $\square$

We will later settle the case $D/\langle z \rangle \cong C_3^2$ in Proposition 8.16 (see Theorem 13.7).

8.2.2   One Family for $p = 3$

We add another result for $p = 3$ and $l = 1$ in Eq. (8.1).

Theorem 8.17 *Let B be a 3-block of a finite group with defect group*

$$D = \langle x, y \mid x^{3^m} = y^{3^n} = 1, \ yxy^{-1} = x^4 \rangle$$

where $2 \leq m \leq n + 1$. Then $k_0(B) = 3^{n+1}$. In particular, the Alperin-McKay Conjecture holds for B.

Proof We may assume that B is non-nilpotent. By Proposition 8.9 we have $k_0(B) \in \{2 \cdot 3^n, 3^{n+1}\}$. By way of contradiction, suppose that $k_0(B) = 2 \cdot 3^n$. Let $P \in \text{Syl}_p(G)$. Since $D/\mathfrak{foc}(B)$ acts freely on $\text{Irr}_0(B)$, there are 3^n characters of degree $a|P : D|$, and 3^n characters of degree $b|P : D|$ in B for some $a, b \geq 1$ such that $3 \nmid a, b$. Hence,

$$\left| \sum_{\chi \in \text{Irr}_0(B)} \chi(1)^2 \right|_3 = 3^n |P : D|^2 (a^2 + b^2)_3 = |P : D|^2 |D : \mathfrak{foc}(B)|.$$

Now Theorem 1.1 in [151] gives a contradiction.   □

A generalization of the argument in the proof shows that in the situation of Proposition 8.9, $k_0(B) = 2p^n$ can only occur if $p \equiv 1 \pmod 4$.

Chapter 9
Products of Metacyclic Groups

After we have handled 2-blocks with metacyclic defect groups completely, there are several ways to proceed. In this chapter we will see that the methods by Brauer and Olsson for dihedral, semidihedral and quaternion groups can be generalized to deal with direct and central products of cyclic groups and 2-groups of maximal class. These results appeared in [247, 249, 250].

Speaking of representation type, the defect groups in this chapter can be roughly described as "finite times tame". We summarize the results of the whole chapter.

Theorem 9.1 *Let M be a 2-group of maximal class, and let C be a cyclic group. Then for every block B with defect group $M \times C$ or $M * C$ the following conjectures are satisfied:*

- *Alperin's Weight Conjecture*
- *Brauer's $k(B)$-Conjecture*
- *Brauer's Height-Zero Conjecture*
- *Olsson's Conjecture*
- *Alperin-McKay Conjecture*
- *Ordinary Weight Conjecture*
- *Gluck's Conjecture*
- *Eaton's Conjecture*
- *Eaton-Moretó Conjecture*
- *Malle-Navarro Conjecture*
- *Robinson's Conjecture*

Moreover, the Gluing Problem for B has a unique solution.

© Springer International Publishing Switzerland 2014
B. Sambale, *Blocks of Finite Groups and Their Invariants*, Lecture Notes
in Mathematics 2127, DOI 10.1007/978-3-319-12006-5_9

Although the proofs in the following four sections are fairly similar, we did not try to condense the matter, since the results build on one another by induction. Most of the conjectures in Theorem 9.1 are immediate consequences of the main Theorems 9.7, 9.18, 9.28, 9.37, and we will omit the details. Observe that Gluck's Conjecture in this setting only applies to defect groups of order at most 16. This will be handled later in Proposition 13.5.

9.1 $D_{2^n} \times C_{2^m}$

Let B be a block of G with defect group

$$D := \langle x, y, z \mid x^{2^{n-1}} = y^2 = z^{2^m} = [x, z] = [y, z] = 1,\ yxy^{-1} = x^{-1}\rangle$$
$$= \langle x, y\rangle \times \langle z\rangle \cong D_{2^n} \times C_{2^m}$$

where $n \geq 2$ and $m \geq 0$. In the case $n = 2$ and $m = 0$ we get a four-group. Then the invariants of B have been known for a long time. If $n = 2$ and $m = 1$, D is elementary abelian of order 8, and the block invariants are also known (see Theorem 13.1 below). Finally, in the case $n = 2 \leq m$ there exists a perfect isometry between B and its Brauer correspondent (see [227, 270]). Thus, also in this case the block invariants are known, and the major conjectures are satisfied. Hence, we assume $n \geq 3$ for the rest of the section. We allow $m = 0$, since the results are completely consistent in this case.

Lemma 9.2 *The automorphism group* $\mathrm{Aut}(D)$ *is a 2-group.*

Proof This is known for $m = 0$. For $m \geq 1$ the subgroups $\Phi(D) < \Phi(D)\mathrm{Z}(D) < \langle x, z\rangle < D$ are characteristic in D. By Theorem 5.3.2 in [94] every automorphism of $\mathrm{Aut}(D)$ of odd order acts trivially on $D/\Phi(D)$. The claim follows from Theorem 5.1.4 in [94]. $\square$

It follows that the inertial index $e(B)$ of B equals 1. Now we investigate the fusion system $\mathscr{F}$ of B. First we compute the $\mathscr{F}$-centric, $\mathscr{F}$-radical subgroups (instead of the $\mathscr{F}$-essential subgroups), since they are needed later for Alperin's Weight Conjecture.

Lemma 9.3 *Let* $Q_1 := \langle x^{2^{n-2}}, y, z\rangle \cong C_2^2 \times C_{2^m}$ *and* $Q_2 := \langle x^{2^{n-2}}, xy, z\rangle \cong C_2^2 \times C_{2^m}$. *Then* Q_1 *and* Q_2 *are the only candidates for proper* $\mathscr{F}$-*centric,* $\mathscr{F}$-*radical subgroups up to conjugation. Moreover, one of the following cases occurs:*

(aa) $\mathrm{Aut}_{\mathscr{F}}(Q_1) \cong \mathrm{Aut}_{\mathscr{F}}(Q_2) \cong S_3$.
(ab) $\mathrm{Aut}_{\mathscr{F}}(Q_1) \cong C_2$ *and* $\mathrm{Aut}_{\mathscr{F}}(Q_2) \cong S_3$.
(ba) $\mathrm{Aut}_{\mathscr{F}}(Q_1) \cong S_3$ *and* $\mathrm{Aut}_{\mathscr{F}}(Q_2) \cong C_2$.
(bb) $\mathrm{Aut}_{\mathscr{F}}(Q_1) \cong \mathrm{Aut}_{\mathscr{F}}(Q_2) \cong C_2$.

In case (bb) the block B *is nilpotent.*

Proof Let $Q < D$ be $\mathscr{F}$-centric and $\mathscr{F}$-radical. Then $z \in Z(D) \subseteq C_D(Q) \subseteq Q$ and $Q = (Q \cap \langle x, y \rangle) \times \langle z \rangle$. Since $\mathrm{Aut}(Q)$ is not a 2-group, $Q \cap \langle x, y \rangle$ and thus Q must be abelian (see Lemma 9.2). Let us consider the case $Q = \langle x, z \rangle$. Then $m = n - 1$ (this is not important here). The group $D \subseteq N_G(Q, b_Q)$ acts trivially on $\Omega(Q) \subseteq Z(D)$, while a non-trivial automorphism of $\mathrm{Aut}(Q)$ of odd order acts non-trivially on $\Omega(Q)$ (see Theorem 5.2.4 in [94]). This contradicts $O_2(\mathrm{Aut}_{\mathscr{F}}(Q)) = 1$. Hence, Q is isomorphic to $C_2^2 \times C_{2^m}$, and contains an element of the form $x^i y$. After conjugation with a suitable power of x we may assume $Q \in \{Q_1, Q_2\}$. This shows the first claim.

Let $S \leq D$ be an arbitrary subgroup isomorphic to $C_2^2 \times C_{2^m}$. If $z \notin S$, the group $\langle S, z \rangle = (\langle S, z \rangle \cap \langle x, y \rangle) \times \langle z \rangle$ is abelian and of order at least 2^{m+3}. Hence, $\langle S, z \rangle \cap \langle x, y \rangle$ would be cyclic. This contradiction shows $z \in S$. Thus, S is conjugate to $Q \in \{Q_1, Q_2\}$. Since $|N_D(Q)| = 2^{m+3}$, we derive that Q is fully $\mathscr{F}$-normalized. In particular, $N_D(Q)/Q \cong C_2$ is a Sylow 2-subgroup of $\mathrm{Aut}_{\mathscr{F}}(Q)$. Hence, $O_{2'}(\mathrm{Aut}_{\mathscr{F}}(Q))$ has index 2 in $\mathrm{Aut}_{\mathscr{F}}(Q)$. Assume $N_D(Q) C_G(Q) < N_G(Q, b_Q)$. Lemma 5.4 in [184] shows $O_2(\mathrm{Aut}_{\mathscr{F}}(Q)) = 1$. If $m \neq 1$, we have $|\mathrm{Aut}(Q)| = 2^k \cdot 3$ for some $k \in \mathbb{N}$, since $\Phi(Q) < \Omega(Q)\Phi(Q) \leq Q$ are characteristic subgroups. Then $\mathrm{Aut}_{\mathscr{F}}(Q) \cong S_3$. Hence, we may assume $m = 1$. Then $\mathrm{Aut}_{\mathscr{F}}(Q) \leq \mathrm{Aut}(Q) \cong \mathrm{GL}(3, 2)$. Since the normalizer of a Sylow 7-subgroup of $\mathrm{GL}(3, 2)$ has order 21, it follows that $|O_{2'}(\mathrm{Aut}_{\mathscr{F}}(Q))| \neq 7$. Since this normalizer is selfnormalizing in $\mathrm{GL}(3, 2)$, we also have $|O_{2'}(\mathrm{Aut}_{\mathscr{F}}(Q))| \neq 21$. This shows $|O_{2'}(\mathrm{Aut}_{\mathscr{F}}(Q))| = 3$ and $\mathrm{Aut}_{\mathscr{F}}(Q) \cong S_3$, because $|\mathrm{GL}(3, 2)| = 2^3 \cdot 3 \cdot 7$ (compare also with Proposition 6.12).

The last claim follows from Alperin's Fusion Theorem and $e(B) = 1$. $\qquad\square$

The naming of these cases is adopted from [41]. Since the cases (ab) and (ba) are symmetric, we ignore case (ba). It is easy to see that Q_1 and Q_2 are not conjugate in D. Hence, by Alperin's Fusion Theorem the subpairs (Q_1, b_{Q_1}) and (Q_2, b_{Q_2}) are not conjugate in G. It is also easy to see that Q_1 and Q_2 are always $\mathscr{F}$-centric.

Lemma 9.4 *Let* $Q \in \{Q_1, Q_2\}$ *such that* $\mathrm{Aut}_{\mathscr{F}}(Q) \cong S_3$. *Then*

$$C_Q(N_G(Q, b_Q)) \in \{\langle z \rangle, \langle x^{2^{n-2}} z \rangle\}.$$

In particular $z^{2j} \in C_Q(N_G(Q, b_Q))$ *and* $x^{2^{n-2}} z^{2j} \notin C_Q(N_G(Q, b_Q))$ *for* $j \in \mathbb{Z}$.

Proof We consider only the case $Q = Q_1$ (the other case is similar). It is easy to see that the elements in $Q \setminus Z(D)$ are not fixed under $N_D(Q) \subseteq N_D(Q, b_Q)$. Since D acts trivially on $Z(D)$, it suffices to determine the fixed points of an automorphism $\alpha \in \mathrm{Aut}_{\mathscr{F}}(Q)$ of order 3 in $Z(D)$. It is easy to see that $C_Q(\alpha) = \langle a \rangle$ has order 2^m. First we show that $a \in Z(D)$. Suppose the contrary. Let $\beta \in \mathrm{Aut}_{\mathscr{F}}(Q)$ be the automorphism induced by $x^{2^{n-3}} \in N_D(Q) \subseteq N_G(Q, b_Q)$. Then we have $\beta(a) \neq a$. Since $\beta \alpha \beta^{-1} = \alpha^{-1}$, we have $\alpha(\beta(a)) = \beta(\alpha^{-1}(a)) = \beta(a)$. Thus, $\beta(a) \in C_Q(\alpha) = \langle a \rangle$. This gives the contradiction $\beta(a) a^{-1} \in D' \cap \langle a \rangle = \langle x^2 \rangle \cap \langle a \rangle = 1$. Now in case $m \neq 1$ the claim is clear. Thus, assume $m = 1$ and $a = x^{2^{n-2}}$.

Then β acts trivially on $Q/\langle a \rangle$ and α acts non-trivially on $Q/\langle a \rangle$. This contradicts $\beta \alpha \beta^{-1} \alpha = 1$. $\square$

It is not possible to decide whether $C_Q(N_G(Q, b_Q))$ is $\langle z \rangle$ or $\langle x^{2^{n-2}} z \rangle$ in Lemma 9.4, since we can replace z by $x^{2^{n-2}} z$.

Lemma 9.5   *A set of representatives $\mathcal{R}$ for the $\mathcal{F}$-conjugacy classes of elements $u \in D$ such that $\langle u \rangle$ is fully $\mathcal{F}$-normalized is given as follows:*

(i) $x^i z^j$ $(i = 0, 1, \ldots, 2^{n-2}, j = 0, 1, \ldots, 2^m - 1)$ *in case (aa)*.
(ii) $x^i z^j$ *and* $y z^j$ $(i = 0, 1, \ldots, 2^{n-2}, j = 0, 1, \ldots, 2^m - 1)$ *in case (ab)*.

Proof By Lemmas 9.3 and 9.4 in any case the elements $x^i z^j$ $(i = 0, 1, \ldots, 2^{n-2}, j = 0, 1, \ldots, 2^m - 1)$ are pairwise non-conjugate in $\mathcal{F}$. Moreover, $\langle x, z \rangle \subseteq C_G(x^i z^j)$ and $|D : N_D(\langle x^i z^j \rangle)| \leq 2$. Suppose that $\langle x^i y z^j \rangle \trianglelefteq D$ for some $i, j \in \mathbb{Z}$. Then we have $x^{i+2} y z^j = x(x^i y z^j) x^{-1} \in \langle x^i y z^j \rangle$ and the contradiction $x^2 \in \langle x^i y z^j \rangle$. This shows that the subgroups $\langle x^i z^j \rangle$ are always fully $\mathcal{F}$-normalized.

Assume that case (aa) occurs. Then the elements of the form $x^{2i} y z^j$ $(i, j \in \mathbb{Z})$ are conjugate to elements of the form $x^{2i} z^j$ under $D \cup N_G(Q_1, b_{Q_1})$. Similarly, the elements of the form $x^{2i+1} y z^j$ $(i, j \in \mathbb{Z})$ are conjugate to elements of the form $x^{2i} z^j$ under $D \cup N_G(Q_2, b_{Q_2})$. The claim follows in this case.

In case (ab) the given elements are pairwise non-conjugate, since no conjugate of $y z^j$ lies in Q_2. As in case (aa) the elements of the form $x^{2i} y z^j$ $(i, j \in \mathbb{Z})$ are conjugate to elements of the form $y z^j$ under D and the elements of the form $x^{2i+1} y z^j$ $(i, j \in \mathbb{Z})$ are conjugate to elements of the form $x^{2i} z^j$ under $D \cup N_G(Q_2, b_{Q_2})$. Finally, the subgroups $\langle y z^j \rangle$ are fully $\mathcal{F}$-normalized, since $y z^j$ is not conjugate to an element in Q_2. $\square$

Lemma 9.6 *Olsson's conjecture $k_0(B) \leq 2^{m+2} = |D : D'|$ is satisfied in all cases.*

Proof We consider the B-subsection (x, b_x). Since $\langle x \rangle$ is fully $\mathcal{F}$-normalized, b_x has defect group $\langle x, z \rangle$. Since $\langle x, z \rangle$ cannot be isomorphic to a subgroup of Q_1 (or Q_2), it follows that $\mathrm{Aut}_{\mathcal{F}}(\langle x, z \rangle)$ is a 2-group. Hence, b_x is nilpotent and $l(b_x) = 1$. Moreover, x is $\mathcal{F}$-conjugate (even D-conjugate) to x^{-1}. Now the claim follows from Theorem 5.3. $\square$

Theorem 9.7 *In all cases we have*

$$k(B) = 2^m(2^{n-2} + 3), \qquad k_0(B) = 2^{m+2}, \qquad k_1(B) = 2^m(2^{n-2} - 1).$$

Moreover,

$$l(B) = \begin{cases} 1 & \text{in case (bb)} \\ 2 & \text{in case (ab)} \\ 3 & \text{in case (aa)} \end{cases}.$$

In particular Brauer's $k(B)$-Conjecture, Brauer's Height Zero Conjecture and the Alperin-McKay Conjecture hold.

Proof Assume first that case (bb) occurs. Then B is nilpotent and $k_i(B)$ is just the number $k_i(D)$ of irreducible characters of D of degree 2^i ($i \geq 0$) and $l(B) = 1$. Since C_{2^m} is abelian, we get $k_i(B) = 2^m k_i(D_{2^n})$. The claim follows in this case. Thus, we assume that case (aa) or case (ab) occurs. We determine the numbers $l(b)$ for the subsections in Lemma 9.5 and apply Theorem 1.35. Let us begin with the non-major subsections. Since $\mathrm{Aut}_{\mathscr{F}}(\langle x, z \rangle)$ is a 2-group, the blocks $b_{x^i z^j}$ for $i = 1, \ldots, 2^{n-2}-1$ and $j = 0, 1, \ldots, 2^m - 1$ are nilpotent by Lemma 1.34. In particular, $l(b_{x^i z^j}) = 1$. The blocks $b_{y z^j}$ ($j = 0, 1, \ldots, 2^m - 1$) have Q_1 as defect group. Since $N_G(Q_1, b_{Q_1}) = N_D(Q_1) C_G(Q_1)$, they are also nilpotent, and it follows that $l(b_{y z^j}) = 1$.

We divide the (non-trivial) major subsections into three sets:

$$U := \{ x^{2^{n-2}} z^{2j} : j = 0, 1, \ldots, 2^{m-1} - 1 \},$$

$$V := \{ z^j : j = 1, \ldots, 2^m - 1 \},$$

$$W := \{ x^{2^{n-2}} z^{2j+1} : j = 0, 1, \ldots, 2^{m-1} - 1 \}.$$

By Lemma 9.4, case (bb) occurs for b_u, and we get $l(b_u) = 1$ for $u \in U$. The blocks b_v with $v \in V$ dominate unique blocks $\overline{b_v}$ of $C_G(v)/\langle v \rangle$ with defect group $D/\langle v \rangle \cong D_{2^n} \times C_{2^m/|\langle v \rangle|}$ such that $l(b_v) = l(\overline{b_v})$. The same argument for $w \in W$ gives blocks $\overline{b_w}$ with defect group $D/\langle w \rangle \cong D_{2^n}$. This allows us to apply induction on m (for the blocks b_v and b_w). The beginning of this induction ($m = 0$) is satisfied by Theorem 8.1. Thus, we may assume $m \geq 1$. By Lemma 1.34 the cases for b_v (resp. b_w) and $\overline{b_v}$ (resp. $\overline{b_w}$) coincide.

Suppose that case (ab) occurs. By Lemma 9.4, case (ab) occurs for exactly $2^m - 1$ blocks in $\{ b_v : v \in V \} \cup \{ b_w : w \in W \}$ and case (bb) occurs for the other 2^{m-1} blocks. Induction gives

$$\sum_{v \in V} l(b_v) + \sum_{w \in W} l(b_w) = \sum_{v \in V} l(\overline{b_v}) + \sum_{w \in W} l(\overline{b_w}) = 2(2^m - 1) + 2^{m-1}.$$

Taking all subsections together, we derive

$$k(B) - l(B) = 2^m(2^{n-2} + 3) - 2.$$

In particular $k(B) \geq 2^m(2^{n-2}+3)-1$. Let $u := x^{2^{n-2}} \in Z(D)$. Then $2^{h(\chi)} \mid d^u_{\chi \varphi_u}$ and $2^{h(\chi)+1} \nmid d^u_{\chi \varphi_u}$ for $\chi \in \mathrm{Irr}(B)$ by Lemma 1.37. In particular $d^u_{\chi \varphi_u} \neq 0$. Lemma 9.6 gives

$$2^{n+m}-4 \leq k_0(B)+4(k(B)-k_0(B)) \leq \sum_{\chi \in \mathrm{Irr}(B)} \left(d^u_{\chi \varphi_u} \right)^2 = (d(u), d(u)) = |D| = 2^{n+m}.$$

$$\tag{9.1}$$

Hence, we have

$$d^u_{\chi\varphi_u} = \begin{cases} \pm 1 & \text{if } h(\chi) = 0 \\ \pm 2 & \text{otherwise} \end{cases},$$

and the claim follows in case (ab).

Now suppose that case (aa) occurs. Then by the same argument as in case (ab) we have

$$\sum_{v \in V} l(b_v) + \sum_{w \in W} l(b_w) = \sum_{v \in V} l(\overline{b_v}) + \sum_{w \in W} l(\overline{b_w}) = 3(2^m - 1) + 2^{m-1}.$$

Observe that this sum does not depend on which case actually occurs for b_z (for example). In fact all three cases for b_z are possible. Taking all subsections together, we derive

$$k(B) - l(B) = 2^m(2^{n-2} + 3) - 3.$$

Here it is not clear a priori whether $l(B) > 1$. Brauer delayed the discussion of the possibility $l(B) = 1$ until Sect. 7 of [41]. Here we argue differently via lower defect groups and centrally controlled blocks. First we consider the case $m \geq 2$. By Lemma 9.4 we have $\langle D, \mathrm{N}_G(Q_1, b_{Q_1}), \mathrm{N}_G(Q_2, b_{Q_2}) \rangle \subseteq \mathrm{C}_G(z^2)$, i.e. B is centrally controlled. By Theorem 1.38 we get $l(B) \geq l(b_{z^2}) = 3$. Hence, the claim follows with Inequality (9.1).

Now consider the case $m = 1$. By Lemma 9.4 there is a (unique) non-trivial fixed point $u \in \mathrm{Z}(D)$ of $\mathrm{N}_G(Q_1, b_{Q_1})$. Then $l(b_u) > 1$. By Theorem 8.1 the Cartan matrix of b_u has 2 as an elementary divisor. Hence, Proposition 1.41 implies $m^{(1)}_{b_u}(Q) > 0$ for some $Q \leq \mathrm{C}_G(u) = \mathrm{N}_G(\langle u \rangle)$ of order 2. Since $\langle u \rangle \leq \mathrm{Z}(\mathrm{C}_G(u))$, we have $Q = \langle u \rangle$ by Lemma 1.44. Now it follows from Lemma 1.42 that $m^{(1)}_B(Q, b_Q) = m^{(1)}_{B_Q}(Q) = m^{(1)}_{b_u}(Q) > 0$. This shows $l(B) \geq 2$ by Proposition 1.41. Now the claim follows again with Inequality (9.1).   □

We add some remarks. For every $n \geq 3$ and each case ((aa), (ab) or (bb)) there is a finite group H with Sylow 2-subgroup D_{2^n} and fusion system $\mathscr{F}$ (see Theorem 10.17 below). Taking the principal block of $H \times C_{2^m}$ we get examples for B for any parameters. Moreover, the principal block of S_6 shows that also $\mathrm{C}_{Q_1}(\mathrm{N}_G(Q_1, b_{Q_1})) \neq \mathrm{C}_{Q_2}(\mathrm{N}_G(Q_2, b_{Q_2}))$ is possible in case (aa). This gives an example, where B is not centrally controlled (and $m = 1$). In particular, the fusion system in case (aa) is not unique. Theorem 9.7 still gives the impression that B should be perfectly isometric (or even Morita equivalent) to a tensor product of a block with defect group D_{2^n} and the group algebra FC_{2^m}. However, we show that this is not always true. This result is new and was suggested by Külshammer.

Proposition 9.8 *The principal 2-block of FS_6 is not perfectly isometric (nor Morita equivalent) to $A \otimes_F FC_2$ where A is a block of a finite group with defect group D_8.*

Proof We have $l(B_0(FS_6)) = 3$ and $k(B_0(FS_6)) = 10$. Since $A \otimes_F FC_2$ can be treated as a block of a direct product of the form $H \times C_2$, we may assume that $l(A) = 3$ (see [278]). Let $Z_1 := Z(B_0(FS_6))$ and $Z_2 := Z(A \otimes_F FC_2) = Z(A) \otimes_F FC_2$. By Theorem 4.11 in [48] it suffices to show that Z_1 and Z_2 are not isomorphic as F-algebras. By the main result of [52] the algebra $Z(A)$ is determined up to isomorphism. Hence, we may assume that $A = B_0(FA_6)$ and $A \otimes_F FC_2 = B_0(F(A_6 \times C_2))$. We compare the kernels of the Frobenius map $\hat{Z}_i := \{a \in Z_i : a^2 = 0\}$ for $i = 1, 2$. The block idempotent of $B_0(FS_6)$ is given by $1 + (1, 2, 3, 4, 5)^+$ where $(1, 2, 3, 4, 5)^+$ is the class sum of the conjugacy class of $(1, 2, 3, 4, 5)$ in S_6. It follows that Z_1 has a basis $b_1, \ldots, b_{10}$ such that each b_i has the form $b_i = \sum_{g \in L_i} g$ for a subset $L_i \subseteq S_6$ which is closed under conjugation (see Proposition 2.2 in [215]). In particular all the structure constants of Z_1 are 0 or 1. An element $a = \sum_{i=1}^{10} a_i b_i$ ($a_i \in F$ for $i = 1, \ldots, 10$) belongs to $\hat{Z}_1$ if and only if $\sum_{i=1}^{10} a_i^2 b_i^2 = 0$. This gives linear equations of the form $0 = a_{i_1}^2 + \ldots + a_{i_j}^2 = (a_{i_1} + \ldots + a_{i_j})^2 = a_{i_1} + \ldots + a_{i_j}$. A computer calculation implies $\dim_F \hat{Z}_1 = 7$. Similarly we obtain $\dim_F \hat{Z}_2 = 8$. Hence, Z_1 and Z_2 are not isomorphic. $\qquad\square$

As another remark we mention that B cannot be a block of maximal defect of a simple group for $m \geq 1$ by the main theorem in [104].

Theorem 9.9 *Alperin's Weight Conjecture holds for B.*

Proof Let $Q \leq D$ be $\mathscr{F}$-centric and $\mathscr{F}$-radical. By Lemma 9.3 we have $\mathrm{Out}_{\mathscr{F}}(Q) \cong S_3$ or $\mathrm{Out}_{\mathscr{F}}(Q) = 1$ (if $Q = D$). In particular $\mathrm{Out}_{\mathscr{F}}(Q)$ has trivial Schur multiplier. Moreover, $F \mathrm{Out}_{\mathscr{F}}(Q)$ has precisely one block of defect 0. Now the claim follows from Theorem 9.7. $\qquad\square$

Theorem 9.10 *The Ordinary Weight Conjecture holds for B.*

Proof Let $Q \leq D$ be $\mathscr{F}$-centric and $\mathscr{F}$-radical. In the case $Q = D$ we have $\mathrm{Out}_{\mathscr{F}}(D) = 1$ and $\mathscr{N}_D$ consists only of the trivial chain. Then it follows easily that $\mathbf{w}(D, d) = k^d(D) = k^d(B)$ for all $d \in \mathbb{N}$. Now let $Q \in \{Q_1, Q_2\}$ such that $\mathrm{Out}_{\mathscr{F}}(Q) = \mathrm{Aut}_{\mathscr{F}}(Q) \cong S_3$. It suffices to show that $\mathbf{w}(Q, d) = 0$ for all $d \in \mathbb{N}$. Since Q is abelian, we have $\mathbf{w}(Q, d) = 0$ unless $d = m + 2$. Thus, let $d = m + 2$. Up to conjugation $\mathscr{N}_Q$ consists of the trivial chain $\sigma : 1$ and the chain $\tau : 1 < C$, where $C \leq \mathrm{Out}_{\mathscr{F}}(Q)$ has order 2.

We consider the chain σ first. Here $I(\sigma) = \mathrm{Out}_{\mathscr{F}}(Q) \cong S_3$ acts faithfully on $\Omega(Q) \cong C_2^3$ and thus fixes a four-group. Hence, the characters in $\mathrm{Irr}(Q)$ split into 2^m orbits of length 3 and 2^m orbits of length 1 under $I(\sigma)$ (see also Lemma 9.4). For a character $\chi \in \mathrm{Irr}(D)$ lying in an orbit of length 3 we have $I(\sigma, \chi) \cong C_2$ and thus $w(Q, \sigma, \chi) = 0$. For the 2^m stable characters $\chi \in \mathrm{Irr}(D)$ we get $w(Q, \sigma, \chi) = 1$, since $I(\sigma, \chi) = \mathrm{Out}_{\mathscr{F}}(Q)$ has precisely one block of defect 0.

Now consider the chain τ. Here $I(\tau) = C$ and the characters in $\mathrm{Irr}(Q)$ split into 2^m orbits of length 2 and 2^{m+1} orbits of length 1 under $I(\tau)$. For a character $\chi \in \mathrm{Irr}(D)$ in an orbit of length 2 we have $I(\tau, \chi) = 1$ and thus $w(Q, \tau, \chi) = 1$. For the 2^{m+1} stable characters $\chi \in \mathrm{Irr}(D)$ we get $I(\tau, \chi) = I(\tau) = C$ and $w(Q, \tau, \chi) = 0$.

Taking both chains together, we derive

$$\mathbf{w}(Q, d) = (-1)^{|\sigma|+1} 2^m + (-1)^{|\tau|+1} 2^m = 2^m - 2^m = 0.$$

This proves the OWC.   □

Finally we show that the Gluing Problem for the block B has a unique solution. This was done for $m = 0$ in [219].

Theorem 9.11 *The Gluing Problem for B has a unique solution.*

Proof We will show that $\mathrm{H}^i(\mathrm{Aut}_{\mathscr{F}}(\sigma), F^\times) = 0$ for $i = 1, 2$ and every chain σ of $\mathscr{F}$-centric subgroups of D. Then it follows that $\mathscr{A}_{\mathscr{F}}^i = 0$ and $\mathrm{H}^0([S(\mathscr{F}^c)], \mathscr{A}_{\mathscr{F}}^2) = \mathrm{H}^1([S(\mathscr{F}^c)], \mathscr{A}_{\mathscr{F}}^1) = 0$. Hence, by Theorem 1.1 in [219] the Gluing Problem has only the trivial solution.

Let $Q \le D$ be the largest ($\mathscr{F}$-centric) subgroup occurring in σ. Then as in the proof of Lemma 9.3 we have $Q = (Q \cap \langle x, y \rangle) \times \langle z \rangle$. If $Q \cap \langle x, y \rangle$ is non-abelian, $\mathrm{Aut}(Q)$ is a 2-group by Lemma 9.2. In this case we get $\mathrm{H}^i(\mathrm{Aut}_{\mathscr{F}}(\sigma), F^\times) = 0$ for $i = 1, 2$ (see proof of Corollary 2.2 in [219]). Hence, we may assume that $Q \in \{Q_1, Q_2\}$ and $\mathrm{Aut}_{\mathscr{F}}(Q) \cong S_3$ (see proof of Lemma 9.6 for the case $Q = \langle x, z \rangle$). Then σ only consists of Q and $\mathrm{Aut}_{\mathscr{F}}(\sigma) = \mathrm{Aut}_{\mathscr{F}}(Q)$. Hence, also in this case we get $\mathrm{H}^i(\mathrm{Aut}_{\mathscr{F}}(\sigma), F^\times) = 0$ for $i = 1, 2$.   □

9.2 $D_{2^n} * C_{2^m}$

It seems natural to proceed with defect groups of type $Q_{2^n} \times C_{2^m}$. However, in order to do so we first need to settle the problem for central products which occur in the induction step. Let

$$D := \langle x, y, z \mid x^{2^{n-1}} = y^2 = z^{2^m} = [x, z] = [y, z] = 1, \ yxy^{-1} = x^{-1}, \ x^{2^{n-2}} = z^{2^{m-1}} \rangle$$

$$= \langle x, y \rangle * \langle z \rangle \cong D_{2^n} * C_{2^m}$$

where $n \ge 2$ and $m \ge 1$. For $m = 1$ we get $D \cong D_{2^n}$. Then the invariants of B are known. Hence, we assume $m \ge 2$. Similarly for $n = 2$ we get $D = \langle y, z \rangle \cong C_2 \times C_{2^m}$. Then B is nilpotent and everything is known. Thus, we also assume $n \ge 3$. Then we have $D = \langle x, yz^{2^{m-2}}, z \rangle \cong Q_{2^n} * C_{2^m}$. For $n \ge 4$ we also have $D = \langle xz^{2^{m-2}}, y, z \rangle \cong SD_{2^n} * C_{2^m}$.

The first lemma shows that the situation splits naturally into two cases according to $n = 3$ or $n \ge 4$.

Lemma 9.12 *The automorphism group $\mathrm{Aut}(D)$ is a 2-group if and only if $n \ge 4$.*

Proof Since $\mathrm{Aut}(Q_8) \cong S_4$, we see that $\mathrm{Aut}(Q_8 \times C_{2^m})$ is not a 2-group. An automorphism of $Q_8 \times C_{2^m}$ of odd order acts trivially on $(Q_8 \times C_{2^m})' \cong C_2$ and on

$Z(Q_8 \times C_{2^m})/(Q_8 \times C_{2^m})' \cong C_{2^m}$ and thus also on $Z(Q_8 \times C_{2^m})$ by Theorem 5.3.2 in [94]. Hence, $\mathrm{Aut}(Q_8 * C_{2^m}) = \mathrm{Aut}(D_8 * C_{2^m})$ is not a 2-group.

Now assume $n \geq 4$. Then $\Phi(D) = \langle x^2, z^2 \rangle < \Phi(D)Z(D) = \langle x^2, z \rangle$ are characteristic subgroups of D. Moreover, $\langle x, z \rangle$ is the only abelian maximal subgroup containing $\Phi(D)Z(D)$. Hence, every automorphism of $\mathrm{Aut}(D)$ of odd order acts trivially on $D/\Phi(D)$. The claim follows from Theorem 5.1.4 in [94]. $\qquad\square$

It follows that the inertial index $e(B)$ of B equals 1 for $n \geq 4$. In case $n = 3$ there are two possibilities $e(B) \in \{1, 3\}$, since $\Phi(D)Z(D)$ is still characteristic in D. Now we investigate the fusion system $\mathscr{F}$ of B.

Lemma 9.13 *Let* $Q_1 := \langle x^{2^{n-3}}, y, z \rangle \cong D_8 * C_{2^m}$ *and* $Q_2 := \langle x^{2^{n-3}}, xy, z \rangle \cong D_8 * C_{2^m}$. *Then* Q_1 *and* Q_2 *are the only candidates for proper* $\mathscr{F}$-*centric,* $\mathscr{F}$-*radical subgroups up to conjugation. Moreover, one of the following cases occurs:*

(aa) $n = e(B) = 3$ *or* $(n \geq 4$ *and* $\mathrm{Out}_{\mathscr{F}}(Q_1) \cong \mathrm{Out}_{\mathscr{F}}(Q_2) \cong S_3)$.
(ab) $n \geq 4$, $\mathrm{Out}_{\mathscr{F}}(Q_1) \cong C_2$, *and* $\mathrm{Out}_{\mathscr{F}}(Q_2) \cong S_3$.
(ba) $n \geq 4$, $\mathrm{Out}_{\mathscr{F}}(Q_1) \cong S_3$, *and* $\mathrm{Out}_{\mathscr{F}}(Q_2) \cong C_2$.
(bb) $\mathrm{Out}_{\mathscr{F}}(Q_1) \cong \mathrm{Out}_{\mathscr{F}}(Q_2) \cong C_2$.

In case (bb) the block B is nilpotent.

Proof Let $Q < D$ be $\mathscr{F}$-centric and $\mathscr{F}$-radical. Then $z \in Z(D) \subseteq C_D(Q) \subseteq Q$ and $Q = (Q \cap \langle x, y \rangle) * \langle z \rangle$. If $Q \cap \langle x, y \rangle$ is abelian, we have

$$Q = \langle x^i y, z \rangle \cong C_2 \times C_{2^m} \quad \text{or}$$

$$Q = \langle x, z \rangle \cong C_{2^n} * C_{2^m} \cong C_{2^{\max(n,m)}} \times C_{2^{\min(n,m)-1}}$$

for some $i \in \mathbb{Z}$. In the first case, $\mathrm{Aut}(Q)$ is a 2-group, since $m \geq 2$. Then

$$O_2(\mathrm{Aut}_{\mathscr{F}}(Q)) \neq 1.$$

Thus, assume $Q = \langle x, z \rangle$. The group $D \subseteq N_G(Q, b_Q)$ acts trivially on $Q/\Phi(Q)$, while a non-trivial automorphism of $\mathrm{Aut}(Q)$ of odd order acts non-trivially on $Q/\Phi(Q)$ (see Theorem 5.1.4 in [94]). This contradicts $O_2(\mathrm{Aut}_{\mathscr{F}}(Q)) = 1$. (Moreover, by Lemma 5.4 in [184] we see that $\mathrm{Aut}_{\mathscr{F}}(Q)$ is a 2-group.)

Hence by Lemma 9.12, Q is isomorphic to $D_8 * C_{2^m}$ and contains an element of the form $x^i y$. After conjugation with a suitable power of x we may assume $Q \in \{Q_1, Q_2\}$. This shows the first claim.

Let $S \leq D$ be an arbitrary subgroup isomorphic to $D_8 * C_{2^m}$. If $z \notin S$, then for $\langle S, z \rangle = (\langle S, z \rangle \cap \langle x, y \rangle)\langle z \rangle$ we have $\langle S, z \rangle' = S' \cong C_2$. However, this is impossible, since $\langle S, z \rangle \cap \langle x, y \rangle$ has at least order 16. This contradiction shows $z \in S$. Thus, S is conjugate to $Q \in \{Q_1, Q_2\}$ under D. In particular Q is fully $\mathscr{F}$-normalized. Hence, $N_D(Q)/Q \cong C_2$ is a Sylow 2-subgroup of

$\text{Out}_{\mathscr{F}}(Q)$. Assume $N_D(Q) C_G(Q) < N_G(Q, b_Q)$. Since $O_2(\text{Out}_{\mathscr{F}}(Q)) = 1$ and $|\text{Aut}(Q)| = 2^k \cdot 3$ for some $k \in \mathbb{N}$, we get $\text{Out}_{\mathscr{F}}(Q) \cong S_3$.

The last claim follows from Alperin's Fusion Theorem and $e(B) = 1$ (for $n \geq 4$). □

Since the cases (ab) and (ba) are symmetric, we ignore case (ba). It is easy to see that Q_1 and Q_2 are not conjugate in D if $n \geq 4$. Hence, by Alperin's Fusion Theorem the subpairs (Q_1, b_{Q_1}) and (Q_2, b_{Q_2}) are not conjugate in G. It is also easy to see that Q_1 and Q_2 are always $\mathscr{F}$-centric.

Lemma 9.14 *Let* $Q \in \{Q_1, Q_2\}$ *such that* $\text{Out}_{\mathscr{F}}(Q) \cong S_3$. *Then*

$$C_Q(N_G(Q, b_Q)) = Z(Q) = \langle z \rangle.$$

Proof Since $Q \subseteq N_D(Q, b_Q)$, we have $C_Q(N_G(Q, b_Q)) \subseteq C_Q(Q) = Z(Q)$. On the other hand, $N_D(Q)$ and every automorphism of $\text{Aut}_{\mathscr{F}}(Q)$ of odd order act trivially on $Z(Q) = Z(D) = \langle z \rangle \cong C_{2^m}$. Hence, the claim follows. □

Lemma 9.15 *A set of representatives* $\mathscr{R}$ *for the* $\mathscr{F}$*-conjugacy classes of elements* $u \in D$ *such that* $\langle u \rangle$ *is fully* $\mathscr{F}$*-normalized is given as follows:*

(i) $x^i z^j$ $(i = 0, 1, \dots, 2^{n-2}, j = 0, 1, \dots, 2^{m-1} - 1)$ *in case (aa)*.
(ii) $x^i z^j$ *and* yz^j $(i = 0, 1, \dots, 2^{n-2}, j = 0, 1, \dots, 2^{m-1} - 1)$ *in case (ab)*.

Proof The proof works exactly as in Lemma 9.5. □

Lemma 9.16 *Olsson's conjecture* $k_0(B) \leq 2^{m+1} = |D : D'|$ *is satisfied in all cases.*

Proof This follows from Theorem 5.3 (cf. Lemma 9.6). □

Lemma 9.17 *Let* v *be the 2-adic valuation, and let* ζ *be a primitive* 2^k*-th root of unity for* $k \geq 2$. *Then* $0 < v(1 + \zeta) < 1$.

Proof We prove this by induction on k. For $k = 2$ we have $\zeta \in \{\pm i\}$, where $i = \sqrt{-1}$. Then $2v(1 + i) = v((1 + i)^2) = v(2i) = 1$ and the claim follows. Now let $k \geq 3$. Then $2v(1 + \zeta) = v((1 + \zeta)^2) = v(1 + \zeta^2 + 2\zeta) = v(1 + \zeta^2)$, since $v(1 + \zeta^2) < 1 = v(2\zeta)$ by induction. □

Theorem 9.18

(i) *In case (aa) and* $n = 3$ *we have* $k(B) = 2^{m-1} \cdot 7$, $k_0(B) = 2^{m+1}$, $k_1(B) = 2^{m-1} \cdot 3$, *and* $l(B) = 3$.
(ii) *In case (aa) and* $n \geq 4$ *we have* $k(B) = 2^{m-1}(2^{n-2} + 5)$, $k_0(B) = 2^{m+1}$, $k_1(B) = 2^{m-1}(2^{n-2} - 1)$, $k_{n-2}(B) = 2^m$, *and* $l(B) = 3$.
(iii) *In case (ab) we have* $k(B) = 2^{m-1}(2^{n-2} + 4)$, $k_0(B) = 2^{m+1}$, $k_1(B) = 2^{m-1}(2^{n-2} - 1)$, $k_{n-2}(B) = 2^{m-1}$, *and* $l(B) = 2$.
(iv) *In case (bb) we have* $k(B) = 2^{m-1}(2^{n-2} + 3)$, $k_0(B) = 2^{m+1}$, $k_1(B) = 2^{m-1}(2^{n-2} - 1)$, *and* $l(B) = 1$.

In particular Brauer's $k(B)$-Conjecture, Brauer's Height Zero Conjecture and the Alperin-McKay Conjecture hold.

Proof Assume first that case (bb) occurs. Then B is nilpotent and $k_i(B)$ is just the number $k_i(D)$ of irreducible characters of D of degree 2^i ($i \geq 0$) and $l(B) = 1$. In particular $k_0(B) = |D : D'| = 2^{m+1}$ and $k(B) = k(D) = 2^{m-1}(2^{n-2} + 3)$. Since $|D|$ is the sum of the squares of the degrees of the irreducible characters, we get $k_1(B) = k_1(D) = 2^{m-1}(2^{n-2} - 1)$.

Now assume that case (aa) or case (ab) occurs. We determine the numbers $l(b)$ for the subsections in Lemma 9.15 and apply Theorem 1.35. Let us begin with the non-major subsections. Since $\mathrm{Aut}_{\mathcal{F}}(\langle x, z \rangle)$ is a 2-group, we have $l(b_{x^i z^j}) = 1$ for all $i = 1, \ldots, 2^{n-2} - 1$ and $j = 0, 1, \ldots, 2^{m-1} - 1$. The blocks b_{yz^j} ($j = 0, 1, \ldots, 2^{m-1} - 1$) have $C_D(yz^j) = \langle yz^j, z \rangle \cong C_2 \times C_{2^m}$ as defect group. Hence, they are also nilpotent, and it follows that $l(b_{yz^j}) = 1$.

The major subsections of B are given by (z^j, b_{z^j}) for $j = 0, 1, \ldots, 2^m - 1$ up to conjugation. By Lemma 9.14, the cases for B and b_{z^j} coincide. As usual, the blocks b_{z^j} dominate blocks $\overline{b_{z^j}}$ of $C_G(z^j)/\langle z^j \rangle$ with defect group $D/\langle z^j \rangle \cong D_{2^{n-1}} \times C_{2^m/|\langle z^j \rangle|}$ for $j \neq 0$. We have $l(b_{z^j}) = l(\overline{b_{z^j}})$. The cases for b_{z^j} and $\overline{b_{z^j}}$ also coincide. Now we discuss the cases (ab) and (aa) separately.

Case (ab):

Then we have $l(b_{z^j}) = l(\overline{b_{z^j}}) = 2$ for $j = 1, \ldots, 2^m - 1$ by Theorem 9.7. Hence, Theorem 1.35 implies

$$k(B) - l(B) = 2^{m-1}(2^{n-2} - 1) + 2^{m-1} + 2(2^m - 1) = 2^{m-1}(2^{n-2} + 4) - 2.$$

Since B is a centrally controlled block, we have $l(B) \geq l(b_z) = 2$ and $k(B) \geq 2^{m-1}(2^{n-2} + 4)$ by Theorem 1.38. In order to bound $k(B)$ from above we study the numbers $d^z_{\chi\varphi}$. Let $D^z := (d^z_{\chi\varphi_i})_{\chi \in \mathrm{Irr}(B), i=1,2}$. Then $(D^z)^{\mathrm{T}} \overline{D^z} = C^z$ is the Cartan matrix of b_z. Since $\overline{b_z}$ has defect group $D_{2^{n-1}}$, the matrix C^z is given by

$$C^z = 2^m \begin{pmatrix} 2^{n-3} + 1 & 2 \\ 2 & 4 \end{pmatrix}$$

up to basic sets (see Theorem 8.1). We consider the generalized decomposition numbers more carefully. As usual we write

$$d^z_{\chi\varphi_i} = \sum_{j=0}^{2^{m-1}-1} a^i_j(\chi)\zeta^j$$

for $i = 1, 2$, where ζ is a primitive 2^m-th root of unity. Since the subsections (z^j, b_{z^j}) are pairwise non-conjugate for $j = 0, \ldots, 2^m - 1$, we get

$$(a^1_i, a^1_j) = (2^{n-2} + 2)\delta_{ij}, \qquad (a^1_i, a^2_j) = 4\delta_{ij}, \qquad (a^2_i, a^2_j) = 8\delta_{ij}.$$

Then

$$m^z_{\chi\psi} = 4d^z_{\chi\varphi_1}\overline{d^z_{\psi\varphi_1}} - 2(d^z_{\chi\varphi_1}\overline{d^z_{\psi\varphi_2}} + d^z_{\chi\varphi_2}\overline{d^z_{\psi\varphi_1}}) + (2^{n-3}+1)d^z_{\chi\varphi_2}\overline{d^z_{\psi\varphi_2}}.$$

It follows from Proposition 1.36 that

$$h(\chi) = 0 \iff m^z_{\chi\chi} \in \mathcal{O}^\times \iff d^z_{\chi\varphi_2} \in \mathcal{O}^\times \iff \sum_{j=0}^{2^{m-1}-1} a^2_j(\chi) \equiv 1 \pmod{2}.$$

$$(9.2)$$

Assume that $k(B)$ is as large as possible. Since (z, b_z) is a major subsection, no row of D^z vanishes. Hence, for $j \in \{0, 1, \ldots, 2^{m-1}-1\}$ we have essentially the following possibilities (where $\epsilon_1, \epsilon_2, \epsilon_3, \epsilon_4 \in \{\pm 1\}$; cf. proof of Theorem 3.15 in [212]):

$$(I): \begin{pmatrix} a^1_j & \pm1 \cdots \pm1 & \epsilon_1 & \epsilon_2 & \epsilon_3 & \epsilon_4 & . & \cdots & \cdots & \cdots & \cdots & \cdots \\ a^2_j & . & \cdots & . & \epsilon_1 & \epsilon_2 & \epsilon_3 & \epsilon_4 & \pm1 & \pm1 & \pm1 & \pm1 & . & \cdots . \end{pmatrix},$$

$$(II): \begin{pmatrix} a^1_j & \pm1 \cdots \pm1 & \epsilon_1 & \epsilon_2 & \epsilon_3 & . & \cdots & \cdots & \cdots \\ a^2_j & . & \cdots & . & 2\epsilon_1 & \epsilon_2 & \epsilon_3 & \pm1 & \pm1 & . & \cdots . \end{pmatrix},$$

$$(III): \begin{pmatrix} a^1_j & \pm1 \cdots \pm1 & \epsilon_1 & \epsilon_2 & \cdots . \\ a^2_j & . & \cdots & . & 2\epsilon_1 & 2\epsilon_2 & \cdots . \end{pmatrix}.$$

The number $k(B)$ would be maximal if case (I) occurs for all j and for every character $\chi \in \mathrm{Irr}(B)$ we have $\sum_{j=0}^{2^{m-1}-1} |a^1_j(\chi)| \le 1$ and $\sum_{j=0}^{2^{m-1}-1} |a^2_j(\chi)| \le 1$. However, this contradicts Lemma 9.16 and Eq. (9.2). This explains why we have to take the cases (II) and (III) also into account. Now let α (resp. γ, δ) be the number of indices $j \in \{0, 1, \ldots, 2^{m-1}-1\}$ such that case (I) (resp. (II), (III)) occurs for a^i_j. Then obviously $\alpha + \beta + \gamma = 2^{m-1}$. It is easy to see that we may assume for all $\chi \in \mathrm{Irr}(B)$ that $\sum_{j=0}^{2^{m-1}-1} |a^1_j(\chi)| \le 1$ in order to maximize $k(B)$. In contrast to that it does make sense to have $a^2_j(\chi) \ne 0 \ne a^2_k(\chi)$ for some $j \ne k$ in order to satisfy Olsson's Conjecture in view of Eq. (9.2). Let δ be the number of pairs $(\chi, j) \in \mathrm{Irr}(B) \times \{0, 1, \ldots, 2^{m-1}-1\}$ such that there exists a $k \ne j$ with $a^2_j(\chi)a^2_k(\chi) \ne 0$. Then it follows that

$$\gamma = 2^{m-1} - \alpha - \beta,$$

$$k(B) \le (2^{n-2}+6)\alpha + (2^{n-2}+4)\beta + (2^{n-2}+2)\gamma - \delta/2$$

$$= 2^{m+n-3} + 6\alpha + 4\beta + 2\gamma - \delta/2$$

$$= 2^{m+n-3} + 2^m + 4\alpha + 2\beta - \delta/2,$$

$$8\alpha + 4\beta - \delta \le k_0(B) \le 2^{m+1}.$$

This gives $k(B) \leq 2^{m+n-3} + 2^{m+1} = 2^{m-1}(2^{n-2} + 4)$. Together with the lower bound above, we have shown that $k(B) = 2^{m-1}(2^{n-2} + 4)$ and $l(B) = 2$. In particular the cases (I), (II) and (III) are really the only possibilities which can occur. The inequalities above imply also $k_0(B) = 2^{m+1}$. However we do not know the precise values of α, β, γ, and δ. We will see in a moment that $\delta = 0$. Assume the contrary. If $\chi \in \mathrm{Irr}(B)$ is a character such that $a_j^2(\chi)a_k^2(\chi) \neq 0$ for some $j \neq k$, then it is easy to see that $a_j^2(\chi)a_k^2(\chi) \in \{\pm 1\}$ and $a_l^2(\chi) = 0$ for all $l \notin \{j, k\}$. For if not, we would have $8\alpha + 4\beta - \delta < k_0(B)$ or $k(B) < 2^{m+n-3} + 2^m + 4\alpha + 2\beta - \delta/2$. Hence, we have to exclude the following types of rows of D^z (where $\epsilon \in \{\pm 1\}$): $(\epsilon\zeta^j, \epsilon\zeta^j + \epsilon\zeta^k)$, $(\epsilon\zeta^j, \epsilon\zeta^j - \epsilon\zeta^k)$, $(0, \epsilon\zeta^j + \epsilon\zeta^k)$, and $(0, \epsilon\zeta^j - \epsilon\zeta^k)$. Let d_χ^z be the row of D^z corresponding to the character $\chi \in \mathrm{Irr}(B)$. If $d_\chi^z = (\epsilon\zeta^j, \epsilon\zeta^j + \epsilon\zeta^k)$ for $j \neq k$ we have

$$m_{\chi\chi}^z = 4 - 2(2 + \zeta^{j-k} + \zeta^{k-j}) + (2^{n-3} + 1)(2 + \zeta^{j-k} + \zeta^{k-j})$$

$$= 4 + (2^{n-3} - 1)(2 + \zeta^{j-k} + \zeta^{k-j}).$$

Since $\nu(\zeta^{j-k} + \zeta^{k-j}) = \nu(\zeta^{j-k}(\zeta^{j-k} + \zeta^{k-j})) = \nu(1 + \zeta^{2(j-k)})$, Lemma 9.17 implies $\nu(2 + \zeta^{j-k} + \zeta^{k-j}) \leq 1$. This yields the contradiction $1 \leq h(\chi) < \nu(m_{\chi\chi}^z) \leq 1$. A very similar calculation works for the other types of rows. Thus, we have shown $\delta = 0$. Then the rows of D^z have the following forms: $(\pm\zeta^j, 0)$, $(\epsilon\zeta^j, \epsilon\zeta^j)$, $(0, \pm\zeta^j)$, and $(\epsilon\zeta^j, 2\epsilon\zeta^j)$. We already know which of these rows correspond to characters of height 0. In order to determine $k_i(B)$ we calculate the contributions for the remaining rows. If $d_\chi^z = (\pm\zeta^j, 0)$, we have $m_{\chi\chi}^z = 4$. Then Proposition 1.36 implies $h(\chi) = 1$. The number of these rows is precisely

$$(2^{n-2} - 2)\alpha + (2^{n-2} - 1)\beta + 2^{n-2}\gamma = 2^{n+m-3} - 2\alpha - \beta = 2^{n+m-3} - 2^{m-1}$$

$$= 2^{m-1}(2^{n-2} - 1).$$

Now assume that $\psi \in \mathrm{Irr}(B)$ is a character of height 0 such that $d_\psi^z = (0, \pm\zeta^j)$ (such characters always exist). Let $\chi \in \mathrm{Irr}(B)$ such that $d_\chi^z = (\epsilon\zeta^k, 2\epsilon\zeta^k)$, where $\epsilon \in \{\pm 1\}$. Then $m_{\chi\psi}^z = -2(\pm\epsilon\zeta^{k-j}) + (2^{n-3} + 1)(\pm\epsilon 2\zeta^{k-j}) = \pm\epsilon 2^{n-2}\zeta^{k-j}$, and Proposition 1.36 implies $h(\chi) = n - 2$. The number of these characters is precisely $k(B) - k_0(B) - 2^{m-1}(2^{n-2} - 1) = 2^{m-1}$. This gives $k_i(B)$ for $i \in \mathbb{N}$ (recall that $n \geq 4$ in case (ab)).

Case (aa):

Here the arguments are similar, so that we will leave out some details. By Theorem 9.7 we have

$$k(B) - l(B) = 2^{m-1}(2^{n-2} - 1) + 3(2^m - 1) = 2^{m-1}(2^{n-2} + 5) - 3.$$

Again B is centrally controlled, and $l(B) \geq 3$ and $k(B) \geq 2^{m-1}(2^{n-2} + 5)$ follow from Theorem 1.38. The Cartan matrix C^z of b_z is given by

$$C^z = 2^m \begin{pmatrix} 2^{n-3} + 1 & 1 & 1 \\ 1 & 2 & 0 \\ 1 & 0 & 2 \end{pmatrix}$$

up to basic sets (see Theorem 8.1). We write $\mathrm{IBr}(b_z) = \{\varphi_1, \varphi_2, \varphi_3\}$ and define the integral columns a^i_j for $i = 1, 2, 3$ and $j = 0, 1, \ldots, 2^{m-1} - 1$ as in case (ab). Then we can calculate the scalar products (a^i_j, a^k_l). In particular the orthogonality relations imply that the columns a^2_j and a^3_j consist of four entries ± 1 and zeros elsewhere. The contributions are given by

$$m^z_{\chi\psi} = 4 d^z_{\chi\varphi_1}\overline{d^z_{\psi\varphi_1}} - 2\big(d^z_{\chi\varphi_1}\overline{d^z_{\psi\varphi_2}} + d^z_{\chi\varphi_2}\overline{d^z_{\psi\varphi_1}} + d^z_{\chi\varphi_1}\overline{d^z_{\psi\varphi_3}} + d^z_{\chi\varphi_3}\overline{d^z_{\psi\varphi_1}}\big)$$
$$+ d^z_{\chi\varphi_2}\overline{d^z_{\psi\varphi_3}} + d^z_{\chi\varphi_3}\overline{d^z_{\psi\varphi_2}} + (2^{n-2} + 1)\big(d^z_{\chi\varphi_2}\overline{d^z_{\psi\varphi_2}} + d^z_{\chi\varphi_3}\overline{d^z_{\psi\varphi_3}}\big)$$

for $\chi, \psi \in \mathrm{Irr}(B)$. As before, Proposition 1.36 implies

$$h(\chi) = 0 \iff m^z_{\chi\chi} \in \mathcal{O}^\times \iff |d^z_{\chi\varphi_2} + d^z_{\chi\varphi_3}|^2 \in \mathcal{O}^\times$$

$$\iff d^z_{\chi\varphi_2} + d^z_{\chi\varphi_3} \in \mathcal{O}^\times \iff \sum_{j=0}^{2^{m-1}-1} \big(a^2_j(\chi) + a^3_j(\chi)\big) \equiv 1 \pmod 2.$$

$$(9.3)$$

In order to search the maximum value for $k(B)$ (in view of Lemma 9.16 and Eq. (9.3)) we have to consider the following possibilities (where $\epsilon_1, \epsilon_2, \epsilon_3, \epsilon_4 \in \{\pm 1\}$):

$$(I): \begin{pmatrix} a^1_j & \pm 1 \cdots \pm 1 & \epsilon_1\ \epsilon_2\ \epsilon_3\ \epsilon_4 & . & \cdots\ \cdots\ \cdots\ \cdots\ . \\ a^2_j & . \quad \cdots \quad . & \epsilon_1\ \epsilon_2\ .\ . & \pm 1\ \pm 1 & . \quad \cdots\ \cdots\ \cdots\ . \\ a^3_j & . \quad \cdots\ \cdots\ \cdots\ . & \epsilon_3\ \epsilon_4 & .\quad . \ \pm 1\ \pm 1 & . \quad \cdots\ . \end{pmatrix},$$

$$(II): \begin{pmatrix} a^1_j & \pm 1 \cdots \pm 1 & \epsilon_1\ \epsilon_2\ \epsilon_3 & . & \cdots\ \cdots\ \cdots\ \cdots\ . \\ a^2_j & . \quad \cdots \quad . & \epsilon_1\ \epsilon_2\ . & \epsilon_4\ \pm 1 & . \quad \cdots\ \cdots\ \cdots\ . \\ a^3_j & . \quad \cdots\ \cdots\ . & \epsilon_2\ \epsilon_3\ -\epsilon_4 & . \quad \pm 1 & . \quad \cdots\ . \end{pmatrix},$$

$$(III): \begin{pmatrix} a^1_j & \pm 1 \cdots \pm 1 & \epsilon_1\ \epsilon_2 & . & \cdots\ \cdots\ \cdots\ . \\ a^2_j & . \quad \cdots \quad . & \epsilon_1\ \epsilon_2\ \epsilon_3 & \epsilon_4 & . \quad \cdots\ . \\ a^3_j & . \quad \cdots \quad . & \epsilon_1\ \epsilon_2\ -\epsilon_3\ -\epsilon_4 & . \quad \cdots\ . \end{pmatrix}.$$

We define α, β and γ as in case (ab). Then we have $\alpha + \beta + \gamma = 2^{m-1}$. Let δ be the number of triples $(\chi, i, j) \in \mathrm{Irr}(B) \times \{2, 3\} \times \{0, 1, \ldots, 2^{m-1} - 1\}$ such that there exists a $k \neq j$ with $a^i_j(\chi)a^2_k(\chi) \neq 0$ or $a^i_j(\chi)a^3_k(\chi) \neq 0$. Then the following holds:

$$\gamma = 2^{m-1} - \alpha - \beta,$$

$$k(B) \leq (2^{n-2} + 6)\alpha + (2^{n-2} + 5)\beta + (2^{n-2} + 4)\gamma - \delta/2$$

$$= 2^{n+m-3} + 2^{m+1} + 2\alpha + \beta - \delta/2,$$

$$8\alpha + 4\beta - \delta \leq k_0(B) \leq 2^{m+1}.$$

This gives $k(B) \leq 2^{n+m-3} + 2^{m+1} + 2^{m-1} = 2^{m-1}(2^{n-2} + 5)$. Together with the lower bound we have shown that $k(B) = 2^{m-1}(2^{n-2} + 5)$, $k_0(B) = 2^{m+1}$, and $l(B) = 3$. In particular the maximal value for $k(B)$ is indeed attended. Moreover, $\delta = 0$. Let $\chi \in \mathrm{Irr}(B)$ such that $d^z_{\chi.} = (\pm\zeta^j, 0, 0)$. Then $m^z_{\chi\chi} = 4$ and $h(\chi) = 1$ by Proposition 1.36. The number of these characters is

$$(2^{n-2} - 2)\alpha + (2^{n-2} - 1)\beta + 2^{n-2}\gamma = 2^{n+m-1} - 2^{m-1} = 2^{m-1}(2^{n-2} - 1).$$

Now let $\psi \in \mathrm{Irr}(B)$ a character of height 0 such that $d^z_{\psi.} = (0, 0, \pm\zeta^j)$, and let $\chi \in \mathrm{Irr}(B)$ such that $d^z_{\chi.} = (\epsilon\zeta^k, \epsilon\zeta^k, \epsilon\zeta^k)$, where $\epsilon \in \{\pm1\}$. Then we have $m^z_{\chi\psi} = -2(\pm\epsilon\zeta^{k-j}) \pm \epsilon\zeta^{k-j} + (2^{n-2} + 1)(\pm\epsilon\zeta^{k-j}) = \pm\epsilon2^{n-2}\zeta^{k-j}$ and $h(\chi) = n - 2$. The same holds if $d^z_{\chi.} = (0, \epsilon\zeta^k, -\epsilon\zeta^k)$. This gives the numbers $k_i(B)$ for $i \in \mathbb{N}$. Observe that we have to add $k_1(B)$ and $k_{n-2}(B)$ in case $n = 3$. $\qquad\square$

If we take $m = 1$ in the formulas for $k_i(B)$ and $l(B)$ we get exactly the invariants for the defect group Q_{2^n}. However, recall that $D_{2^n} * C_2 \cong D_{2^n}$. Using Theorem 10.17 below it is easy to construct examples for B in all cases.

Theorem 9.19 *Alperin's Weight Conjecture holds for B.*

Proof Let $Q \leq D$ be $\mathscr{F}$-centric and $\mathscr{F}$-radical. By Lemma 9.13 we have $\mathrm{Out}_\mathscr{F}(Q) \cong S_3$, $\mathrm{Out}_\mathscr{F}(Q) \cong C_3$, or $\mathrm{Out}_\mathscr{F}(Q) = 1$ (in the last two cases we have $Q = D$). In particular $\mathrm{Out}_\mathscr{F}(Q)$ has trivial Schur multiplier. Moreover, the group algebras $F1$ and FS_3 have precisely one block of defect 0, while FC_3 has three blocks of defect 0. Now the claim follows from Theorem 9.18. $\qquad\square$

Lemma 9.20 *Let ζ be a primitive 2^m-th root of unity. Then for $n = 3$ the (ordinary) character table of D is given as follows:*

1	x	y	z
1	1	1	ζ^{2r}
1	-1	1	ζ^{2r}
1	1	-1	ζ^{2r}
1	-1	-1	ζ^{2r}
2	0	0	ζ^{2r+1}

where $r = 0, 1, \ldots, 2^{m-1} - 1$.

Proof We just take the characters $\chi \in \mathrm{Irr}(D_8 \times C_{2^m})$ with $\chi(x^2 z^{2^{m-1}}) = \chi(1)$. $\square$

Theorem 9.21 *The Ordinary Weight Conjecture holds for B.*

Proof We may assume that B is not nilpotent, and thus case (bb) does not occur. Suppose that $n = 3$ and case (aa) occurs. Then D is the only $\mathscr{F}$-centric, $\mathscr{F}$-radical subgroup of D. Since $\mathrm{Out}_{\mathscr{F}}(D) \cong C_3$, the set $\mathscr{N}_D$ consists only of the trivial chain. We have $\mathbf{w}(D, d) = 0$ for $d \notin \{m + 1, m + 2\}$, since then $k^d(D) = 0$. For $d = m + 1$ we get $\mathbf{w}(D, d) = 3 \cdot 2^{m-1}$ by Lemma 9.20. In case $d = m + 2$ it follows that $\mathbf{w}(D, d) = 3 \cdot 2^{m-1} + 2^{m-1} = 2^{m+1}$. Hence, the OWC follows from Theorem 9.18.

Now let $n \geq 4$ and assume that case (aa) occurs. Then there are three $\mathscr{F}$-centric, $\mathscr{F}$-radical subgroups up to conjugation: Q_1, Q_2 and D. Since $\mathrm{Out}_{\mathscr{F}}(D) = 1$, it follows easily that $\mathbf{w}(D, d) = k^d(D)$ for all $d \in \mathbb{N}$. By Theorem 9.18 it suffices to show

$$\mathbf{w}(Q, d) = \begin{cases} 2^{m-1} & \text{if } d = m + 1 \\ 0 & \text{otherwise} \end{cases}$$

for $Q \in \{Q_1, Q_2\}$, because $k^{m+1}(B) = k_{n-2}(B) = 2^m$. We already have $\mathbf{w}(Q, d) = 0$ unless $d \in \{m + 1, m + 2\}$. Without loss of generality let $Q = Q_1$.

Let $d = m + 1$. Up to conjugation $\mathscr{N}_Q$ consists of the trivial chain $\sigma : 1$ and the chain $\tau : 1 < C$, where $C \leq \mathrm{Out}_{\mathscr{F}}(Q)$ has order 2. We consider the chain σ first. Here $I(\sigma) = \mathrm{Out}_{\mathscr{F}}(Q) \cong S_3$ acts trivially on the characters of Q or defect $m + 1$ by Lemma 9.20. This contributes 2^{m-1} to the alternating sum of $\mathbf{w}(Q, d)$. Now consider the chain τ. Here $I(\tau) = C$ and $z(FC) = 0$ where $z(FC)$ is the number of blocks of defect 0 in FC. Hence, the contribution of τ vanishes and we get $\mathbf{w}(Q, d) = 2^{m-1}$ as desired.

Let $d = m + 2$. Then we have $I(\sigma, \mu) \cong S_3$ for every character $\mu \in \mathrm{Irr}(Q)$ with $\mu(x^{2^{n-3}}) = \mu(y) = 1$. For the other characters of Q with defect d we have $I(\sigma, \mu) \cong C_2$. Hence, the chain σ contributes 2^{m-1} to the alternating sum. There are 2^m characters $\mu \in \mathrm{Irr}(D)$ which are not fixed under $I(\tau) = C$. Hence, they split into 2^{m-1} orbits of length 2. For these characters we have $I(\tau, \mu) = 1$. For the other irreducible characters μ of D of defect d we have $I(\tau, \mu) = C$. Thus, the contribution of τ to the alternating sum is -2^{m-1}. This shows $\mathbf{w}(Q, d) = 0$.

In case (ab) we have only two $\mathscr{F}$-centric, $\mathscr{F}$-radical subgroups: Q_2 and D. Since $k_{n-2}(B) = 2^{m-1}$ in this case, the calculations above imply the result. $\square$

Theorem 9.22 *The Gluing Problem for B has a unique solution.*

Proof Assume first that $n \geq 4$. Let σ be a chain of $\mathscr{F}$-centric subgroups of D, and let $Q \leq D$ be the largest subgroup occurring in σ. Then as in the proof of Lemma 9.13 we have $Q = (Q \cap \langle x, y \rangle) * \langle z \rangle$. If $Q \cap \langle x, y \rangle$ is abelian or $Q = D$, then

$\mathrm{Aut}_{\mathscr{F}}(Q)$ and $\mathrm{Aut}_{\mathscr{F}}(\sigma)$ are 2-groups. In this case we get $\mathrm{H}^i(\mathrm{Aut}_{\mathscr{F}}(\sigma), F^{\times}) = 0$ for $i = 1, 2$ (see proof of Corollary 2.2 in [219]). Now assume that $Q \in \{Q_1, Q_2\}$ and $\mathrm{Aut}_{\mathscr{F}}(Q) \cong S_4$. Then it is easy to see that Q does not contain a proper $\mathscr{F}$-centric subgroup. Hence, σ consists only of Q and $\mathrm{Aut}_{\mathscr{F}}(\sigma) = \mathrm{Aut}_{\mathscr{F}}(Q)$. Thus, also in this case we get $\mathrm{H}^i(\mathrm{Aut}_{\mathscr{F}}(\sigma), F^{\times}) = 0$ for $i = 1, 2$. It follows that $\mathscr{A}_{\mathscr{F}}^i = 0$ and $\mathrm{H}^0([S(\mathscr{F}^c)], \mathscr{A}_{\mathscr{F}}^2) = \mathrm{H}^1([S(\mathscr{F}^c)], \mathscr{A}_{\mathscr{F}}^1) = 0$. Hence, by Theorem 1.1 in [219] the Gluing Problem has only the trivial solution.

Now let $n = 3$. Then we have $\mathrm{H}^i(\mathrm{Aut}_{\mathscr{F}}(\sigma), F^{\times}) = 0$ for $i = 1, 2$ unless $\sigma = D$ and case (aa) occurs. In this case $\mathrm{Aut}_{\mathscr{F}}(\sigma) = \mathrm{Aut}_{\mathscr{F}}(D) \cong A_4$. Here $\mathrm{H}^2(\mathrm{Aut}_{\mathscr{F}}(\sigma), F^{\times}) = 0$, but $\mathrm{H}^1(\mathrm{Aut}_{\mathscr{F}}(\sigma), F^{\times}) \cong \mathrm{H}^1(A_4, F^{\times}) \cong \mathrm{H}^1(C_3, F^{\times}) \cong C_3$. Hence, we have to consider the situation more closely. Up to conjugation there are three chains of $\mathscr{F}$-centric subgroups: $Q := \langle x, z \rangle$, D, and $Q < D$. Since $[S(\mathscr{F}^c)]$ is partially ordered by taking subchains, one can view $[S(\mathscr{F}^c)]$ as a category, where the morphisms are given by the pairs of ordered chains. In our case $[S(\mathscr{F}^c)]$ has precisely five morphisms. With the notations of [284] the functor $\mathscr{A}_{\mathscr{F}}^1$ is a representation of $[S(\mathscr{F}^c)]$ over $\mathbb{Z}$. Hence, we can view $\mathscr{A}_{\mathscr{F}}^1$ as a module $\mathscr{M}$ over the incidence algebra of $[S(\mathscr{F}^c)]$. More precisely, we have

$$\mathscr{M} := \bigoplus_{a \in \mathrm{Ob}[S(\mathscr{F}^c)]} \mathscr{A}_{\mathscr{F}}^1(a) = \mathscr{A}_{\mathscr{F}}^1(D) \cong C_3.$$

Now we can determine $\mathrm{H}^1([S(\mathscr{F}^c)], \mathscr{A}_{\mathscr{F}}^1)$ using Lemma 6.2(2) in [284]. For this let $d : \mathrm{Hom}[S(\mathscr{F}^c)] \to \mathscr{M}$ a derivation. Then we have $d(\alpha) = 0$ for all $\alpha \in \mathrm{Hom}[S(\mathscr{F}^c)]$ with $\alpha \neq (D, D) =: \alpha_1$. Moreover,

$$d(\alpha_1) = d(\alpha_1 \alpha_1) = (\mathscr{A}_{\mathscr{F}}^1(\alpha_1))(d(\alpha_1)) + d(\alpha_1) = 2d(\alpha_1) = 0.$$

Hence, $\mathrm{H}^1([S(\mathscr{F}^c)], \mathscr{A}_{\mathscr{F}}^1) = 0$. $\qquad\square$

9.3 $Q_{2^n} \times C_{2^m}$

We write

$$D := \langle x, y, z \mid x^{2^{n-1}} = z^{2^m} = [x, z] = [y, z] = 1, \ y^2 = x^{2^{n-2}}, \ yxy^{-1} = x^{-1} \rangle$$
$$= \langle x, y \rangle \times \langle z \rangle \cong Q_{2^n} \times C_{2^m}$$

where $n \geq 3$ and $m \geq 0$. We allow $m = 0$, since the results are completely consistent in this case.

The first lemma shows that the situation splits naturally into two cases according to $n = 3$ or $n \geq 4$.

Lemma 9.23 *The automorphism group $\mathrm{Aut}(D)$ is a 2-group if and only if $n \geq 4$.*

Proof Since Aut(Q_8) $\cong$ S_4, the "only if"-part is easy to see. Now let $n \geq 4$. Then the subgroups $\Phi(D) < \Phi(D)Z(D) < \langle x, z \rangle < D$ are characteristic in D. By Theorem 5.3.2 in [94] every automorphism of Aut(D) of odd order acts trivially on $D/\Phi(D)$. The claim follows from Theorem 5.1.4 in [94].   □

It follows that the inertial index $e(B)$ of B equals 1 for $n \geq 4$. In case $n = 3$ there are two possibilities $e(B) \in \{1, 3\}$, since $\Phi(D)Z(D)$ is still characteristic in D. Now we investigate the fusion system $\mathscr{F}$ of B.

Lemma 9.24 *Let* $Q_1 := \langle x^{2^{n-3}}, y, z \rangle \cong Q_8 \times C_{2^m}$ *and* $Q_2 := \langle x^{2^{n-3}}, xy, z \rangle \cong Q_8 \times C_{2^m}$. *Then* Q_1 *and* Q_2 *are the only candidates for proper* $\mathscr{F}$-*centric,* $\mathscr{F}$-*radical subgroups up to conjugation. Moreover, one of the following cases occurs:*

(aa) $n = e(B) = 3$ *or* ($n \geq 4$ *and* $\mathrm{Out}_{\mathscr{F}}(Q_1) \cong \mathrm{Out}_{\mathscr{F}}(Q_2) \cong S_3$).
(ab) $n \geq 4$, $\mathrm{Out}_{\mathscr{F}}(Q_1) \cong C_2$ *and* $\mathrm{Out}_{\mathscr{F}}(Q_2) \cong S_3$.
(ba) $n \geq 4$, $\mathrm{Out}_{\mathscr{F}}(Q_1) \cong S_3$ *and* $\mathrm{Out}_{\mathscr{F}}(Q_2) \cong C_2$.
(bb) $\mathrm{Out}_{\mathscr{F}}(Q_1) \cong \mathrm{Out}_{\mathscr{F}}(Q_2) \cong C_2$.

In case (bb) the block B is nilpotent.

Proof Let $Q < D$ be $\mathscr{F}$-centric and $\mathscr{F}$-radical. Then $z \in Z(D) \subseteq C_D(Q) \subseteq Q$ and $Q = (Q \cap \langle x, y \rangle) \times \langle z \rangle$. Let us consider the case $Q = \langle x, z \rangle$. Then $m = n - 1$ (this is not important here). The group $D \subseteq N_G(Q, b_Q)$ acts trivially on $\Omega(Q) \subseteq Z(D)$, while a non-trivial automorphism of Aut(Q) of odd order acts non-trivially on $\Omega(Q)$ (see Theorem 5.2.4 in [94]). This contradicts $O_2(\mathrm{Aut}_{\mathscr{F}}(Q)) = 1$. Moreover, by Lemma 5.4 in [184] we see that $\mathrm{Aut}_{\mathscr{F}}(Q)$ is a 2-group (this will be needed later).

Now let $Q = \langle x^i y, z \rangle$ for some $i \in \mathbb{Z}$. Then we have $m = 2$, and the same argument as before leads to a contradiction.

Hence by Lemma 9.23, Q is isomorphic to $Q_8 \times C_{2^m}$, and contains an element of the form $x^i y$. After conjugation with a suitable power of x we may assume $Q \in \{Q_1, Q_2\}$. This shows the first claim.

Let $S \leq D$ be an arbitrary subgroup isomorphic to $Q_8 \times C_{2^m}$. If $z \notin S$, then for $\langle S, z \rangle = (\langle S, z \rangle \cap \langle x, y \rangle) \times \langle z \rangle$ we have $\langle S, z \rangle' = S' \cong C_2$. However, this is impossible, since $\langle S, z \rangle \cap \langle x, y \rangle$ has at least order 16. This contradiction shows $z \in S$. Thus, S is conjugate to $Q \in \{Q_1, Q_2\}$ under D. In particular, Q is fully $\mathscr{F}$-normalized. Hence, $N_D(Q)/Q \cong C_2$ is a Sylow 2-subgroup of $\mathrm{Out}_{\mathscr{F}}(Q)$. Assume $N_D(Q)C_G(Q) < N_G(Q, b_Q)$. Since $O_2(\mathrm{Out}_{\mathscr{F}}(Q)) = 1$ and $|\mathrm{Aut}(Q)| = 2^k \cdot 3$ for some $k \in \mathbb{N}$, we get $\mathrm{Out}_{\mathscr{F}}(Q) \cong S_3$.

The last claim follows from Alperin's Fusion Theorem and $e(B) = 1$ (for $n \geq 4$).   □

The naming of these cases is adopted from [212]. Since the cases (ab) and (ba) are symmetric, we ignore case (ba). It is easy to see that Q_1 and Q_2 are not conjugate in D if $n \geq 4$. Hence, by Alperin's Fusion Theorem the subpairs (Q_1, b_{Q_1}) and (Q_2, b_{Q_2}) are not conjugate in G. It is also easy to see that Q_1 and Q_2 are always $\mathscr{F}$-centric.

Lemma 9.25 *Let* $Q \in \{Q_1, Q_2\}$ *such that* $\mathrm{Out}_{\mathscr{F}}(Q) \cong S_3$. *Then*

$$\mathrm{C}_Q(\mathrm{N}_G(Q, b_Q)) = Z(Q) = \langle x^{2^{n-2}}, z \rangle.$$

Proof Since $Q \subseteq \mathrm{N}_D(Q, b_Q)$, we have $\mathrm{C}_Q(\mathrm{N}_G(Q, b_Q)) \subseteq \mathrm{C}_Q(Q) = Z(Q)$. On the other hand, $\mathrm{N}_D(Q)$ acts trivially on $Z(Q) = Z(D)$. Hence, it suffices to determine the fixed points of an automorphism $\alpha \in \mathrm{Aut}(Q)$ of order 3 in $Z(Q)$. Since α acts trivially on $Q' \cong C_2$ and on $Z(Q)/Q' \cong C_{2^m}$, the claim follows from Theorem 5.3.2 in [94]. □

Lemma 9.26 *A set of representatives $\mathscr{R}$ for the $\mathscr{F}$-conjugacy classes of elements* $u \in D$ *such that* $\langle u \rangle$ *is fully $\mathscr{F}$-normalized is given as follows:*

(i) $x^i z^j$ $(i = 0, 1, \ldots, 2^{n-2}, j = 0, 1, \ldots, 2^m - 1)$ *in case (aa).*
(ii) $x^i z^j$ *and* $y z^j$ $(i = 0, 1, \ldots, 2^{n-2}, j = 0, 1, \ldots, 2^m - 1)$ *in case (ab).*

Proof By Lemma 9.25, in any case the elements $x^i z^j$ $(i = 0, 1, \ldots, 2^{n-2}, j = 0, 1, \ldots, 2^m - 1)$ are pairwise non-conjugate in $\mathscr{F}$. If $n = 3$, the block B is controlled and every subgroup is fully $\mathscr{F}$-normalized. Thus, assume for the moment that $n \geq 4$. Then $\langle x, z \rangle \subseteq \mathrm{C}_G(x^i z^j)$ and $|D : \mathrm{N}_D(\langle x^i z^j \rangle)| \leq 2$. Suppose that $\langle x^i y z^j \rangle \trianglelefteq D$ for some $i, j \in \mathbb{Z}$. Then we have $x^{i+2} y z^j = x(x^i y z^j)x^{-1} \in \langle x^i y z^j \rangle$ and the contradiction $x^2 \in \langle x^i y z^j \rangle$. This shows that the subgroups $\langle x^i z^j \rangle$ are always fully $\mathscr{F}$-normalized.

Assume that case (aa) occurs. Then the elements of the form $x^{2i} y z^j$ $(i, j \in \mathbb{Z})$ are conjugate to elements of the form $x^{2i} z^j$ under $D \cup \mathrm{N}_G(Q_1, b_{Q_1})$. Similarly, the elements of the form $x^{2i+1} y z^j$ $(i, j \in \mathbb{Z})$ are conjugate to elements of the form $x^{2i} z^j$ under $D \cup \mathrm{N}_G(Q_2, b_{Q_2})$. The claim follows in this case.

In case (ab) the given elements are pairwise non-conjugate, since no conjugate of $y z^j$ lies in Q_2. As in case (aa), the elements of the form $x^{2i} y z^j$ $(i, j \in \mathbb{Z})$ are conjugate to elements of the form $y z^j$ under D and the elements of the form $x^{2i+1} y z^j$ $(i, j \in \mathbb{Z})$ are conjugate to elements of the form $x^{2i} z^j$ under $D \cup \mathrm{N}_G(Q_2, b_{Q_2})$. Finally, the subgroups $\langle y z^j \rangle$ are fully $\mathscr{F}$-normalized, since $y z^j$ is not conjugate to an element in Q_2. □

Lemma 9.27 *Olsson's Conjecture $k_0(B) \leq 2^{m+2} = |D : D'|$ is satisfied in all cases.*

Proof This follows from Theorem 5.3 (cf. Lemma 9.6). □

Theorem 9.28

(i) *In case (aa) and $n = 3$ we have $k(B) = 2^m \cdot 7$, $k_0(B) = 2^{m+2}$, $k_1(B) = 2^m \cdot 3$ and $l(B) = 3$.*
(ii) *In case (aa) and $n \geq 4$ we have $k(B) = 2^m(2^{n-2} + 5)$, $k_0(B) = 2^{m+2}$, $k_1(B) = 2^m(2^{n-2} - 1)$, $k_{n-2}(B) = 2^{m+1}$ and $l(B) = 3$.*

(iii) *In case (ab) we have* $k(B) = 2^m(2^{n-2} + 4)$, $k_0(B) = 2^{m+2}$, $k_1(B) = 2^m(2^{n-2} - 1)$, $k_{n-2}(B) = 2^m$ *and* $l(B) = 2$.

(iv) *In case (bb) we have* $k(B) = 2^m(2^{n-2} + 3)$, $k_0(B) = 2^{m+2}$, $k_1(B) = 2^m(2^{n-2} - 1)$ *and* $l(B) = 1$.

In particular Brauer's $k(B)$-Conjecture, Brauer's Height-Zero Conjecture and the Alperin-McKay Conjecture hold.

Proof Assume first that case (bb) occurs. Then B is nilpotent and $k_i(B)$ is just the number $k_i(D)$ of irreducible characters of D of degree 2^i ($i \geq 0$) and $l(B) = 1$. Since C_{2^m} is abelian, we get $k_i(B) = 2^m k_i(Q_{2^n})$. The claim follows in this case.

Now assume that case (aa) or case (ab) occurs. We determine the numbers $l(b)$ for the subsections in Lemma 9.26 and apply Theorem 1.35. Let us begin with the non-major subsections. Since $\mathrm{Aut}_{\mathscr{F}}(\langle x, z \rangle)$ is a 2-group, we have $l(b_{x^i z^j}) = 1$ for all $i = 1, \ldots, 2^{n-2} - 1$ and $j = 0, 1, \ldots, 2^m - 1$. The blocks b_{yz^j} ($j = 0, 1, \ldots, 2^{m-1} - 1$) have $C_D(yz^j) = \langle y, z \rangle \cong C_4 \times C_{2^m}$ as defect group. In case (ab), $\mathrm{Aut}_{\mathscr{F}}(N_D(\langle y, z \rangle)) = \mathrm{Aut}_{\mathscr{F}}(Q_1)$ is a 2-group. Thus, by Lemma 5.4 in [184] also $\mathrm{Aut}_{\mathscr{F}}(\langle y, z \rangle)$ is a 2-group. Hence, it follows that $l(b_{yz^j}) = 1$ for $j = 0, 1, \ldots, 2^{m-1} - 1$.

Now let (u, b_u) be a major subsection. By Lemma 9.25, the cases for B and b_u coincide. As usual, the blocks b_u dominate blocks $\overline{b_u}$ of $C_G(u)/\langle u \rangle$ with defect group $D/\langle u \rangle$. In case $u = z^j$ for some $j \in \mathbb{Z}$ we have $D/\langle u \rangle \cong Q_{2^n} \times C_{2^m/|\langle z^j \rangle|}$. Of course the cases for b_u and $\overline{b_u}$ coincide, and we have $l(b_{z^j}) = l(\overline{b_{z^j}})$. Thus, we can apply induction on m. The beginning of this induction ($m = 0$) is satisfied by Theorem 8.1.

In case $u = x^{2^{n-2}}$ we have $D/\langle u \rangle \cong D_{2^{n-1}} \times C_{2^m}$. Then we can apply Theorem 9.7. Observe again that the cases for b_u and $\overline{b_u}$ coincide.

Finally, if $u = x^{2^{n-2}} z^j$ for some $j \in \{1, \ldots, 2^m - 1\}$, we have

$$D/\langle u \rangle \cong (D/\langle z^{2j} \rangle)/(\langle x^{2^{n-2}} z^j \rangle / \langle z^{2j} \rangle) \cong Q_{2^n} * C_{2^m/|\langle z^{2j} \rangle|}.$$

For $\langle z^j \rangle = \langle z \rangle$ we get $D/\langle u \rangle \cong Q_{2^n}$. Otherwise, $Q_{2^n} * C_{2^m/|\langle z^{2j} \rangle|} \cong D_{2^n} * C_{2^m/|\langle z^{2j} \rangle|}$. Here we can apply Theorem 9.18. Now we discuss the cases (ab) and (aa) separately.

Case (ab):

Then we have $l(b_u) = l(\overline{b_u}) = 2$ for $1 \neq u \in Z(D)$. Hence, Theorem 1.35 implies

$$k(B) - l(B) = 2^m(2^{n-2} - 1) + 2^m + 2(2^{m+1} - 1) = 2^m(2^{n-2} + 4) - 2.$$

Since B is a centrally controlled block, we have $l(B) \geq l(b_z) = 2$ and $k(B) \geq 2^m(2^{n-2} + 4)$ by Theorem 1.38. In orderto bound $k(B)$ from above we study the numbers $d_{\chi\varphi}^z$. Let $D^z := (d_{\chi\varphi_i}^z)_{\substack{\chi\in\mathrm{Irr}(B),\\ i=1,2}}$. Then $(D^z)^{\mathrm{T}}\overline{D^z} = C^z$ is the Cartan matrix

of b_z. Since $\overline{b_z}$ has defect group Q_{2^n}, it follows that

$$C^z = 2^m \begin{pmatrix} 2^{n-2}+2 & 4 \\ 4 & 8 \end{pmatrix}$$

up to basic sets (see Theorem 8.1). We consider the generalized decomposition numbers more carefully. Here the proof follows the lines of Theorem 9.18. However, we have to consider more cases. As in the previous section we write

$$d_{\chi\varphi_i}^z = \sum_{j=0}^{2^{m-1}-1} a_j^i(\chi)\zeta^j$$

for $i = 1, 2$, where ζ is a primitive 2^m-th root of unity. Since the subsections (z^j, b_{z^j}) are pairwise non-conjugate for $j = 0, \ldots, 2^m - 1$, we get

$$(a_i^1, a_j^1) = (2^{n-1} + 4)\delta_{ij}, \qquad (a_i^1, a_j^2) = 8\delta_{ij}, \qquad (a_i^2, a_j^2) = 16\delta_{ij}.$$

Since C^z is just twice as large as in the proof of Theorem 9.18, the contributions remain the same in terms of $d_{\chi\varphi}^z$. In particular we get

$$h(\chi) = 0 \iff \sum_{j=0}^{2^{m-1}-1} a_j^2(\chi) \equiv 1 \pmod{2}. \tag{9.4}$$

Assume that $k(B)$ is as large as possible. Since (z, b_z) is a major subsection, no row of D^z vanishes. Hence, for $j \in \{0, 1, \ldots, 2^{m-1} - 1\}$ we have essentially the following possibilities (where $\epsilon_1, \ldots, \epsilon_8 \in \{\pm 1\}$):

$$(I) : \begin{pmatrix} a_j^1 & \pm 1 \cdots \pm 1 & \epsilon_1 \cdots \epsilon_8 & . & \cdots\cdots\cdots\cdots. \\ a_j^2 & . & \cdots & . & \epsilon_1 \cdots \epsilon_8 & \pm 1 \cdots \pm 1 & . & \cdots. \end{pmatrix},$$

$$(II) : \begin{pmatrix} a_j^1 & \pm 1 \cdots \pm 1 & \epsilon_1 & \epsilon_2 \cdots \epsilon_7 & . & \cdots\cdots\cdots\cdots. \\ a_j^2 & . & \cdots & . & 2\epsilon_1 & \epsilon_2 \cdots \epsilon_7 & \pm 1 \cdots \pm 1 & . & \cdots. \end{pmatrix},$$

$$(III) : \begin{pmatrix} a_j^1 & \pm 1 \cdots \pm 1 & \epsilon_1 & \epsilon_2 & \epsilon_3 \cdots \epsilon_6 & . & \cdots\cdots\cdots\cdots. \\ a_j^2 & . & \cdots & . & 2\epsilon_1 & 2\epsilon_2 & \epsilon_3 \cdots \epsilon_6 & \pm 1 & \pm 1 & \pm 1 & \pm 1 & . & \cdots. \end{pmatrix},$$

$$(IV): \begin{pmatrix} a_j^1 & \pm 1 \cdots \pm 1 & \epsilon_1 & \epsilon_2 & \epsilon_3 & \epsilon_4 & \epsilon_5 & . & \cdots \cdots \cdots \\ a_j^2 & . & \cdots & . & 2\epsilon_1 & 2\epsilon_2 & 2\epsilon_3 & \epsilon_4 & \epsilon_5 & \pm 1 & \pm 1 & . & \cdots \end{pmatrix},$$

$$(V): \begin{pmatrix} a_j^1 & \pm 1 \cdots \pm 1 & \epsilon_1 & \epsilon_2 & \epsilon_3 & \epsilon_4 & \cdots \cdot \\ a_j^2 & . & \cdots & . & 2\epsilon_1 & 2\epsilon_2 & 2\epsilon_3 & 2\epsilon_4 & \cdots \cdot \end{pmatrix}.$$

The number $k(B)$ would be maximal if case (I) occurs for all j and for every character $\chi \in \mathrm{Irr}(B)$ we have $\sum_{j=0}^{2^{m-1}-1} |a_j^1(\chi)| \leq 1$ and $\sum_{j=0}^{2^{m-1}-1} |a_j^2(\chi)| \leq 1$. However, this contradicts Lemma 9.27 and Eq. (9.4). This explains why we have to allow other possibilities. We illustrate with two example that the given forms (I) to (V) are the only possibilities we need. For that consider

$$(IIa): \begin{pmatrix} a_j^1 & \pm 1 \cdots \pm 1 & 2\epsilon_1 \cdots \epsilon_7 & . & \cdots \cdots \cdots \cdots \cdot \\ a_j^2 & . & \cdots & . & \epsilon_1 \cdots \epsilon_7 & \pm 1 \cdots \pm 1 & . & \cdots \cdot \end{pmatrix},$$

$$(IVa): \begin{pmatrix} a_j^1 & \pm 1 \cdots \pm 1 & \epsilon_1 & \epsilon_2 \cdots \epsilon_6 & . & \cdots \cdots \cdots \cdot \\ a_j^2 & . & \cdots & . & 3\epsilon_1 & \epsilon_2 \cdots \epsilon_6 & \pm 1 & \pm 1 & . & \cdots \cdot \end{pmatrix}.$$

Then both (II) and (IIa) contribute $2^{n-1} + 10$ to $k(B)$. However, (II) contributes 12 to $k_0(B)$, while (IIa) contributes 16 to $k_0(B)$. Hence, (II) is "better" than (IIa). In the same way (IV) is "better" than (IVa). Now let α_1 (resp. $\alpha_2, \ldots, \alpha_5$) be the number of indices $j \in \{0, 1, \ldots, 2^{m-1} - 1\}$ such that case (I) (resp. (II), ... ,(V)) occurs for a_j^i. Then obviously $\alpha_1 + \ldots + \alpha_5 = 2^{m-1}$. It is easy to see that we may assume for all $\chi \in \mathrm{Irr}(B)$ that $\sum_{j=0}^{2^{m-1}-1} |a_j^1(\chi)| \leq 1$ in order to maximize $k(B)$. In contrast to that it does make sense to have $a_j^2(\chi) \neq 0 \neq a_k^2(\chi)$ for some $j \neq k$ in order to satisfy Olsson's Conjecture in view of Eq. (9.4). Let δ be the number of pairs $(\chi, j) \in \mathrm{Irr}(B) \times \{0, 1, \ldots, 2^{m-1} - 1\}$ such that there exists a $k \neq j$ with $a_j^2(\chi)a_k^2(\chi) \neq 0$. Then it follows that

$$\alpha_5 = 2^{m-1} - \alpha_1 - \alpha_2 - \alpha_3 - \alpha_4,$$

$$k(B) \leq (2^{n-1} + 12)\alpha_1 + (2^{n-1} + 10)\alpha_2 + (2^{n-1} + 8)\alpha_3$$
$$+ (2^{n-1} + 6)\alpha_4 + (2^{n-1} + 4)\alpha_5 - \delta/2$$
$$= 2^{m+n-2} + 12\alpha_1 + 10\alpha_2 + 8\alpha_3 + 6\alpha_4 + 4\alpha_5 - \delta/2$$
$$= 2^{m+n-2} + 2^{m+1} + 8\alpha_1 + 6\alpha_2 + 4\alpha_3 + 2\alpha_4 - \delta/2,$$

$$16\alpha_1 + 12\alpha_2 + 8\alpha_3 + 4\alpha_4 - \delta \leq k_0(B) \leq 2^{m+2}.$$

This gives $k(B) \leq 2^{m+n-2} + 2^{m+2} = 2^m(2^{n-2} + 4)$. Together with the lower bound above, we have shown that $k(B) = 2^{m-1}(2^{n-2} + 4)$ and $l(B) = 2$. In particular the cases (I),...,(V) are really the only possibilities which can occur. The inequalities above imply also $k_0(B) = 2^{m+2}$. As in theprevious section we can show that $\delta = 0$. Moreover, as there we see that the rows of type $(\pm \zeta^j, 0)$ of D^z correspond to characters of height 1. The number of these rows is

$$(2^{n-1} - 4)\alpha_1 + (2^{n-1} - 3)\alpha_2 + (2^{n-1} - 2)\alpha_3 + (2^{n-1} - 1)\alpha_4 + 2^{n-1}\alpha_5$$
$$= 2^m(2^{n-2} - 1).$$

The remaining rows of D^z correspond to characters of height 0 or $n - 2$. This gives $k_i(B)$ for $i \in \mathbb{N}$ (recall that $n \geq 4$ in case (ab)).

Case (aa):

Here we have $l(b_u) = l(\overline{b_u}) = 3$ for $1 \neq u \in Z(D)$. Hence, Theorem 1.35 implies

$$k(B) - l(B) = 2^m(2^{n-2} - 1) + 3(2^{m+1} - 1) = 2^m(2^{n-2} + 5) - 3.$$

Again B is a centrally controlled, $l(B) \geq l(b_z) = 3$ and $k(B) \geq 2^m(2^{n-2} + 5)$ by Theorem 1.38. The Cartan matrix of b_z is

$$C^z = 2^m \begin{pmatrix} 2^{n-2} + 2 & 2 & 2 \\ 2 & 4 & . \\ 2 & . & 4 \end{pmatrix}$$

up to basic sets. We write $\mathrm{IBr}(b_z) = \{\varphi_1, \varphi_2, \varphi_3\}$ and define the integral columns a^i_j for $i = 1, 2, 3$ and $j = 0, 1, \ldots, 2^{m-1} - 1$ as in case (ab). Then we can calculate the scalar products (a^i_j, a^k_l). Again C^z is just twice as large as in the proof of Theorem 9.18 and we get

$$h(\chi) = 0 \iff \sum_{j=0}^{2^{m-1}-1} \left(a^2_j(\chi) + a^3_j(\chi)\right) \equiv 1 \pmod{2}. \tag{9.5}$$

In order to search the maximum value for $k(B)$ (in view of Lemma 9.27 and Eq. (9.5)) we have to consider the following possibilities (where $\epsilon_1, \ldots, \epsilon_8 \in \{\pm 1\}$):

(I)

a_j^1	a_j^2	a_j^3
±1	.	.
$\vdots$	$\vdots$	$\vdots$
±1	.	.
ϵ_1	ϵ_1	.
$\vdots$	$\vdots$	$\vdots$
ϵ_4	ϵ_4	.
ϵ_5	.	ϵ_5
$\vdots$	$\vdots$	$\vdots$
ϵ_8	.	ϵ_8
.	±1	.
$\vdots$	$\vdots$	$\vdots$
.	±1	.
.	.	±1
$\vdots$	$\vdots$	$\vdots$
.	.	±1
$\vdots$	$\vdots$	$\vdots$
.	.	.

(II)

a_j^1	a_j^2	a_j^3
±1	.	.
$\vdots$	$\vdots$	$\vdots$
±1	.	.
ϵ_1	ϵ_1	.
ϵ_2	ϵ_2	.
ϵ_3	ϵ_3	.
ϵ_4	ϵ_4	ϵ_4
ϵ_5	.	ϵ_5
ϵ_6	.	ϵ_6
ϵ_7	.	ϵ_7
.	ϵ_8	$-\epsilon_8$
.	±1	.
.	±1	.
.	±1	.
.	.	±1
.	.	±1
.	.	±1
$\vdots$	$\vdots$	$\vdots$
.	.	.

(III)

a_j^1	a_j^2	a_j^3
±1	.	.
$\vdots$	$\vdots$	$\vdots$
±1	.	.
ϵ_1	ϵ_1	.
ϵ_2	ϵ_2	.
ϵ_3	ϵ_3	ϵ_3
ϵ_4	ϵ_4	ϵ_4
ϵ_5	.	ϵ_5
ϵ_6	.	ϵ_6
.	ϵ_7	$-\epsilon_7$
.	ϵ_8	$-\epsilon_8$
.	±1	.
.	±1	.
.	.	±1
.	.	±1
$\vdots$	$\vdots$	$\vdots$
.	.	.

(IV)

a_j^1	a_j^2	a_j^3
±1	.	.
$\vdots$	$\vdots$	$\vdots$
±1	.	.
ϵ_1	ϵ_1	.
ϵ_2	ϵ_2	ϵ_2
ϵ_3	ϵ_3	ϵ_3
ϵ_4	ϵ_4	ϵ_4
ϵ_5	.	ϵ_5
.	ϵ_6	$-\epsilon_6$
.	ϵ_7	$-\epsilon_7$
.	ϵ_8	$-\epsilon_8$
.	±1	.
.	.	±1
$\vdots$	$\vdots$	$\vdots$
.	.	.

(V)

a_j^1	a_j^2	a_j^3
±1	.	.
$\vdots$	$\vdots$	$\vdots$
±1	.	.
ϵ_1	ϵ_1	ϵ_1
ϵ_2	ϵ_2	ϵ_2
ϵ_3	ϵ_3	ϵ_3
ϵ_4	ϵ_4	ϵ_4
.	ϵ_5	$-\epsilon_5$
.	ϵ_6	$-\epsilon_6$
.	ϵ_7	$-\epsilon_7$
.	ϵ_8	$-\epsilon_8$
.	.	.
$\vdots$	$\vdots$	$\vdots$
.	.	.

Define $\alpha_1, \ldots, \alpha_5$ as before. Let δ be the number of triples $(\chi, i, j) \in \mathrm{Irr}(B) \times \{2, 3\} \times \{0, 1, \ldots, 2^{m-1} - 1\}$ such that there exists a $k \neq j$ with $a_j^i(\chi) a_k^2(\chi) \neq 0$ or $a_j^i(\chi) a_k^3(\chi) \neq 0$. Then the following holds:

$$\alpha_5 = 2^{m-1} - \alpha_1 - \alpha_2 - \alpha_3 - \alpha_4,$$

$$k(B) \leq (2^{n-1} + 12)\alpha_1 + (2^{n-1} + 11)\alpha_2 + (2^{n-1} + 10)\alpha_3$$
$$+ (2^{n-1} + 9)\alpha_4 + (2^{n-1} + 8)\alpha_5 - \delta/2$$
$$= 2^{m+n-2} + 12\alpha_1 + 11\alpha_2 + 10\alpha_3 + 9\alpha_4 + 8\alpha_5 - \delta/2$$
$$= 2^{m+n-2} + 2^{m+2} + 4\alpha_1 + 3\alpha_2 + 2\alpha_3 + \alpha_4 - \delta/2,$$

$$16\alpha_1 + 12\alpha_2 + 8\alpha_3 + 4\alpha_4 - \delta \leq k_0(B) \leq 2^{m+2}.$$

This gives $k(B) \leq 2^{n+m-2} + 2^{m+2} + 2^m = 2^m(2^{n-2} + 5)$. Together with the lower bound we have shown that $k(B) = 2^m(2^{n-2} + 5)$, $k_0(B) = 2^{m+2}$, and $l(B) = 3$. In particular the maximal value for $k(B)$ is indeed attended. Moreover, $\delta = 0$. As in the previous section we see that the rows of D^z of type $(\pm\zeta^j, 0, 0)$ correspond to characters of height 1. The number of these rows is

$$(2^{n-1} - 4)\alpha_1 + (2^{n-1} - 3)\alpha_2 + (2^{n-1} - 2)\alpha_3 + (2^{n-1} - 1)\alpha_4 + 2^{n-1}\alpha_5$$
$$= 2^m(2^{n-2} - 1).$$

The remaining rows of D^z correspond to characters of height 0 or $n - 2$. This gives $k_i(B)$ for $i \in \mathbb{N}$. Observe that we have to add $k_1(B)$ and $k_{n-2}(B)$ in case $n = 3$. $\square$

We add some remarks. Using Theorem 10.17 below it is easy to construct examples for B in all cases. If $\tilde{B}$ is a block with defect group $Q_{2^n} * C_{2^{m+1}}$, then the invariants of B and $\tilde{B}$ coincide in the corresponding cases. However, it was shown in [246] (for $n = 3$ and $m = 1$) that the numbers of 2-rational characters of B resp. $\tilde{B}$ are different.

Theorem 9.29 *Alperin's Weight Conjecture holds for B.*

Proof Just copy the proof of Theorem 9.19. $\square$

Theorem 9.30 *The Ordinary Weight Conjecture holds for B.*

Proof We may assume that B is not nilpotent, and thus case (bb) does not occur. Suppose that $n = 3$ and case (aa) occurs. Then D is the only $\mathcal{F}$-centric, $\mathcal{F}$-radical subgroup of D. Since $\text{Out}_{\mathcal{F}}(D) \cong C_3$, the set $\mathcal{N}_D$ consists only of the trivial chain. We have $\mathbf{w}(D, d) = 0$ for $d \notin \{m+2, m+3\}$, since then $k^d(D) = 0$. For $d = m+2$ we get $\mathbf{w}(D, d) = 3 \cdot 2^m$, since the irreducible characters of D of degree 2 are stable under $\text{Out}_{\mathcal{F}}(D)$. In case $d = m + 3$ it follows that $\mathbf{w}(D, d) = 3 \cdot 2^m + 2^m = 2^{m+2}$. Hence, the OWC follows from Theorem 9.28.

Now let $n \geq 4$ and assume that case (aa) occurs. Then there are three $\mathcal{F}$-centric, $\mathcal{F}$-radical subgroups up to conjugation: Q_1, Q_2 and D. Since $\text{Out}_{\mathcal{F}}(D) = 1$, it follows easily that $\mathbf{w}(D, d) = k^d(D)$ for all $d \in \mathbb{N}$. By Theorem 9.28 it suffices to show

$$\mathbf{w}(Q, d) = \begin{cases} 2^m & \text{if } d = m + 2 \\ 0 & \text{otherwise} \end{cases}$$

for $Q \in \{Q_1, Q_2\}$, because $k^{m+2}(B) = k_{n-2}(B) = 2^{m+1}$. We already have $\mathbf{w}(Q, d) = 0$ unless $d \in \{m + 2, m + 3\}$. Without loss of generality let $Q = Q_1$.

Let $d = m + 2$. Up to conjugation $\mathcal{N}_Q$ consists of the trivial chain $\sigma : 1$ and the chain $\tau : 1 < C$, where $C \leq \text{Out}_{\mathcal{F}}(Q)$ has order 2. We consider the chain σ first. Here $I(\sigma) = \text{Out}_{\mathcal{F}}(Q) \cong S_3$ acts trivially on the characters of Q or defect $m + 2$. This contributes 2^m to the alternating sum of $\mathbf{w}(Q, d)$. Now consider the chain τ.

Here $I(\tau) = C$ and $z(FC) = 0$. Hence, the contribution of τ vanishes and we get $\mathbf{w}(Q, d) = 2^m$ as desired.

Let $d = m + 3$. Then we have $I(\sigma, \mu) \cong S_3$ for every character $\mu \in \mathrm{Irr}(Q)$ with $\mu(x^{2^{n-3}}) = \mu(y) = 1$. For the other characters of Q with defect d we have $I(\sigma, \mu) \cong C_2$. Hence, the chain σ contributes 2^m to the alternating sum. There are 2^{m+1} characters $\mu \in \mathrm{Irr}(D)$ which are not fixed under $I(\tau) = C$. Hence, they split into 2^m orbits of length 2. For these characters we have $I(\tau, \mu) = 1$. For the other irreducible characters μ of D of defect d we have $I(\tau, \mu) = C$. Thus, the contribution of τ to the alternating sum is -2^m. This shows $\mathbf{w}(Q, d) = 0$.

In case (ab) we have only two $\mathscr{F}$-centric, $\mathscr{F}$-radical subgroups: Q_2 and D. Since $k_{n-2}(B) = 2^m$ in this case, the calculations above imply the result.   □

Finally we show that the Gluing Problem for the block B has a unique solution. This was done for $m = 0$ in [219].

Theorem 9.31 *The Gluing Problem for B has a unique solution.*

Proof Let σ be a chain of $\mathscr{F}$-centric subgroups of D, and let Q be the largest subgroup occurring in σ. Then $Q = (Q \cap \langle x, y \rangle) \times \langle z \rangle$. If $Q \cap \langle x, y \rangle$ is abelian, then $\mathrm{Aut}_{\mathscr{F}}(Q)$ and $\mathrm{Aut}_{\mathscr{F}}(\sigma)$ are 2-groups. So we have $\mathrm{H}^i(\mathrm{Aut}_{\mathscr{F}}(\sigma), F^\times) = 0$ for $i = 1, 2$.

Now assume that $Q \cap \langle x, y \rangle$ is non-abelian. Again $\mathrm{Aut}_{\mathscr{F}}(\sigma)$ is a 2-group unless $Q \in \{Q_1, Q_2\}$ (up to conjugation). Without loss of generality assume $Q = Q_1$ and $\mathrm{Aut}_{\mathscr{F}}(Q) \cong S_4$. If Q is the only subgroup occurring in σ, we get $\mathrm{Aut}_{\mathscr{F}}(\sigma) = \mathrm{Aut}_{\mathscr{F}}(Q) \cong S_4$. If σ consists of another subgroup, $\mathrm{Aut}_{\mathscr{F}}(\sigma)$ must be a 2-group, since an automorphism of $\mathrm{Aut}_{\mathscr{F}}(Q)$ of order 3 permutes the three maximal subgroups of $\langle x^{2^{n-3}}, y \rangle$ transitively. So in both cases we have $\mathrm{H}^i(\mathrm{Aut}_{\mathscr{F}}(\sigma), F^\times) = 0$ for $i = 1, 2$.

Hence, $\mathscr{A}_{\mathscr{F}}^2 = 0$ and $\mathrm{H}^0([S(\mathscr{F}^c)], \mathscr{A}_{\mathscr{F}}^2) = \mathrm{H}^1([S(\mathscr{F}^c)], \mathscr{A}_{\mathscr{F}}^1) = 0$. Now by Theorem 1.1 in [219] the Gluing Problem has only the trivial solution.   □

9.4 $SD_{2^n} \times C_{2^m}$

Let

$$D := \langle x, y, z \mid x^{2^{n-1}} = y^2 = z^{2^m} = [x, z] = [y, z] = 1, \; yxy^{-1} = x^{-1+2^{n-2}} \rangle$$

$$= \langle x, y \rangle \times \langle z \rangle \cong SD_{2^n} \times C_{2^m}$$

with $n \geq 4$ and $m \geq 0$.

Lemma 9.32 *The automorphism group* $\mathrm{Aut}(D)$ *is a 2-group.*

Proof This follows as in Lemma 9.23, because the maximal subgroups of the semidihedral group are pairwise non-isomorphic.   □

The last lemma implies that the inertial index of B is $e(B) = 1$.

Lemma 9.33 *Let* $Q_1 := \langle x^{2^{n-2}}, y, z \rangle \cong C_2^2 \times C_{2^m}$ *and* $Q_2 := \langle x^{2^{n-3}}, xy, z \rangle \cong$ $Q_8 \times C_{2^m}$. *Then* Q_1 *and* Q_2 *are the only candidates for proper* $\mathscr{F}$-*centric,* $\mathscr{F}$-*radical subgroups up to conjugation. Moreover, one of the following cases occurs:*

(aa) $\mathrm{Aut}_{\mathscr{F}}(Q_1) \cong \mathrm{Out}_{\mathscr{F}}(Q_2) \cong S_3$.
(ab) $\mathrm{Aut}_{\mathscr{F}}(Q_1) \cong S_3$ *and* $\mathrm{Out}_{\mathscr{F}}(Q_2) \cong C_2$.
(ba) $\mathrm{Aut}_{\mathscr{F}}(Q_1) \cong C_2$ *and* $\mathrm{Out}_{\mathscr{F}}(Q_2) \cong S_3$.
(bb) $\mathrm{Aut}_{\mathscr{F}}(Q_1) \cong \mathrm{Out}_{\mathscr{F}}(Q_2) \cong C_2$.

In case (bb) the block B is nilpotent.

Proof Let $Q < D$ be $\mathscr{F}$-centric and $\mathscr{F}$-radical. Then $z \in Z(D) \subseteq C_D(Q) \subseteq Q$ and $Q = (Q \cap \langle x, y \rangle) \times \langle z \rangle$. Since $\mathrm{Aut}(Q)$ is not a 2-group, only the following cases are possible: $Q \cong C_{2^m}^2$, $C_2^2 \times C_{2^m}$, $Q_8 \times C_{2^m}$. In the first case we have $Q = \langle x, z \rangle$ or $Q = \langle x^i y, z \rangle$ for some odd i. Then $m = n - 1$ or $m = 2$ respectively (this is not important here). The group $D \subseteq \mathrm{N}_G(Q, b_Q)$ (resp. $\langle x^{2^{n-3}} \rangle Q$) acts trivially on $\Omega(Q) \subseteq Z(D)$, while a non-trivial automorphism of $\mathrm{Aut}(Q)$ of odd order acts non-trivially on $\Omega(Q)$ (see Theorem 5.2.4 in [94]). This contradicts $O_2(\mathrm{Aut}_{\mathscr{F}}(Q)) = 1$. Moreover, by Lemma 5.4 in [184] we see that $\mathrm{Aut}_{\mathscr{F}}(\langle x, z \rangle)$ is a 2-group (this will be needed later).

If $Q \cong C_2^2 \times C_{2^m}$, then Q contains an element of the form $x^{2^i} y$. After conjugation with a suitable power of x we may assume $Q = Q_1$. Similarly, Q is conjugate to Q_2 if $Q \cong Q_8 \times C_{2^m}$. This shows the first claim.

It remains to show that one of the given cases occurs. For the subgroup Q_1 this can be done as in Lemma 9.3. For the subgroup Q_2 we can copy the proof of Lemma 9.24. In particular both Q_1 and Q_2 are fully $\mathscr{F}$-normalized. The last claim follows from Alperin's Fusion Theorem and $e(B) = 1$. □

Again the naming of these cases is adopted from Olsson's paper [212], but in contrast to the dihedral and quaternion case, the cases (ab) and (ba) are *not* symmetric, since $Q_1 \not\cong Q_2$. Moreover, it is easy to see that Q_1 and Q_2 are always $\mathscr{F}$-centric.

Lemma 9.34 *Let* $Q \in \{Q_1, Q_2\}$ *such that* $\mathrm{Out}_{\mathscr{F}}(Q) \cong S_3$. *Then*

$$C_Q(\mathrm{N}_G(Q, b_Q)) = \begin{cases} \langle z \rangle & \text{if } Q = Q_1, \\ \langle x^{2^{n-2}}, z \rangle & \text{if } Q = Q_2. \end{cases}$$

Proof For Q_2 this follows as in the quaternion case. For Q_1 we can consult Sect. 9.2. Observe that we may have to replace z by $x^{2^{n-2}} z$ here. However, this does not affect $C_{Q_2}(\mathrm{N}_G(Q_2, b_{Q_2}))$. □

Lemma 9.35 *A set of representatives* $\mathscr{R}$ *for the* $\mathscr{F}$-*conjugacy classes of elements* $u \in D$ *such that* $\langle u \rangle$ *is fully* $\mathscr{F}$-*normalized is given as follows:*

(i) $x^i z^j$ $(i = 0, 1, \ldots, 2^{n-2}, \ j = 0, 1, \ldots, 2^m - 1)$ *in case (aa).*

(ii) $x^i z^j$ and xyz^j $(i = 0, 1, \ldots, 2^{n-2}, j = 0, 1, \ldots, 2^m - 1)$ in case (ab).

(iii) $x^i z^j$ and yz^j $(i = 0, 1, \ldots, 2^{n-2}, j = 0, 1, \ldots, 2^m - 1)$ in case (ab).

Proof By Lemma 9.34, in any case the elements $x^i z^j$ $(i = 0, 1, \ldots, 2^{n-2}, j = 0, 1, \ldots, 2^m - 1)$ are pairwise non-conjugate in $\mathscr{F}$. Moreover, $\langle x, z \rangle \subseteq C_G(x^i z^j)$ and $|D : N_D(\langle x^i z^j \rangle)| \le 2$. Suppose that $\langle x^i yz^j \rangle \trianglelefteq D$ for some $i, j \in \mathbb{Z}$. Then we have $x^{i+2+2^{n-2}} yz^j = x(x^i yz^j)x^{-1} \in \langle x^i yz^j \rangle$ and the contradiction $x^{2+2^{n-2}} \in \langle x^i yz^j \rangle$. This shows that the subgroups $\langle x^i z^j \rangle$ are always fully $\mathscr{F}$-normalized.

Assume that case (aa) occurs. Then the elements of the form $x^{2i} yz^j$ $(i, j \in \mathbb{Z})$ are conjugate to elements of the form $x^{2i} z^j$ under $D \cup N_G(Q_1, b_{Q_1})$. Similarly, the elements of the form $x^{2i+1} yz^j$ $(i, j \in \mathbb{Z})$ are conjugate to elements of the form $x^{2i} z^j$ under $D \cup N_G(Q_2, b_{Q_2})$. The claim follows in this case.

In case (ab) the given elements are pairwise non-conjugate, since no conjugate of xyz^j lies in Q_1. As in case (aa), the elements of the form $x^{2i} yz^j$ $(i, j \in \mathbb{Z})$ are conjugate to elements of the form $x^{2i} z^j$ under $D \cup N_G(Q_1, b_{Q_1})$, and the elements of the form $x^{2i+1} yz^j$ $(i, j \in \mathbb{Z})$ are conjugate to elements of the form xyz^j under D. Finally, the subgroups $\langle xyz^j \rangle$ are fully $\mathscr{F}$-normalized, since xyz^j is not conjugate to an element in Q_1.

The situation in case (ba) is very similar. We omit the details.   □

Lemma 9.36 *Olsson's Conjecture* $k_0(B) \le 2^{m+2} = |D : D'|$ *is satisfied in all cases.*

Proof This follows from Theorem 5.3 (cf. Lemma 9.6).   □

Theorem 9.37

  (i) *In case (aa) we have* $k(B) = 2^m(2^{n-2} + 4)$, $k_0(B) = 2^{m+2}$, $k_1(B) = 2^m(2^{n-2} - 1)$, $k_{n-2}(B) = 2^m$ *and* $l(B) = 3$.

  (ii) *In case (ab) we have* $k(B) = 2^m(2^{n-2} + 3)$, $k_0(B) = 2^{m+2}$, $k_1(B) = 2^m(2^{n-2} - 1)$ *and* $l(B) = 2$.

  (iii) *In case (ba) we have* $k(B) = 2^m(2^{n-2} + 4)$, $k_0(B) = 2^{m+2}$, $k_1(B) = 2^m(2^{n-2} - 1)$, $k_{n-2}(B) = 2^m$ *and* $l(B) = 2$.

  (iv) *In case (bb) we have* $k(B) = 2^m(2^{n-2} + 3)$, $k_0(B) = 2^{m+2}$, $k_1(B) = 2^m(2^{n-2} - 1)$ *and* $l(B) = 1$.

In particular Brauer's $k(B)$-Conjecture, Brauer's Height-Zero Conjecture and the Alperin-McKay Conjecture hold.

Proof Assume first that case (bb) occurs. Then B is nilpotent, and the result follows.

Now assume that case (aa), (ab) or (ba) occurs. We determine the numbers $l(b)$ for the subsections in Lemma 9.35 and apply Theorem 1.35. Let us begin with the non-major subsections. Since $\text{Aut}_{\mathscr{F}}(\langle x, z \rangle)$ is a 2-group, we have $l(b_{x^i z^j}) = 1$ for all $i = 1, \ldots, 2^{n-2} - 1$ and $j = 0, 1, \ldots, 2^m - 1$. The blocks b_{xyz^j} $(j = 0, 1, \ldots, 2^{m-1} - 1)$ have $C_D(xyz^j) = \langle xy, z \rangle \cong C_4 \times C_{2^m}$ as defect group. In case (ab), $\text{Aut}_{\mathscr{F}}(N_D(\langle xy, z \rangle)) = \text{Aut}_{\mathscr{F}}(Q_2)$ is a 2-group. Hence, Lemma 5.4 in [184] implies that also $\text{Aut}_{\mathscr{F}}(\langle xy, z \rangle)$ is a 2-group. This gives $l(b_{xyz^j}) = 1$ for $j = 0, 1, \ldots, 2^m - 1$. Similarly, in case (ba) we have $l(b_{yz^j}) = 1$.

Now we consider the major subsections. By Lemma 9.34, the cases for B and b_{z^j} coincide. As usual, the blocks b_{z^j} dominate blocks $\overline{b_{z^j}}$ of $C_G(z^j)/\langle z^j \rangle$ with defect group $D/\langle z^j \rangle \cong SD_{2^n} \times C_{2^m/|\langle z^j \rangle|}$. Of course the cases for b_{z^j} and $\overline{b_{z^j}}$ coincide, and we have $l(b_{z^j}) = l(\overline{b_{z^j}})$. Thus, we can apply induction on m. The beginning of this induction ($m = 0$) is satisfied by Theorem 8.1.

Let $u := x^{2^{n-1}} z^j$ for a $j \in \{0, 1, \ldots, 2^m - 1\}$. If case (ab) occurs for B, then case (bb) occurs for b_u by Lemma 9.34. Thus, $l(b_u) = 1$ in this case. If case (ba) or (aa) occurs for B, then case (ba) occurs for b_u. In case $j = 0$, b_u dominates a block $\overline{b_u}$ with defect group $D/\langle u \rangle \cong D_{2^{n-1}} \times C_{2^m}$. Then we can apply Theorem 9.7. Observe again that the cases for b_u and $\overline{b_u}$ coincide.

Finally, if $j \in \{1, \ldots, 2^m - 1\}$, we have

$$D/\langle u \rangle \cong (D/\langle z^{2j} \rangle)/((\langle x^{2^{n-2}} z^j \rangle / \langle z^{2j} \rangle) \cong SD_{2^n} * C_{2^m/|\langle z^{2j} \rangle|}.$$

For $\langle z^j \rangle = \langle z \rangle$ we get $D/\langle u \rangle \cong SD_{2^n}$. Otherwise, $SD_{2^n} * C_{2^m/|\langle z^{2j} \rangle|} \cong D_{2^n} * C_{2^m/|\langle z^{2j} \rangle|}$. Here we can apply Theorem 9.18. Now we discuss the cases (ab), (ba) and (aa) separately.

Case (ab):

Then we have $l(b_{z^j}) = l(\overline{b_{z^j}}) = 2$ for $j = 1, \ldots, 2^m - 1$ by induction on m. As explained above, we also have $l(b_u) = 1$ for $u = x^{2^{n-1}} z^j$ and $j = 0, 1, \ldots, 2^m - 1$. Hence, Theorem 1.35 implies

$$k(B) - l(B) = 2^m(2^{n-2} - 1) + 2^m + 2(2^m - 1) + 2^m = 2^m(2^{n-2} + 3) - 2.$$

Since B is a centrally controlled block, we have $l(B) \geq l(b_z) = 2$ and $k(B) \geq 2^m(2^{n-2} + 3)$ by Theorem 1.38.

Let $u := x^{2^{n-2}} \in Z(D)$. Lemma 1.37 implies $2^{h(\chi)} \mid d^u_{\chi\varphi_u}$ and $2^{h(\chi)+1} \nmid d^u_{\chi\varphi_u}$ for $\chi \in \mathrm{Irr}(B)$. In particular, $d^u_{\chi\varphi_u} \neq 0$. Lemma 9.36 gives

$$2^{n+m} \leq k_0(B) + 4(k(B) - k_0(B)) \leq \sum_{\chi \in \mathrm{Irr}(B)} \left(d^u_{\chi\varphi_u}\right)^2 = (d(u), d(u)) = |D| = 2^{n+m}.$$

Hence, we have

$$d^u_{\chi\varphi_u} = \begin{cases} \pm 1 & \text{if } h(\chi) = 0 \\ \pm 2 & \text{otherwise} \end{cases},$$

and the claim follows in case (ab).

Case (ba):

Here we have $l(b_u) = 2$ for all $1 \neq u \in Z(D)$ by induction on m. This gives

$$k(B) - l(B) = 2^m(2^{n-2} - 1) + 2^m + 2(2^{m+1} - 1) = 2^m(2^{n-2} + 4) - 2.$$

Since B is a centrally controlled block, we have $l(B) \geq l(b_z) = 2$ and $k(B) \geq 2^m(2^{n-2} + 4)$ by Theorem 1.38. Now the proof works as in the quaternion case by studying the numbers $d_{\chi\varphi}^z$. Since $\overline{b}_z$ has defect group SD_{2^n}, the Cartan matrix of b_z is given by

$$2^m \begin{pmatrix} 2^{n-2} + 2 & 4 \\ 4 & 8 \end{pmatrix}$$

up to basic sets. This is exactly the same matrix as in the quaternion case. So we omit the details.

Case (aa):

We have $l(b_{z^j}) = 3$ for $j = 1, \ldots, 2^m - 1$ by induction on m. Moreover, for $u = x^{2^{n-1}} z^j$ we get $l(b_u) = 2$. Hence,

$$k(B) - l(B) = 2^m(2^{n-2} - 1) + 3(2^m - 1) + 2^{m+1} = 2^m(2^{n-2} + 4) - 3.$$

Again B is centrally controlled which implies $l(B) \geq l(b_z) = 3$ and $k(B) \geq 2^m(2^{n-2} + 4)$. In contrast to case (ba) we study the generalized decomposition numbers of the element $u := x^{2^{n-2}} z$. Then case (ba) occurs for b_u and the Cartan matrix of b_u is given by

$$2^m \begin{pmatrix} 2^{n-2} + 2 & 4 \\ 4 & 8 \end{pmatrix}$$

up to basic sets. Hence, the proof works as above. $\square$

Using Theorem 10.17 below it is easy to construct examples for B in all cases.

Theorem 9.38 *Alperin's Weight Conjecture holds for B.*

Proof Just copy the proof of Theorem 9.19. $\square$

Theorem 9.39 *The Ordinary Weight Conjecture holds for B.*

Proof We may assume that B is not nilpotent, and thus case (bb) does not occur.

Assume first that case (aa) occurs. Then there are three $\mathscr{F}$-centric, $\mathscr{F}$-radical subgroups up to conjugation: Q_1, Q_2 and D. Since $\text{Out}_{\mathscr{F}}(D) = 1$, it follows easily that $\mathbf{w}(D, d) = k^d(D)$ for all $d \in \mathbb{N}$. By Theorem 9.37, it suffices to show $\mathbf{w}(Q_1, d) = 0$ for all d and

$$\mathbf{w}(Q_2, d) = \begin{cases} 2^m & \text{if } d = m + 2, \\ 0 & \text{otherwise}, \end{cases}$$

because $k^{m+2}(B) = k_{n-2}(B) = 2^m$. For the group Q_1 this works exactly as in Sect. 9.1 and for Q_2 we can copy the proof of Theorem 9.30.

In the cases (ab) and (ba) we have only two $\mathscr{F}$-centric, $\mathscr{F}$-radical subgroups: Q_1 (resp. Q_2) and D. In case (ab), Theorem 9.37 implies $k^d(B) = k^d(D)$ for all $d \in \mathbb{N}$ while in case (ba) we still have $k^{m+2}(B) = 2^m$. So the calculations above imply the result. $\square$

Theorem 9.40 *The Gluing Problem for B has a unique solution.*

Proof Let σ be a chain of $\mathscr{F}$-centric subgroups of D, and let Q be the largest subgroup occurring in σ. Then $Q = (Q \cap \langle x, y \rangle) \times \langle z \rangle$. If $Q \cap \langle x, y \rangle$ is abelian, then $\text{Aut}_{\mathscr{F}}(Q)$ and $\text{Aut}_{\mathscr{F}}(\sigma)$ are 2-groups unless $Q = Q_1$ (up to conjugation). In case $Q = Q_1$, σ only consists of Q, and we can also have $\text{Aut}_{\mathscr{F}}(\sigma) = \text{Aut}_{\mathscr{F}}(Q) \cong S_3$. So in all these cases we have $\text{H}^i(\text{Aut}_{\mathscr{F}}(\sigma), F^\times) = 0$ for $i = 1, 2$.

Now assume that $Q \cap \langle x, y \rangle$ is non-abelian. Again $\text{Aut}_{\mathscr{F}}(\sigma)$ is a 2-group unless $Q = Q_2$ (up to conjugation). Now the claim follows as in Theorem 9.31. $\square$

Chapter 10
Bicyclic Groups

Another interesting generalization of metacyclic groups are bicyclic groups. Here a group G is called *bicyclic* if there exist $x, y \in G$ such that $G = \langle x \rangle \langle y \rangle$. For odd primes p, Huppert showed in [126] that a bicyclic p-group is metacyclic and conversely (see also Satz III.11.5 [128]). This shifts again the focus to the case $p = 2$ where the class of bicyclic p-groups is strictly larger than the class of metacyclic p-groups. Apart from Huppert's work, there are many other contributions to the theory of bicyclic 2-groups. We mention some of them: [29, 70, 135–137]. One of these early results is the following: Let P be a non-metacyclic, bicyclic 2-group. Then the commutator subgroup P' is abelian of rank at most 2 and P/P' contains a cyclic maximal subgroup. Moreover, if P/P' has exponent at least 8, then also P' contains a cyclic maximal subgroup.

Here we are primarily interested in the classification of the corresponding fusion systems. Later we give corollaries for blocks with bicyclic defect groups. The material comes from [252, 257]. We will use the following notation: A group P is called *minimal non-abelian of type* (r, s) if

$$P \cong \langle x, y \mid x^{p^r} = y^{p^s} = [x, y]^p = [x, x, y] = [y, x, y] = 1 \rangle$$

for $r \geq s \geq 1$ (see Chap. 12 for more details).

10.1 Fusion Systems

Janko gave the following characterization of bicyclic 2-groups (see [140] or alternatively Sect. 87 in [24]). Notice that Janko defines commutators in [140] differently than we do.

Theorem 10.1 (Janko) *A non-metacyclic 2-group P is bicyclic if and only if P has rank 2 and contains exactly one non-metacyclic maximal subgroup.*

© Springer International Publishing Switzerland 2014

B. Sambale, *Blocks of Finite Groups and Their Invariants*, Lecture Notes in Mathematics 2127, DOI 10.1007/978-3-319-12006-5_10

Using this, he classified all bicyclic 2-groups in terms of generators and relations. However, it is not clear if different parameters in his paper give non-isomorphic groups. In particular the number of isomorphism types of bicyclic 2-groups is unknown.

As a corollary, we obtain the structure of the automorphism group of a bicyclic 2-group.

Proposition 10.2 *Let P be a bicyclic 2-group such that* $\mathrm{Aut}(P)$ *is not a 2-group. Then P is homocyclic or a quaternion group of order* 8. *In particular, P is metacyclic.*

Proof By Lemma 1 in [190] we may assume that P is non-metacyclic. Since P has rank 2, every non-trivial automorphism of odd order permutes the maximal subgroups of P transitively. By Theorem 10.1 such an automorphism cannot exist.□

As another corollary, we see that every subgroup of a bicyclic 2-group contains a metacyclic maximal subgroup. Since quotients of bicyclic groups are also bicyclic, it follows that every section of a bicyclic 2-group has rank at most 3. This will be used in the following without an explicit comment. Since here and in the following the arguments are very specific (i.e. not of general interest), we will sometimes apply computer calculations in order to handle small cases.

Proposition 10.3 *Let $\mathscr{F}$ be a fusion system on a bicyclic, non-metacyclic 2-group P. Suppose that P contains an $\mathscr{F}$-essential subgroup Q of rank* 2. *Then $Q \cong C_{2^m}^2$ and $P \cong C_{2^m} \wr C_2$ for some $m \geq 2$. Moreover, $\mathscr{F} = \mathscr{F}_P(C_{2^m}^2 \rtimes S_3)$ or $\mathscr{F} = \mathscr{F}_P(\mathrm{PSL}(3,q))$ for some $q \equiv 1$* (mod 4).

Proof By Proposition 6.11 it suffices for the first claim to show that Q is metacyclic, since minimal non-abelian groups of type (m, m) for $m \geq 2$ are non-metacyclic (see Proposition 2.8 in [140]). Let $M \leq P$ be a metacyclic maximal subgroup of P. We may assume $Q \nsubseteq M$. Then $M \cap Q$ is a maximal subgroup of Q. Since Q admits an automorphism of order 3, the maximal subgroups of Q are isomorphic. Now the first claim follows from Proposition 2.2 in [140]. The fusion systems on $C_{2^m} \wr C_2$ are given by Theorem 5.3 in [63]. Two of them have $C_{2^m}^2$ as essential subgroup.   □

It can be seen that the group $C_{2^m} \wr C_2$ is in fact bicyclic. Observe that Theorem 5.3 in [63] provides another non-nilpotent fusion system on $C_{2^m} \wr C_2$ which we will discover later. For the rest of this section we consider the case where the bicyclic, non-metacyclic 2-group P has no $\mathscr{F}$-essential subgroup of rank 2.

In the following we consider fusion systems only up to isomorphism (see Definition 1.25).

Proposition 10.4 *Let $\mathscr{F}$ be a non-nilpotent fusion system on a bicyclic 2-group P. Suppose that P contains an elementary abelian normal subgroup of order* 8. *Then P is minimal non-abelian of type $(n, 1)$ for some $n \geq 2$ and $C_{2^{n-1}} \times C_2^2$ is the only $\mathscr{F}$-essential subgroup of P. Moreover, $\mathscr{F} = \mathscr{F}_P(A_4 \rtimes C_{2^n})$ where C_{2^n} acts as a transposition in $\mathrm{Aut}(A_4) \cong S_4$ (thus $A_4 \rtimes C_{2^n}$ is unique up to isomorphism).*

Proof Suppose first $|P'| = 2$. Then P is minimal non-abelian of type $(n, 1)$ for some $n \geq 2$ by Theorem 4.1 in [140]. We show that P contains exactly one $\mathscr{F}$-essential subgroup Q. Since P is minimal non-abelian, every selfcentralizing subgroup is maximal. Moreover, Q has rank 3 by Proposition 10.3. Hence, $Q = \langle x^2, y, z \rangle \cong C_{2^{n-1}} \times C_2^2$ is the unique non-metacyclic maximal subgroup of P. We prove that $\mathscr{F}$ is unique up to isomorphism. By Alperin's Fusion Theorem it suffices to describe the action of $\mathrm{Aut}_{\mathscr{F}}(Q)$ on Q. First of all $P = \mathrm{N}_P(Q)$ acts on only two four-subgroups $\langle y, z \rangle$ and $\langle x^{2^{n-1}} y, z \rangle$ of Q non-trivially. Let $\alpha \in \mathrm{Aut}_{\mathscr{F}}(Q)$ be of order 3. Then α is unique up to conjugation in $\mathrm{Aut}(Q)$. Hence, α acts on only one four-subgroup R of Q. Let $\beta \in P/Q \leq \mathrm{Aut}_{\mathscr{F}}(Q)$. Then $(\alpha\beta)(R) = (\beta\alpha^{-1})(R) = \beta(R) = R$, since $\mathrm{Aut}_{\mathscr{F}}(Q) \cong S_3$. Thus, $\mathrm{Aut}_{\mathscr{F}}(Q)$ acts (non-trivially) on $\langle y, z \rangle$ or on $\langle x^{2^{n-1}} y, z \rangle$. It can be seen easily that the elements x and $x^{2^{n-1}} y$ satisfy the same relations as x and y. Hence, after replacing y by $x^{2^{n-1}} y$ if necessary, we may assume that $\mathrm{Aut}_{\mathscr{F}}(Q)$ acts on $\langle y, z \rangle$. Since $\mathrm{C}_Q(\alpha) \cong C_{2^{n-1}}$, we see that $x^2 y \notin \mathrm{C}_Q(\alpha)$ or $x^2 yz \notin \mathrm{C}_Q(\alpha)$. But then both $x^2 y, x^2 yz \notin \mathrm{C}_Q(\alpha)$, because $\beta(x^2 y) = x^2 yz$. Hence, $\mathrm{C}_Q(\alpha) = \mathrm{C}_Q(\mathrm{Aut}_{\mathscr{F}}(Q)) \in \{\langle x^2 \rangle, \langle x^2 z \rangle\}$. However, xy and y fulfill the same relations as x and y. Hence, after replacing x by xy if necessary, we have $\mathrm{C}_Q(\mathrm{Aut}_{\mathscr{F}}(Q)) = \langle x^2 \rangle$. This determines the action of $\mathrm{Aut}_{\mathscr{F}}(Q)$ on Q completely. In particular, $\mathscr{F}$ is uniquely determined (up to isomorphism). The group $G = A_4 \rtimes C_{2^n}$ as described in the proposition has a minimal non-abelian Sylow 2-subgroup of type $(n, 1)$. Since A_4 is not 2-nilpotent, $\mathscr{F}_P(G)$ is not nilpotent. It follows that $\mathscr{F} = \mathscr{F}_P(G)$.

Now suppose $|P'| > 2$. Then Theorem 4.2 in [140] describes the structure of P. We use the notation of this theorem. Let $Q < P$ be $\mathscr{F}$-essential. By Proposition 10.3, Q has rank 3. In particular Q is contained in the unique non-metacyclic maximal subgroup $M := E\langle a^2 \rangle$ of P. Since $\langle a^4, u \rangle = \mathrm{Z}(M) < Q$, it follows that $Q \in \{\langle a^4, u, v \rangle, \langle a^4, a^2 v, u \rangle, M\}$. In the first two cases we have $P' = \langle u, z \rangle \subseteq Q \trianglelefteq P$ which contradicts Proposition 6.12. Hence, $Q = M$. Every automorphism of M of order 3 acts freely on $M/\mathrm{Z}(M) \cong C_2^2$. However, the subgroups $L \leq M$ such that $\mathrm{Z}(M) < L < M$ are non-isomorphic. Contradiction. $\square$

It remains to deal with the case where P does not contain an elementary abelian normal subgroup of order 8. In particular Theorem 4.3 in [140] applies.

Lemma 10.5 *Let $\mathscr{F}$ be a fusion system on a bicyclic 2-group P. If $Q \leq P$ is $\mathscr{F}$-essential of rank 3, then one of the following holds:*

(i) $Q \trianglelefteq P$ and $P/\Phi(Q)$ is minimal non-abelian of type $(2, 1)$.
(ii) $Q \ntrianglelefteq P$ and $P/\Phi(Q) \cong D_8 \times C_2$.

Proof By Proposition 6.12 we have $|\mathrm{N}_P(Q) : Q| = 2$. Since $\mathrm{N}_P(Q)$ acts non-trivially on $Q/\Phi(Q)$, we conclude that $\mathrm{N}_P(Q)/\Phi(Q)$ is non-abelian. Then $\mathrm{N}_P(Q)/\Phi(Q)$ is minimal non-abelian of type $(2, 1)$ or $\mathrm{N}_P(Q)/\Phi(Q) \cong D_8 \times C_2$, because the group $\mathrm{N}_P(Q)/\Phi(Q)$ contains an elementary abelian subgroup of order 8. In case $\mathrm{N}_P(Q) = P$ only the first possibility can apply, since P has rank 2. Now assume that $Q \ntrianglelefteq P$ and $\mathrm{N}_P(Q)/\Phi(Q)$ is minimal non-abelian of type $(2, 1)$.

Take $g \in N_P(N_P(Q)) \setminus N_P(Q)$ such that $g^2 \in N_P(Q)$. Then $Q_1 := {}^g Q \neq Q$ and $Q_1 \cap Q$ is $\langle g \rangle$-invariant. Moreover, $\Phi(Q) \subseteq \Phi(N_P(Q)) \subseteq Q_1$ and

$$|\Phi(Q) : \Phi(Q) \cap \Phi(Q_1)| = |\Phi(Q_1) : \Phi(Q) \cap \Phi(Q_1)| = |\Phi(Q_1)\Phi(Q) : \Phi(Q)|$$

$$= |\Phi(Q_1/\Phi(Q))| = 2,$$

since $Q_1/\Phi(Q)$ $(\neq Q/\Phi(Q))$ is abelian of rank 2. Hence, $N_P(Q)/\Phi(Q) \cap \Phi(Q_1)$ is a group of order 32 of rank 2 with two distinct normal subgroups of order 2 such that their quotients are isomorphic to the minimal non-abelian group of type $(2, 1)$. It follows that $N_P(Q)/\Phi(Q) \cap \Phi(Q_1)$ is the minimal non-abelian group of type $(2, 2)$ (this can be checked by computer). However, then all maximal subgroups of $N_P(Q)/\Phi(Q) \cap \Phi(Q_1)$ have rank 3 which contradicts Theorem 10.1. Thus, we have proved that $N_P(Q)/\Phi(Q) \cong D_8 \times C_2$. $\square$

Now we are in a position to determine all $\mathscr{F}$-essential subgroups on a bicyclic 2-group. This is a key result for the remainder of the section.

Proposition 10.6 *Let $\mathscr{F}$ be a fusion system on a bicyclic 2-group P. If $Q \leq P$ is $\mathscr{F}$-essential of rank 3, then one of the following holds:*

(i) $Q \cong C_{2^m} \times C_2^2$ for some $m \geq 1$.
(ii) $Q \cong C_{2^m} \times Q_8$ for some $m \geq 1$.
(iii) $Q \cong C_{2^m} * Q_8$ for some $m \geq 2$.

Proof If P contains an elementary abelian normal subgroup of order 8, then the conclusion holds by Proposition 10.4. Hence, we will assume that there is no such normal subgroup. Let $\alpha \in \text{Out}_{\mathscr{F}}(Q)$ be of order 3 (see Proposition 6.12). Since $|\text{Aut}(Q)|$ is not divisible by 9, we can regard α as an element of $\text{Aut}(Q)$ by choosing a suitable preimage. We apply [267] to the group Q (observe that the rank in [267] is the p-rank in our setting). Let $C := C_Q(\alpha)$. Suppose first that C has 2-rank 3, i.e. $m(C) = 3$ with the notation of [267]. Since $[Q, \alpha]$ is generated by at most three elements, only the first part of Theorem B in [267] can occur. In particular $Q \cong Q_8 * C$. However, this implies that Q contains a subgroup of rank at least 4. Contradiction.

Now assume $m(C) = 2$. Then Theorem A in [267] gives $Q \cong Q_8 * C$. Let $Z \leq Z(Q_8 \times C) = \Phi(Q_8) \times Z(C)$ such that $Q \cong (Q_8 \times C)/Z$. Then $|Z| = 2$ and C has rank at most 2, since Q has rank 3. Moreover, it follows that $\Omega(Z(C)) \nsubseteq \Phi(C)$ (otherwise: $Z \leq \Phi(Q_8) \times \Phi(C) = \Phi(Q_8 \times C)$). By Burnside's Basis Theorem, $C \cong C_2 \times C_{2^m}$ is abelian and $Q \cong Q_8 \times C_{2^m}$ for some $m \geq 1$.

Finally suppose that $m(C) = 1$, i.e. C is cyclic or quaternion. By Theorem 10.1, $\Phi(P)$ is metacyclic. Since $\Phi(Q) \subseteq \Phi(P)$ (Satz III.3.14 in [128]), also $\Phi(Q)$ is metacyclic. According to the action of α on $\Phi(Q)$ one of the following holds (see Proposition 10.2):

(a) $\Phi(Q) \leq C \trianglelefteq Q$.
(b) $\Phi(Q) \cong Q_8$.
(c) $\Phi(Q) \cap C = 1$ and $\Phi(Q) \cong C_{2^n}^2$ for some $n \geq 1$.

We handle these cases separately. First assume case (a). By 8.2.2(a) in [159] we have $|Q : C| = 4$ and α acts freely on Q/C. On the other hand α acts trivially on $Q/C_Q(C)$ by 8.1.2(b) in [159]. This shows $Q = C\,C_Q(C)$. If C is quaternion, then $Q = Q_{2^n} * C_Q(C)$. In particular, $C_Q(C)$ has rank at most 2. Thus, a similar argument as above yields $Q \cong Q_{2^n} \times C_{2^m}$. However, this is impossible here, because α would act trivially on $Q/\Phi(Q)$ by the definition of C. Hence, C is cyclic and central of index 4 in Q. Since, Q has rank 3, the exponents of C and Q coincide. If Q is abelian, we must have $Q \cong C_{2^m} \times C_2^2$ for some $m \geq 1$. Now assume that Q is non-abelian. Write $C = \langle a \rangle$ and choose $b, c \in Q$ such that $Q/C = \langle bC, cC \rangle$. Since $\langle b \rangle C$ is abelian and non-cyclic, we may assume $b^2 = 1$. Similarly $c^2 = 1$. Since Q is non-abelian, ${}^c b \neq b$. Let $|C| = 2^m$ where $m \geq 2$. Then $a \in Z(Q)$ implies ${}^c b = a^{2^{m-1}} b$. Thus, Q is uniquely determined as

$$Q = \langle a, b, c \mid a^{2^m} = b^2 = c^2 = [a, b] = [a, c] = 1,\ {}^c b = a^{2^{m-1}} b \rangle.$$

Since the group $Q_8 * C_{2^m} \cong D_8 * C_{2^m}$ has the same properties, we get $Q \cong Q_8 * C_{2^m}$.

Next we will show that case (b) cannot occur for any finite group Q. On the one hand we have $Q/C_Q(\Phi(Q)) \leq \mathrm{Aut}(Q_8) \cong S_4$. On the other hand $C_2^2 \cong \Phi(Q)C_Q(\Phi(Q))/C_Q(\Phi(Q)) \leq \Phi(Q/C_Q(\Phi(Q)))$. Contradiction.

It remains to deal with case (c). Again we will derive a contradiction. By Theorem D in [267], $C \neq 1$ (U_{64} has rank 4). The action of α on $Q/\Phi(Q)$ shows $|P : C\Phi(Q)| \geq 4$. Now $\Phi(Q) \cap C = 1$ implies $|C| = 2$. There exists an α-invariant maximal subgroup $N \trianglelefteq Q$. Thus, $N \cap C \subseteq N \cap C\Phi(Q) \cap C = \Phi(Q) \cap C = 1$. In particular we can apply Theorem D in [267] which gives $N \cong C_{2^{n+1}}^2$. Hence, $Q \cong N \rtimes C = C_{2^{n+1}}^2 \rtimes C_2$ (here $\rtimes$ can also mean $\times$). Choose $x, y \in N$ such that $\alpha(x) = y$ and $\alpha(y) = x^{-1} y^{-1}$. Let $C = \langle c \rangle$. Since Q has rank 3, c acts trivially on $N/\Phi(N)$. Hence, we find integers i, j such that ${}^z x = x^i y^j$ and $i \equiv 1 \pmod{2}$ and $j \equiv 0 \pmod{2}$. Then ${}^c y = \alpha({}^z x) = x^{-j} y^{i-j}$. In particular, the isomorphism type of Q does only depend on i, j. Since $c^2 = 1$, we obtain $i^2 - j^2 \equiv 1 \pmod{2^{n+1}}$ and $j(2i - j) \equiv 0 \pmod{2^{n+1}}$. We will show that $j \equiv 0 \pmod{2^n}$. This is true for $n = 1$. Thus, assume $n \geq 2$. Then $1 - j^2 \equiv i^2 - j^2 \equiv 1 \pmod{8}$. Therefore, $j \equiv 0 \pmod{4}$. Now $j(2i - j) \equiv 0 \pmod{2^{n+1}}$ implies $j \equiv 0 \pmod{2^n}$. In particular $i^2 \equiv i^2 - j^2 \equiv 1 \pmod{2^{n+1}}$. Hence, we have two possibilities for j and at most four possibilities for i. This gives at most eight isomorphism types for Q. Now we split the proof into the cases $Q \trianglelefteq P$ and $Q \ntrianglelefteq P$.

Suppose $Q \trianglelefteq P$. Then $|P : Q| = 2$ by Proposition 6.12. Moreover, $\Omega(Q) \trianglelefteq P$. Since P does not contain an elementary abelian normal subgroup of order 8, it follows that Q contains more than seven involutions. With the notation above, let $x^r y^s c$ be an involution such that $x^r y^s \notin \Omega(N)$. Then $1 = x^r y^s c x^r y^s c = x^{r+ir-js} y^{s+jr+(i-j)s}$ and $r(1 + i) - js \equiv s(1 + i) + jr - js \equiv 0 \pmod{2^{n+1}}$. In case $n = 1$ we have $|P| = 64$. Here it can be shown by computer that P does not exist. Hence, suppose $n \geq 2$ in the following. Suppose further that $i \equiv 1 \pmod{2^n}$. Then we obtain $2r \equiv 2s \equiv 0 \pmod{2^n}$. Since $x^r y^s \notin \Omega(N)$ we may assume that $r \equiv \pm 2^{n-1} \pmod{2^{n+1}}$ (the case $s \equiv \pm 2^{n-1} \pmod{2^{n+1}}$ is similar). However, this leads to the contradiction $0 \equiv r(1 + i) - js \equiv 2^n \pmod{2^{n+1}}$. This shows that $i \equiv -1 \pmod{2^n}$. In particular, $x^{i-1}y^i = {}^c x x^{-1} = [c, x] \in Q'$ and $x^{-j}y^{i-j-1} = [c, y] \in Q'$. This shows $C_{2^n}^2 \cong Q' = \Phi(Q)$. By Lemma 10.5, $P/\Phi(Q)$ is minimal non-abelian of type $(2, 1)$. Since $Q' \subseteq P'$, we conclude that $P/P' \cong C_4 \times C_2$. Then P is described in Theorem 4.11 in [140]. In particular $\Phi(P)$ is abelian. Choose $g \in P \setminus Q$. Then g acts non-trivially on $N/\Phi(Q)$, because α does as well. This shows $N \trianglelefteq P$ and $C_2^2 \cong N/\Phi(Q) \neq Z(P/\Phi(Q)) = \Phi(P/\Phi(Q))$. Hence, P/N is cyclic and $\Phi(P) \neq N$. Therefore, Q contains two abelian maximal subgroups and $N \cap \Phi(P) \subseteq Z(Q)$. Now a result of Knoche (see Aufgabe III.7.24) gives the contradiction $|Q'| = 2$.

Now assume $Q \ntrianglelefteq P$. We will derive the contradiction that $N_P(Q)$ does not contain a metacyclic maximal subgroup. By Lemma 10.5, $N_P(Q)/\Phi(Q) \cong D_8 \times C_2$. Choose $g \in N_P(Q) \setminus Q$. Then g acts non-trivially on $N/\Phi(N)$, because α does as well. In particular $N \trianglelefteq N_P(Q)$. This implies

$$g^2 \Phi(Q) \in \mho(N_P(Q)/\Phi(Q)) = (N_P(Q)/\Phi(Q))' \subseteq N/\Phi(Q)$$

and $g^2 \in N$. As above, we may choose $x, y \in N$ such that ${}^g x = y$ and ${}^g y = x$. Since g centralizes g^2, we can write $g^2 = (xy)^i$ for some $i \in \mathbb{Z}$. Then gx^{-i} has order 2. Hence, we may assume that $g^2 = 1$ and $\langle N, g \rangle \cong C_{2^{n+1}} \wr C_2$. In case $n = 1$ we have $|N_P(Q)| = 64$. Here one can show by computer that $N_P(Q)$ does not exist. Hence, $n \geq 2$. Let M be a metacyclic maximal subgroup of $N_P(Q)$. Since $\langle \Phi(Q), g \rangle \cong C_{2^n} \wr C_2$ is not metacyclic, we conclude that $g \notin M$. Let $C = \langle c \rangle$. Then $\langle \Phi(Q), c \rangle$ has rank 3. In particular, $c \notin M$. This leaves two possibilities for M. It is easy to see that $\langle N, gc \rangle \cong C_{2^{n+1}} \wr C_2$. Thus, $M = \langle \Phi(Q), xc, gc \rangle$. Assume $(gc)^2 \in \Phi(Q)$. Then it is easy to see that $\langle \Phi(Q), gc \rangle \cong C_{2^n} \wr C_2$ is not metacyclic. This contradiction shows $(gc)^2 \equiv xy \pmod{\Phi(Q)}$. Moreover, $c(gc)^2 c = (cg)^2 = (gc)^{-2}$. Since $N = \langle gc, \alpha(gc) \rangle$, c acts as inversion on N. In particular, $(xc)^2 = 1$. Hence $\langle \Omega(Q), xc \rangle \subseteq M$ is elementary abelian of order 8. Contradiction. $\square$

Let Q be one of the groups in Proposition 10.6. Then it can be seen that there is an automorphism $\alpha \in \mathrm{Aut}(Q)$ of order 3. Since the kernel of the canonical map $\mathrm{Aut}(Q) \to \mathrm{Aut}(Q/\Phi(Q)) \cong \mathrm{GL}(3, 2)$ is a 2-group, we have $\langle \alpha \rangle \in \mathrm{Syl}_3(\mathrm{Aut}(Q))$. If α is not conjugate to α^{-1} in $\mathrm{Aut}(Q)$, then Burnside's Transfer Theorem implies that $\mathrm{Aut}(Q)$ is 3-nilpotent. But then also $\mathrm{Out}_{\mathscr{F}}(Q) \cong S_3$ would be 3-nilpotent which is not the case. Hence, α is unique up to conjugation in $\mathrm{Aut}(Q)$. In particular the isomorphism type of $C_Q(\alpha)$ is uniquely determined.

Proposition 10.7 *Let $\mathscr{F}$ be a fusion system on a bicyclic 2-group P. If $Q \trianglelefteq P$ is $\mathscr{F}$-essential of rank 3, then one of the following holds:*

(i) *P is minimal non-abelian of type $(n, 1)$ for some $n \geq 2$.*

(ii) *$P \cong Q_8 \rtimes C_{2^n}$ for some $n \geq 2$. Here C_{2^n} acts as a transposition in $\mathrm{Aut}(Q_8) \cong S_4$.*

(iii) *$P \cong Q_8.C_{2^n}$ for some $n \geq 2$.*

In particular P' is cyclic.

Proof We use Proposition 10.6. If Q is abelian, then $C_2^3 \cong \Omega(Q) \trianglelefteq P$. By Proposition 10.4, P is minimal non-abelian of type $(n, 1)$ for some $n \geq 2$. Now assume $Q \cong Q_8 \times C_{2^{n-1}}$ for some $n \geq 2$. We write $Q = \langle x, y, z \rangle$ such that $\langle x, y \rangle \cong Q_8$ and $\langle z \rangle \cong C_{2^{n-1}}$. Moreover, choose $g \in P \setminus Q$. Let $\alpha \in \mathrm{Out}_{\mathscr{F}}(Q)$ as usual. Then α acts non-trivially on $Q/Z(Q) \cong C_2^2$ and so does g. Hence, we may assume ${}^g x = y$. Since $g^2 \in Q$, it follows that ${}^g y = {}^{g^2} x \in \{x, x^{-1}\}$. By replacing g with gx if necessary, we may assume that ${}^g y = x$. Hence, $g^2 \in Z(Q)$. By Lemma 10.5, $P/\Phi(Q)$ is minimal non-abelian of type $(2, 1)$. In particular, $Q/\Phi(Q) = \Omega(P/\Phi(Q))$. This gives $g^2 \notin \Phi(Q)$ and $g^2 \in z\langle x^2, z^2 \rangle$. Since ${}^g(x^2) = x^2$, we get ${}^g z = z$. After replacing g by gz^i for a suitable integer i, it turns out that $g^2 \in \{z, zx^2\}$. In the latter case we replace z by $x^2 z$ and obtain $g^2 = z$. Hence, $P = Q_8 \rtimes C_{2^n}$ as stated. Moreover, g acts on $\langle x, y \rangle$ as an involution in $\mathrm{Aut}(Q_8) \cong S_4$. Since an involution which is a square in $\mathrm{Aut}(Q_8)$ cannot act non-trivially on $Q_8/\Phi(Q_8)$, g must correspond to a transposition in S_4. This describes P up to isomorphism. Since $P = \langle gx \rangle \langle g \rangle$, P is bicyclic. In particular $P' \subseteq \langle x, y \rangle$ is abelian and thus cyclic.

Finally suppose that $Q = Q_8 * C_{2^n}$ for some $n \geq 2$. We use the same notation as before. In particular $x^2 = z^{2^{n-1}}$. The same arguments as above give $g^2 = z$ and

$$P = \langle x, y, g \mid x^4 = 1, \, x^2 = y^2 = g^{2^n}, \, {}^y x = x^{-1}, \, {}^g x = y, \, {}^g y = x \rangle \cong Q_8.C_{2^n}.$$

Then $P = \langle gx \rangle \langle g \rangle$ is bicyclic and P' cyclic. $\qquad\square$

We will construct the groups and fusion systems in the last proposition systematically in our main Theorem 10.17.

The following result is useful to reduce the search for essential subgroups. Notice that the centerfree fusion systems on metacyclic 2-groups are determined in [63].

Proposition 10.8 *Let $\mathscr{F}$ be a centerfree fusion system on a bicyclic, non-metacyclic 2-group P. Then there exists an abelian $\mathscr{F}$-essential subgroup $Q \leq P$ isomorphic to $C_{2^m}^2$ or to $C_{2^m} \times C_2^2$ for some $m \geq 1$.*

Proof By way of contradiction assume that all $\mathscr{F}$-essential subgroups are isomorphic to $C_{2^m} \times Q_8$ or to $C_{2^m} * Q_8$ (use Propositions 10.3 and 10.6). Let $z \in Z(P)$ be an involution. Since $Z(\mathscr{F}) = 1$, Alperin's Fusion Theorem in connection with Theorem 10.1 implies that there exists an $\mathscr{F}$-essential subgroup $Q \leq P$ such that $z \in Z(Q)$. Moreover, there is an automorphism $\alpha \in \mathrm{Aut}(Q)$ such that $\alpha(z) \neq z$. Of course α restricts to an automorphism of $Z(Q)$. In case $Q \cong C_{2^m} * Q_8$ this is not possible, since $Z(Q)$ is cyclic. Now assume $Q \cong C_{2^m} \times Q_8$. Observe that we can assume that α has order 3, because the automorphisms in $\mathrm{Aut}_P(Q)$ fix z anyway. But then α acts trivially on Q' and on $\Omega(Q)/Q'$ and thus also on $\Omega(Q) \ni z$. Contradiction. $\qquad\square$

10.1.1   The Case P' Non-cyclic

The aim of this section is to prove that there are only nilpotent fusion system provided P' is non-cyclic. We do this by a case by case analysis corresponding to the theorems in [140]. By Proposition 10.7 we may assume that there are no normal $\mathscr{F}$-essential subgroups. Let $\mathscr{F}$ be a non-nilpotent fusion system on the bicyclic 2-group P. Assume for the moment that $P' \cong C_2^2$. Then P does not contain an elementary abelian subgroup of order 8 by Proposition 10.4. Hence, Theorem 4.6 in [140] shows that P is unique of order 32. In this case we can prove with computer that there are no candidates for $\mathscr{F}$-essential subgroups. Hence, we may assume $\Phi(P') \neq 1$ in the following.

We introduce a few notation from Theorem 4.3 in [140] that will be used for the rest of the paper:

$$\Phi(P) = P'\langle a^2 \rangle = \langle a^2 \rangle \langle v \rangle, \qquad M = E\langle a^2 \rangle = \langle x \rangle \langle a^2 \rangle \langle v \rangle.$$

Here, M is the unique non-metacyclic maximal subgroup of P.

Proposition 10.9 *Let P be a bicyclic 2-group such that P' is non-cyclic and $P/\Phi(P')$ contains no elementary abelian normal subgroup of order 8. Then every fusion system on P is nilpotent.*

Proof The case $\Phi(P') = 1$ was already handled. So we may assume $\Phi(P') \neq 1$. In particular Theorem 4.7 in [140] applies. Let $\mathscr{F}$ be a non-nilpotent fusion system on P. Assume first that there exists an $\mathscr{F}$-essential subgroup $Q \in \{C_{2^m} \times C_2^2, C_{2^m} * Q_8 \cong C_{2^m} * D_8\}$ (the letter m is not used in Theorem 4.7 of [140]). Theorem 4.7 of [140] also shows that $\Phi(P)$ is metacyclic and abelian. Since Q contains more than three involutions, there is an involution $\beta \in M \setminus \Phi(P)$. Hence, we can write $\beta = xa^{2i}v^j$ for some $i, j \in \mathbb{Z}$. Now in case (a) of Theorem 4.7 of [140] we derive the following contradiction:

$$\beta^2 = xa^{2i}v^j xa^{2i}v^j = xa^{2i}xa^{2i} = x^2(av)^{2i}a^{2i} = x^2 a^{2i} u^i z^{\xi i} a^{2i} = u z^{(\eta + \xi)i} \neq 1.$$

Similarly in case (b) we get:

$$\beta^2 = xa^{2i}v^j xa^{2i}v^j = xa^{2i}xz^j a^{2i} = x^2(av)^{2i}z^j a^{2i} = x^2 a^{2i} u^i v^{2^{n-2}i} z^{\xi i} z^j a^{2i}$$

$$= x^2 v^{2^{n-2}i} z^{\eta i} v^{2^{n-2}i} z^{\xi i} z^j = u z^{i(1+\eta+\xi)+j} \neq 1.$$

Next assume that there is an $\mathscr{F}$-essential subgroup $C_{2^m} \times Q_8 \cong Q \leq P$ for some $m \geq 1$. Suppose $m \geq 3$ for the moment. Since $Q \subseteq M$, it is easy to see that $M \setminus \Phi(P)$ contains an element of order at least 8. However, we have seen above that this is impossible. Hence, $m \leq 2$. By Proposition 10.7, Q is not normal in P. Since $Q < N_M(Q) \leq N_P(Q)$, we have $N_P(Q) \leq M = N_P(Q)\Phi(P)$. A computer calculation shows that $N_P(Q) \cong Q_{16} \times C_{2^m}$. Thus, $N_P(Q) \cap \Phi(P) \cong C_8 \times C_{2^m}$, because $\Phi(P)$ is abelian. Hence, there exist $\beta = xa^{2^i}y^j \in N_P(Q) \setminus \Phi(P) \subseteq M \setminus \Phi(P)$ and $\delta \in N_P(Q) \cap \Phi(P)$ such that $\beta^2 = \delta^4$. As above we always have $\beta^2 \in u\langle z \rangle$. However, in both cases (a) and (b) we have $\delta^4 \in \mho_2(\Phi(P)) \cap \Omega(\Phi(P)) = \langle a^8 \rangle \langle v^{2^{n-1}} \rangle = \langle z \rangle$. Contradiction. $\square$

If P' is cyclic, $P/\Phi(P')$ is minimal non-abelian and thus contains an elementary abelian normal subgroup of order 8. Hence, it remains to deal with the case where $P/\Phi(P')$ has a normal subgroup isomorphic to C_2^3.

Our next goal is to show that P' requires a cyclic maximal subgroup in order to admit a non-nilpotent fusion system.

Proposition 10.10 *Let P be a bicyclic 2-group such that $P' \cong C_{2^r} \times C_{2^{r+s}}$ for some $r \geq 2$ and $s \in \{1, 2\}$. Then every fusion system on P is nilpotent.*

Proof We apply Theorems 4.11 and 4.12 in [140] simultaneously. As usual assume first that P contains an $\mathscr{F}$-essential subgroup $Q \cong C_{2^m} \times C_2^2$ for some $m \geq 1$ (m is not used in the statement of Theorem 4.11 in [140]). Then $Q \cap \Phi(P) \cong C_{2^m} \times C_2$, since $\Phi(P)$ is abelian and metacyclic. We choose $\beta := xa^{2^i}v^j \in Q \setminus \Phi(P)$. In case $m \geq 2$, β fixes an element of order 4 in $Q \cap \Phi(P)$. Since $\Phi(P)$ is abelian, all elements of $\Phi(P)$ of order 4 are contained in

$$\Omega_2(\Phi(P)) = \begin{cases} \langle b^{2^{r-2}}, v^{2^{r-1}} \rangle & \text{if Theorem 4.11 applies,} \\ \langle b^{2^{r-1}}, v^{2^{r-1}} \rangle & \text{if Theorem 4.12 applies.} \end{cases}$$

However, the relations in Theorem 4.11/12 in [140] show that x and thus β acts as inversion on $\Omega_2(\Phi(P))$. Hence, $m = 1$. Then $N_P(Q) \cap \Phi(P) \cong C_4 \times C_2$ by Lemma 10.5. In particular there exists an element $\rho \in \Omega_2(\Phi(P)) \setminus (N_P(Q) \cap \Phi(P))$. Then $^\rho\beta = \beta\rho^{-2} \in Q$. Since $Q = \langle \beta \rangle (Q \cap \Phi(P))$, we derive the contradiction $\rho \in N_P(Q)$.

Next suppose that $Q \cong C_{2^m} \times Q_8$ for some $m \geq 1$. Here we can repeat the argument word by word. Finally the case $Q \cong C_{2^m} * Q_8$ cannot occur, since $Z(P)$ is non-cyclic. $\square$

The next lemma is useful in a more general context.

Lemma 10.11 *Let P be a metacyclic 2-group which does not have maximal class. Then every homocyclic subgroup of P is given by $\Omega_i(P)$ for some $i \geq 0$.*

Proof Let $C_{2^k}^2 \cong Q \leq P$ with $k \in \mathbb{N}$. We argue by induction on k. By Exercise 1.85 in [23], $C_2^2 \cong \Omega(P)$. Hence, we may assume $k \geq 2$. By induction it suffices to show that $P/\Omega(P)$ does not have maximal class. Let us assume the contrary. Since $P/\Omega(P)$ contains more than one involution, $P/\Omega(P)$ is a dihedral group or a semidihedral group. Let $\langle x \rangle \trianglelefteq P$ such that $P/\langle x \rangle$ is cyclic. Then $\langle x \rangle \Omega(P)/\Omega(P)$ and $(P/\Omega(P))/(\langle x \rangle \Omega(P)/\Omega(P)) \cong P/\langle x \rangle \Omega(P)$ are also cyclic. This yields

$|P/\langle x \rangle \Omega(P)| = 2$ and $|P/\langle x \rangle| = 4$. Since $P/\Omega(P)$ is a dihedral group or a semidihedral group, there exists an element $y \in P$ such that the following holds:

(i) $P/\Omega(P) = \langle x\Omega(P), y\Omega(P) \rangle$,
(ii) $y^2 \in \Omega(P)$,
(iii) $^y x \equiv x^{-1} \pmod{\Omega(P)}$ or $^y x \equiv x^{-1+2^{n-2}} \pmod{\Omega(P)}$ with $|P/\Omega(P)| = 2^n$ and without loss of generality, $n \geq 4$.

Since $P = \langle x, y \rangle \Omega(P) \subseteq \langle x, y \rangle \Phi(P) = \langle x, y \rangle$, we have shown that P is the semidirect product of $\langle x \rangle$ with $\langle y \rangle$. Moreover

$$^y x \in \{x^{-1},\ x^{-1+2^{n-1}},\ x^{-1+2^{n-2}},\ x^{-1-2^{n-2}}\}.$$

Since $Q \cap \langle x \rangle$ and $Q/Q \cap \langle x \rangle \cong Q\langle x \rangle/\langle x \rangle$ are cyclic, we get $k = 2$ and $x^{2^{n-2}} \in Q$. But then, Q cannot be abelian, since $n \geq 4$. Contradiction. $\qquad \square$

Note that in general for a metacyclic 2-group P which does not have maximal class it can happen that $P/\Omega(P)$ has maximal class.

Proposition 10.12 *Let P be a bicyclic 2-group such that $P' \cong C_{2^r}^2$ for some $r \geq 2$. Then every fusion system on P is nilpotent.*

Proof We apply Theorem 4.9 in [140]. The general argument is quite similar as in Proposition 10.10, but we need more details. Assume first that $Q \cong C_{2^m} \times C_2^2$ for some $m \geq 1$ is $\mathscr{F}$-essential in P (m is not used in the statement of Theorem 4.9 in [140]). Since $\Phi(P)$ has rank 2, we get $Q \cap \Phi(P) \cong C_{2^m} \times C_2$. We choose $\beta := xa^{2i}v^j \in Q \setminus \Phi(P)$. Suppose first that $m \geq 2$. Then β fixes an element $\delta \in Q \cap \Phi(P)$ of order 4. Now $\Phi(P)$ is a metacyclic group with $\Omega(\Phi(P)) \cong C_2^2$ and $C_4^2 \cong \Omega_2(P') \leq \Phi(P)$. So Lemma 10.11 implies $\Omega_2(\Phi(P)) = \langle v^{2^{r-2}}, b^{2^{r-2}} \rangle \cong C_4^2$. In case $r = 2$ we have $|P| = 2^7$, and the claim follows by a computer verification. Thus, we may assume $r \geq 3$. Then $x^{-1} v^{2^{r-2}} x = v^{-2^{r-2}}$. Moreover, $\Omega_2(\Phi(P)) \subseteq \mho(\Phi(P)) = \Phi(\Phi(P)) \subseteq Z(\Phi(P))$, since $\Phi(P)$ is abelian or minimal non-abelian depending on η. This shows that β acts as inversion on $\Omega_2(\Phi(P))$ and thus cannot fix δ. It follows that $m = 1$. Then $|N_P(Q) \cap \Phi(P)| \leq 8$. In particular there exists an element $\rho \in \Omega_2(\Phi(P)) \setminus (N_P(Q) \cap \Phi(P))$. Then $^\rho \beta = \beta \rho^{-2} \in Q$. Since $Q = \langle \beta \rangle (Q \cap \Phi(P))$, we derive the contradiction $\rho \in N_P(Q)$.

Now assume $Q \cong C_{2^m} \times Q_8$ for some $m \geq 1$. We choose again $\beta := xa^{2i}v^j \in Q \setminus \Phi(P)$. If $\Phi(P)$ contains a subgroup isomorphic to Q_8, then $\Omega_2(\Phi(P))$ cannot be abelian. So, in case $m = 1$ we have $N_P(Q) \cap \Phi(P) \cong C_8 \times C_2$. Then the argument above reveals a contradiction (using $r \geq 3$). Now let $m \geq 2$. We write $Q = \langle q_1 \rangle \times \langle q_2, q_3 \rangle$ where $\langle q_1 \rangle \cong C_{2^m}$ and $\langle q_2, q_3 \rangle \cong Q_8$. In case $q_1 \notin \Phi(P)$ we can choose $\beta = q_1$. In any case it follows that β fixes an element of order 4 in $Q \cap \Phi(P)$. This leads to a contradiction as above.

Finally suppose that $Q \cong C_{2^m} * Q_8 \cong C_{2^m} * D_8$ for some $m \geq 2$. Here we can choose $\beta \in Q \setminus \Phi(P)$ as an involution. Then there is always an element of order 4 in $Q \cap \Phi(P)$ which is fixed by β. The contradiction follows as before. $\qquad \square$

Proposition 10.13 *Let P be a bicyclic 2-group such that $P' \cong C_{2^r} \times C_{2^{r+s+1}}$ for some $r, s \geq 2$. Then every fusion system on P is nilpotent.*

Proof Here Theorem 4.13 in [140] applies. The proof is a combination of the proofs of Propositions 10.10 and 10.12. In fact for part (a) of Theorem 4.13 we can copy the proof of Proposition 10.10. Similarly the arguments of Proposition 10.12 remain correct for case (b). Here observe that there is no need to discuss the case $r = 2$ separately, since $x^{-1}v^{2^{r+s-1}}x = v^{-2^{r+s-1}}$. $\qquad\square$

Now it suffices to consider the case where P' contains a cyclic maximal subgroup. If P' is non-cyclic, Theorem 4.8 in [140] applies. This case is more complicated, since $|P/P'|$ is not bounded anymore.

Proposition 10.14 *Let P be a bicyclic 2-group such that $P' \cong C_{2^n} \times C_2$ for some $n \geq 2$, and $P/\Phi(P')$ has a normal elementary abelian subgroup of order 8. Then every fusion system on P is nilpotent.*

Proof There are two possibilities for P according to if $Z(P)$ is cyclic or not. We handle them separately.

Case 1: $Z(P)$ non-cyclic.

Then $a^{2^m} = uz^\eta$. Moreover,

$$a^{-2}va^2 = a^{-1}vuv^{2+4s}a = a^{-1}uv^{3+4s}a = u(uv^{3+4s})^{3+4s} = v^{(3+4s)^2} \in v\langle v^8\rangle. \tag{10.1}$$

Using this we see that $\langle a^{2^{m-1}}, v^{2^{n-2}}\rangle \cong C_4^2$. Thus, Lemma 10.11 implies $\Omega_2(\Phi(P)) = \langle a^{2^{m-1}}, v^{2^{n-2}}\rangle$. As usual we assume that there is an $\mathscr{F}$-essential subgroup $Q \cong C_{2^t} \times C_2^2$ for some $t \geq 1$. Then $Q \cap \Phi(P) \cong C_{2^t} \times C_2$, since $\Phi(P)$ has rank 2. For $t = 1$ we obtain $Q \cap \Phi(P) = \Omega(\Phi(P)) \subseteq Z(P)$. Write $\overline{P} := P/\Omega(\Phi(P))$, $\overline{Q} := Q/\Omega(\Phi(P))$ and so on. Then $C_{\overline{P}}(\overline{Q}) \subseteq N_P(Q)$. So by Satz III.14.23 in [128], $\overline{P}$ has maximal class. Hence, $P' = \Phi(P)$ and $m = 1$. Contradiction. Thus, we may assume $t \geq 2$. Then as usual we can find an element $\delta \in Q \cap \Phi(P)$ of order 4 which is fixed by some involution $\beta \in Q \setminus \Phi(P)$. We write $\delta = a^{2^{m-1}d_1}v^{2^{n-2}d_2}$ and $\beta = xv^j a^{2i}$. Assume first that $2 \mid d_1$. Then $2 \nmid d_2$. Since $a^{2^m}v^{2^{n-2}} \in Z(\Phi(P))$, it follows that $\delta = {}^\beta\delta = {}^x\delta = \delta^{-1}$. This contradiction shows $2 \nmid d_1$. After replacing δ with its inverse if necessary, we can assume $d_1 = 1$. Now we consider β. We have

$$1 = \beta^2 = (xv^j a^{2i})^2 \equiv x^2 v^{2j} a^{4i} \equiv a^{4i} \pmod{P'}.$$

Since

$$2^{n+m} = |\Phi(P)| = \frac{|\langle a^2\rangle||P'|}{|\langle a^2\rangle \cap P'|} = \frac{2^{n+m+1}}{|\langle a^2\rangle \cap P'|},$$

we get $2^{m-2} \mid i$. In case $i = 2^{m-2}$ we get the contradiction

$$\langle z \rangle \ni x^2 = xv^{j-2^{m-2}d_2} xv^{j-2^{m-2}d_2} = (\beta\delta^{-1})^2 = \delta^2 \in u\langle z \rangle.$$

Hence, $2^{m-1} \mid i$. So, after multiplying β with δ^2 if necessary, we may assume $i = 0$, i.e. $\beta = xv^j$. Then $1 = xv^j xv^j = x^2$. Conjugation with a^{-1} gives $\beta = a^{-1}xv^j a = xv^{-1}a^{-1}v^j a = xu^j v^{(3+4s)j-1}$. Since $u \in Q$, we may assume that $\beta = xv^{2j}$. After we conjugate Q by v^j, we even obtain $\beta = x$. Since $x(a^2v^i)x^{-1} = a^2uv^{4(1+s)-i}$, no element of the form a^2v^i is fixed by x. On the other hand

$$x(a^4v^i)x^{-1} = (a^2uv^{4(1+s)})^2v^{-i} = a^4v^{4(1+s)(3+4s)^2+4(1+s)-i}.$$

This shows that there is an i such that $a^4v^i =: \lambda$ is fixed by x. Assume there is another element $\lambda_1 := a^4v^j$ which is also fixed by x. Then $\lambda^{-1}\lambda_1 = v^{j-i} \in \langle z \rangle$. This holds in a similar way for elements containing higher powers of a. In particular $u = a^{2^m}z^\eta \in \langle \lambda, z \rangle$. Recall that $\Phi(P) = \langle v \rangle \rtimes \langle a^2 \rangle$. This shows $C_{\Phi(P)}(x) = \langle \lambda \rangle \times \langle z \rangle \cong C_{2^{m-1}} \times C_2$. Since $Q \cap \Phi(P) \subseteq C_{\Phi(P)}(x)$ and $Q = (Q \cap \Phi(P))\langle x \rangle$, we deduce $C_{\Phi(P)}(x) \subseteq C_P(Q) \subseteq Q$. Moreover, $Q \cap \Phi(P) = C_{\Phi(P)}(x)$ and $t = m - 1$. Therefore, $Q = \langle \lambda, x, z \rangle$. The calculation above shows that there is an element $\mu := a^2v^j$ such that $^\mu x = ux \in Q$. Now $\mu^2 \in C_{\Phi(P)}(x)$ implies $C_{\Phi(P)}(x) = \langle \mu^2, z \rangle$ and $\mu \in N_P(Q) = Q\langle v^{2^{n-2}} \rangle$. Contradiction.

Now assume $Q \cong C_{2^t} \times Q_8$ for some $t \geq 1$. Since $\Phi(P)$ does not contain a subgroup isomorphic to Q_8, we see that $\Omega(\Phi(P)) \subseteq Q$. First assume $t = 1$. Then we look again at the quotients $\overline{P} := P/\Omega(\Phi(P))$ and $\overline{Q} := Q/\Omega(\Phi(P)) \cong C_2^2$. Since $N_P(Q)$ acts non-trivially on $\overline{Q}$, we get $C_{\overline{P}}(\overline{Q}) \subseteq \overline{Q}$. In particular Proposition 1.8 in [23] implies that $\overline{P}$ has maximal class. This leads to a contradiction as in the first part of the proof. Thus, we may assume $t \geq 2$ from now on. Then $\Omega_2(\Phi(P)) \subseteq Q$. Since Q contains more elements of order 4 than $\Phi(P)$, we can choose $\beta \in Q \setminus \Phi(P)$ of order 4. Write $\beta = xa^{2i}v^j$. Then $\beta^2 \in \Omega(\Phi(P)) \subseteq P'$. So the same discussion as above shows that we can assume $\beta = x$. In particular $|\langle x \rangle| = 4$. Since $C_{\Phi(P)}(x)$ is abelian, λ centralizes $(C_Q(x) \cap \Phi(P))\langle x \rangle\langle v^{2^{n-2}} \rangle = C_Q(x)\langle v^{2^{n-2}} \rangle = Q$. This shows $\lambda \in Q$ and $t = m - 1$ again. More precisely we have $Q = \langle \lambda \rangle \times \langle v^{2^{n-2}}, x \rangle$. Equation (10.1) shows that $v^{2^{n-3}}$ still lies in the center of $\Phi(P)$. It follows easily that $N_P(Q) = Q\langle v^{2^{n-3}} \rangle$. However, as above we also have $\mu \in N_P(Q)$. Contradiction.

Finally, the case $Q \cong C_{2^t} * Q_8$ cannot occur, since $Z(P)$ is non-cyclic.

Case 2: $Z(P)$ cyclic.

Here we have $a^{2^m} = uv^{2^{n-2}}z^\eta$, $n \geq m + 2 \geq 4$ and $1 + s \not\equiv 0 \pmod{2^{n-3}}$. Again we begin with $Q \cong C_{2^t} \times C_2^2$ for some $t \geq 1$. By Theorem 4.3(b) in [140] we still have $\langle u, z \rangle = \Omega(Z(\Phi(P)))$. Since $\Phi(P)$ does not have maximal class, also $\langle u, z \rangle = \Omega(\Phi(P))$ holds. In particular $\Omega(\Phi(P)) \subseteq Q$. In case $t = 1$ we see that $P/\Omega(\Phi(P))$ has maximal class which leads to a contradiction as before.

Thus, $t \geq 2$. Since $u \in Z(\Phi(P))$, Eq. (10.1) is still true. Hence, $\Omega_2(\Phi(P)) = \langle a^{2^{m-1}} v^{2^{n-3}}, v^{2^{n-2}} \rangle \cong C_4^2$. We choose an involution $\beta = x v^j a^{2i} \in Q \setminus \Phi(P)$. Then as usual $v^{2^{n-2}} \in N_P(Q) \setminus Q$. Since $a^{2^m} \in \langle u \rangle \times \langle v^{2^{n-2}} \rangle$, we find an element $\delta = a^{2^{m-1}} v^{d_1} \in Q \cap \Omega_2(\Phi(P))$ of order 4 fixed by β. Now exactly the same argument as in Case 1 shows that $\beta = x$ after changing the representative of β and conjugation of Q if necessary. Similarly we get $\lambda := a^4 v^j \in C_{\Phi(P)}(x)$. Moreover, $u = a^{2^m} v^{-2^{n-2}} z^\eta \in \{\lambda^{2^{m-2}}, \lambda^{2^{m-2}} z\}$. Therefore, $C_{\Phi(P)}(x) = \langle \lambda \rangle \times \langle z \rangle \cong C_{2^{m-1}} \times C_2$. The contradiction follows as before.

Now assume that $Q \cong C_{2^t} \times Q_8$ or $Q \cong C_{2^t+1} * Q_8$ for some $t \geq 1$. Proposition 10.8 shows that $\mathscr{F} = C_{\mathscr{F}}(\langle z \rangle)$. Theorem 6.3 in [184] implies that $\overline{Q} := Q/\langle z \rangle$ is an $\mathscr{F}/\langle z \rangle$-essential subgroup of $\overline{P} := P/\langle z \rangle$. Now $\overline{P}$ is bicyclic and has commutator subgroup isomorphic to $C_{2^{n-1}} \times C_2$. Hence the result follows by induction on t. $\square$

Combining these propositions we deduce one of the main results of this section.

Theorem 10.15 *Every fusion system on a bicyclic 2-group P is nilpotent unless P' is cyclic.*

It seems that there is no general reason for Theorem 10.15. For example there are non-nilpotent fusion systems on 2-groups of rank 2 with non-cyclic commutator subgroup.

For the convenience of the reader we state a consequence for finite groups.

Corollary 10.16 *Let G be a finite group with bicyclic Sylow 2-subgroup P. If P' is non-cyclic, then P has a normal complement in G.*

10.1.2 The Case P' Cyclic

In this section we consider the remaining case where the bicyclic 2-group P has cyclic commutator subgroup. Here Theorem 4.4 in [140] plays an important role. The following theorem classifies all fusion systems on bicyclic 2-groups together with some more information.

Theorem 10.17 *Let $\mathscr{F}$ be a fusion system on a bicyclic 2-group P. Then one of the following holds:*

(1) *$\mathscr{F}$ is nilpotent, i.e. $\mathscr{F} = \mathscr{F}_P(P)$.*
(2) *$P \cong C_{2^n}^2$ and $\mathscr{F} = \mathscr{F}_P(P \rtimes C_3)$ for some $n \geq 1$.*
(3) *$P \cong D_{2^n}$ for some $n \geq 3$ and $\mathscr{F} = \mathscr{F}_P(\mathrm{PGL}(2, 5^{2^{n-3}}))$ or $\mathscr{F}_P(\mathrm{PSL}(2, 5^{2^{n-2}}))$. Moreover, $\mathscr{F}$ provides one respectively two essential subgroups isomorphic to C_2^2 up to conjugation.*
(4) *$P \cong Q_8$ and $\mathscr{F} = \mathscr{F}_P(\mathrm{SL}(2, 3))$ is controlled, i.e. there are no $\mathscr{F}$-essential subgroups.*

(5) $P \cong Q_{2^n}$ for some $n \geq 4$ and $\mathcal{F} = \mathcal{F}_P(\mathrm{SL}(2, 5^{2^{n-4}}).C_2)$ or $\mathcal{F}_P(\mathrm{SL}(2, 5^{2^{n-3}}))$. Moreover, $\mathcal{F}$ provides one respectively two essential subgroups isomorphic to Q_8 up to conjugation.

(6) $P \cong SD_{2^n}$ for some $n \geq 4$ and $\mathcal{F} = \mathcal{F}_P(\mathrm{PSL}(2, 5^{2^{n-3}}) \rtimes C_2)$, $\mathcal{F}_P(\mathrm{GL}(2, q))$ or $\mathcal{F}_P(\mathrm{PSL}(3, q))$ where in the last two cases q is a suitable prime power such that $q \equiv 3 \pmod 4$. Moreover, in the first (resp. second) case C_2^2 (resp. Q_8) is the only $\mathcal{F}$-essential subgroup up to conjugation, in the last case both are $\mathcal{F}$-essential and these are the only ones up to conjugation.

(7) $P \cong C_{2^n} \wr C_2$ for some $n \geq 2$ and $\mathcal{F} = \mathcal{F}_P(C_{2^n}^2 \rtimes S_3)$, $\mathcal{F}_P(\mathrm{GL}(2, q))$ or $\mathcal{F}_P(\mathrm{PSL}(3, q))$ where in the last two cases $q \equiv 1 \pmod 4$. Moreover, in the first (resp. second) case $C_{2^n}^2$ (resp. $C_{2^n} * Q_8$) is the only $\mathcal{F}$-essential subgroup up to conjugation, in the last case both are $\mathcal{F}$-essential and these are the only ones up to conjugation.

(8) $P \cong C_2^2 \rtimes C_{2^n}$ is minimal non-abelian of type $(n, 1)$ for some $n \geq 2$ and $\mathcal{F} = \mathcal{F}_P(A_4 \rtimes C_{2^n})$. Moreover, $C_{2^{n-1}} \times C_2^2$ is the only $\mathcal{F}$-essential subgroup of P.

(9) $P \cong \langle v, x, a \mid v^{2^n} = x^2 = 1, {}^x v = v^{-1}, a^{2^m} = v^{2^{n-1}}, {}^a v = v^{-1+2^{n-m+1}}, {}^a x = vx \rangle \cong D_{2^{n+1}}.C_{2^m}$ for $n > m > 1$ and $\mathcal{F} = \mathcal{F}_P(\mathrm{PSL}(2, 5^{2^{n-1}}).C_{2^m})$. Moreover, $C_{2^{m-1}} \times C_2^2$ is the only $\mathcal{F}$-essential subgroup up to conjugation.

(10) $P \cong \langle v, x, a \mid v^{2^n} = x^2 = a^{2^m} = 1, {}^x v = v^{-1}, {}^a v = v^{-1+2^i}, {}^a x = vx \rangle \cong D_{2^{n+1}} \rtimes C_{2^m}$ for $\max(2, n - m + 2) \leq i \leq n$ and $n, m \geq 2$. Moreover, $\mathcal{F} = \mathcal{F}_P(\mathrm{PSL}(2, 5^{2^{n-1}}) \rtimes C_{2^m})$ and $C_{2^{m-1}} \times C_2^2$ is the only $\mathcal{F}$-essential subgroup up to conjugation. In case $i = n$ there are two possibilities for $\mathcal{F}$ which differ by $Z(\mathcal{F}) \in \{\langle a^2 \rangle, \langle a^2 v^{2^{n-1}} \rangle\}$.

(11) $P \cong \langle v, x, a \mid v^{2^n} = 1, x^2 = a^{2^m} = v^{2^{n-1}}, {}^x v = v^{-1}, {}^a v = v^{-1+2^{n-m+1}}, {}^a x = vx \rangle \cong Q_{2^{n+1}}.C_{2^m}$ for $n > m > 1$ and $\mathcal{F} = \mathcal{F}_P(\mathrm{SL}(2, 5^{2^{n-2}}).C_{2^m})$. Moreover, $C_{2^{m-1}} \times Q_8$ is the only $\mathcal{F}$-essential subgroup up to conjugation.

(12) $P \cong \langle v, x, a \mid v^{2^n} = a^{2^m} = 1, x^2 = v^{2^{n-1}}, {}^x v = v^{-1}, {}^a v = v^{-1+2^i}, {}^a x = vx \rangle \cong Q_{2^{n+1}} \rtimes C_{2^m}$ for $\max(2, n - m + 2) \leq i \leq n$ and $n, m \geq 2$. Moreover, $\mathcal{F} = \mathcal{F}_P(\mathrm{SL}(2, 5^{2^{n-2}}) \rtimes C_{2^m})$ and $C_{2^{m-1}} \times Q_8$ is the only $\mathcal{F}$-essential subgroup up to conjugation.

(13) $P \cong \langle v, x, a \mid v^{2^n} = a^{2^m} = 1, x^2 = v^{2^{n-1}}, {}^x v = v^{-1}, {}^a v = v^{-1+2^{n-m+1}}, {}^a x = vx \rangle \cong Q_{2^{n+1}} \rtimes C_{2^m}$ for $n > m > 1$ and $\mathcal{F} = \mathcal{F}_P(\mathrm{SL}(2, 5^{2^{n-2}}) \rtimes C_{2^m})$. Moreover, $C_{2^m} * Q_8$ is the only $\mathcal{F}$-essential subgroup up to conjugation.

(14) $P \cong \langle v, x, a \mid v^{2^n} = 1, x^2 = a^{2^m} = v^{2^{n-1}}, {}^x v = v^{-1}, {}^a v = v^{-1+2^i}, {}^a x = vx \rangle \cong Q_{2^{n+1}}.C_{2^m}$ for $\max(2, n - m + 2) \leq i \leq n$ and $n, m \geq 2$. In case $m = n$, we have $i \neq n$. Moreover, $\mathcal{F} = \mathcal{F}_P(\mathrm{SL}(2, 5^{2^{n-2}}).C_{2^m})$ and $C_{2^m} * Q_8$ is the only $\mathcal{F}$-essential subgroup up to conjugation.

In particular, $\mathcal{F}$ is non-exotic. Conversely, for every group described in these cases there exists a fusion system with the given properties. Moreover, different parameters give non-isomorphic groups.

Proof Assume that $\mathcal{F}$ is non-nilpotent. By Theorem 10.15, P' is cyclic. The case $P \cong Q_8$ is easy. For the other metacyclic cases and the case $P \cong C_{2^n} \wr C_2$ we refer to Theorem 5.3 in [63]. Here we add a few additional information. An induction on $i \geq 2$ shows $5^{2^{i-2}} \equiv 1 + 2^i \pmod{2^{i+1}}$. This implies that the Sylow 2-subgroups of $SL(2, 5^{2^{n-3}})$, $PSL(2, 5^{2^{n-2}})$ and so on have the right order. For the groups SD_{2^n} and $C_{2^n} \wr C_2$ it is a priori not clear if for every n an odd prime power q can be found. However, this can be shown using Dirichlet's Prime Number Theorem (compare with Theorem 6.2 in [275]). Hence, for a given n all these fusion systems can be constructed.

Using Proposition 10.3 we can assume that every $\mathcal{F}$-essential subgroup has rank 3. Finally by Proposition 10.4 it remains to consider $|P'| > 2$. Hence, let P be as in Theorem 4.4 in [140]. We adapt our notation slightly as follows. We replace a by a^{-1} in order to write $^a v$ instead of v^a. Then we have $^a x = vx$. After replacing v by a suitable power, we may assume that i is a 2-power (accordingly we need to change x to $v^\eta x$ for a suitable number η). Then we can also replace i by $2 + \log i$. This gives

$$P \cong \langle v, x, a \mid v^{2^n} = 1, \ x^2, a^{2^m} \in \langle v^{2^{n-1}} \rangle, \ ^x v = v^{-1}, \ ^a v = v^{-1+2^i}, \ ^a x = vx \rangle.$$
(10.2)

Since Theorem 4.4 in [140] also states that v and $a^{2^{m-1}}$ commute, we obtain $i \in \{\max(n - m + 1, 2), \dots, n\}$. We set $z := v^{2^{n-1}}$ as in [140]. Moreover, let $\lambda := v^{-2^{i-1}} a^2$. Then

$$x\lambda x^{-1} = v^{2^{i-1}} (v^{-1} a)^2 = v^{-2^{i-1}} a^2 = \lambda$$

and $\lambda \in C_{\Phi(P)}(x)$. Assume that also $v^j a^2 \in C_{\Phi(P)}(x)$. Then we get $v^j a^2 \in \{\lambda, \lambda z\}$. Hence, $C_{\Phi(P)}(x) \in \{\langle \lambda \rangle, \langle \lambda \rangle \times \langle z \rangle\}$. It should be pointed out that it was not shown in [140] that these presentations really give groups of order 2^{n+m+1} (although some evidence by computer results is stated). However, we assume in the first part of the proof that these groups with the "right" order exist. Later we construct $\mathcal{F}$ as a fusion of a finite group and it will be clear that P shows up as a Sylow 2-subgroup of order 2^{n+m+1}. Now we distinguish between the three different types of essential subgroups.

Case (1): $Q \cong C_{2^t} \times C_2^2$ is $\mathcal{F}$-essential in P for some $t \geq 1$.

As usual, $Q \leq M = E\langle a^2 \rangle$. Since $Q \cap E$ is abelian and $Q/Q \cap E \cong QE/E \leq P/E$ is cyclic, it follows that E is dihedral and $Q \cap E \cong C_2^2$. After conjugation of Q we may assume $Q \cap E \in \{\langle z, x \rangle, \langle z, vx \rangle\}$. Further conjugation with a gives $Q \cap E = \langle z, x \rangle$. Since $C_Q(x) \cap \Phi(P)$ is non-cyclic, it follows that $C_{\Phi(P)}(x) = \langle \lambda \rangle \times \langle z \rangle \cong C_{2^{m-1}} \times C_2$. As usual we obtain $Q = \langle \lambda, z, x \rangle$ and $t = m - 1$. Moreover, $a^2 v a^{-2} \equiv v \pmod{\langle v^8 \rangle}$. Hence, $N_P(Q) = \langle \lambda, v^{2^{n-2}}, x \rangle$.

We prove that Q is the only $\mathscr{F}$-essential subgroup of P up to conjugation. If there is an $\mathscr{F}$-essential subgroup of rank 2, then Proposition 10.3 implies that P is a wreath product. However, by the proof of Theorem 5.3 in [63] all the other $\mathscr{F}$-essential subgroups are of type $C_{2^r} * Q_8$. Hence, this case cannot occur. Thus, by construction it is clear that Q is the only abelian $\mathscr{F}$-essential subgroup up to conjugation. Now assume that $Q_1 \cong C_{2^s} \times Q_8$ is also $\mathscr{F}$-essential. Since Q_1 has three involutions, $Q_1 \cap E$ is cyclic or isomorphic to C_2^2. In either case $Q/Q \cap E \cong QE/E \le P/E$ cannot be cyclic. Contradiction. Suppose now that $Q_1 \cong C_{2^s} * Q_8 \cong C_{2^s} * D_8$ for some $s \ge 2$. Then $Q_1 \cap E$ cannot be cyclic, since Q_1 has rank 3. Suppose $Q_1 \cap E \cong C_2^2$. Then $\Omega(Z(Q_1)) \subseteq Q_1 \cap E$ and $\exp Q_1/Q_1 \cap E \le 2^{s-1}$. On the other hand, $|Q_1/Q_1 \cap E| = 2^s$. In particular, $Q_1/Q_1 \cap E \cong Q_1 E/E \le P/E$ cannot be cyclic. It follows that $Q_1 \cap E$ must be a (non-abelian) dihedral group. Hence, $2^{s-1}|Q_1 \cap E| = |(Q_1 \cap E)\,Z(Q_1)| \le |Q_1| = 2^{s+2}$ and $Q_1 \cap E \cong D_8$. After conjugation of Q_1 we have $Q_1 \cap E = \langle v^{2^{n-2}}, x\rangle$. Let $\lambda_1 \in Z(Q_1) \setminus E$ be an element of order 2^s such that $\lambda_1^{2^{s-1}} = z$. Since $x \in Q_1$, we have $\lambda_1^2 \in C_{\Phi(P)}(x) = \langle \lambda \rangle \times \langle z \rangle$. This implies $s = 2$ and $\lambda_1 \notin \Phi(P)$. Since $Q_1 = (Q_1 \cap \Phi(P))\langle x\rangle$, we obtain $\lambda_1 x \in C_{\Phi(P)}(x)$. But this contradicts $z = \lambda_1^2 = (\lambda_1 x)^2$. Hence, we have proved that Q is in fact the only $\mathscr{F}$-essential subgroup of P up to conjugation.

Now we try to pin down the structure of P more precisely. We show by induction on $j \ge 0$ that $\lambda^{2^j} = v^{2^{i+j-1}v}a^{2^{j+1}}$ for an odd number v. This is clear for $j = 0$. For arbitrary $j \ge 1$ we have

$$\lambda^{2^j} = \lambda^{2^{j-1}}\lambda^{2^{j-1}} = v^{2^{i+j-2}v}a^{2^j}v^{2^{i+j-2}v}a^{2^j} = v^{2^{i+j-2}v(-1+2^i)^{2^j} + 2^{i+j-2}v}a^{2^{j+1}}$$

$$= v^{2^{i+j-2}v((-1+2^i)^{2^j}+1)}a^{2^{j+1}},$$

and the claim follows. In particular we obtain

$$1 = \lambda^{2^{m-1}} = v^{2^{i+m-2}v}a^{2^m}. \tag{10.3}$$

We distinguish whether P splits or not.

Case (1a): $a^{2^m} = z$.

Here Eq. (10.3) shows $i = n - m + 1$. Then $n > m > 1$, and the isomorphism type of P is completely determined by m and n. We show next that $\mathscr{F}$ is uniquely determined (up to isomorphism). For this we need to describe the action of $\mathrm{Aut}_{\mathscr{F}}(Q)$ in order to apply Alperin's Fusion Theorem. As in the proof of Proposition 10.4, $\mathrm{Aut}_{\mathscr{F}}(Q)$ acts on $\langle x, z\rangle$ or on $\langle x\lambda^{2^{m-2}}, z\rangle$ non-trivially (recall $\mathrm{N}_P(Q) \cong D_8 \times C_{2^{m-1}}$). Set $\tilde{x} := x\lambda^{2^{m-2}}$ and $\tilde{a} := av^{2^{n-2}}$. Then as above $\tilde{x} = xv^{\pm 2^{n-2}}a^{2^{m-1}}$. Hence, $\tilde{x}^2 = 1$ and $\tilde{x}v = v^{-1}$. Moreover, $\tilde{a}^2 = a^2$ and thus $\tilde{a}^{2^m} = z$. Finally, $\tilde{a}v = \tilde{a}v$ and $\tilde{a}\tilde{x} = \tilde{a}(xzv^{\pm 2^{n-2}}a^{2^{m-1}}) = vxzv^{\mp 2^{n-2}}a^{2^{m-1}} = v\tilde{x}$. Hence, $v, \tilde{x}$ and $\tilde{a}$ satisfy the same relations as v, x and a. Obviously, $P = \langle v, \tilde{x}, \tilde{a}\rangle$. Therefore, we may replace x by $\tilde{x}$ and a by $\tilde{a}$. After doing this if necessary, we see that $\mathrm{Aut}_{\mathscr{F}}(Q)$ acts non-trivially on $\langle x, z\rangle$ (observe that Q remains fixed under this transformation). As usual it follows that $C_Q(\text{Aut}_{\mathscr{F}}(Q)) \in \{\langle\lambda\rangle, \langle\lambda z\rangle\}$ (compare with proof of Proposition 10.4). Define $\tilde{a} := a^{1+2^{m-1}}$ and $\tilde{v} := v^{1+2^{n-1}} = vz$. Then $\tilde{a}^2 = a^2 z$, $\tilde{a}^{2^m} = z$, $\tilde{v}^{2^n} = 1$, $^x\tilde{v} = \tilde{v}^{-1}$ and $^{\tilde{a}}\tilde{v} = \tilde{v}^{-1+2^{n-m+1}}$. Now we show by induction on $j \geq 1$ that $a^{2^j} x a^{-2^j} = v^{2^{n-m+j}v}x$ for an odd integer v. For $j = 1$ we have $a^2 x a^{-2} = {}^a(vx) = v^{2^{n-m+1}}x$. For arbitrary $j \geq 1$ induction gives

$$a^{2^{j+1}} x a^{-2^{j+1}} = a^{2^j}(a^{2^j} x a^{-2^j})a^{-2^j} = a^{2^j}(v^{2^{n-m+j}v}x)a^{-2^j}$$

$$= v^{2^{n-m+j}v((-1+2^{n-m+1})^{2^j}+1)}x,$$

and the claim follows. In particular $a^{2^{m-1}} x a^{-2^{m-1}} = zvx$ and $^{\tilde{a}}x = \tilde{v}x$. Obviously, $P = \langle\tilde{v}, \tilde{a}, x\rangle$. Hence, we may replace v, a, x by $\tilde{v}, \tilde{a}, x$ if necessary. Under this transformation Q and $\langle x, z\rangle$ remain fixed as sets and λ goes to λz. So, we may assume $C_Q(\text{Aut}_{\mathscr{F}}(Q)) = \langle\lambda\rangle$. Then the action of $\text{Aut}_{\mathscr{F}}(Q)$ on Q is completely described. In particular $\mathscr{F}$ is uniquely determined.

It remains to prove that P and $\mathscr{F}$ really exist. Let $q := 5^{2^{n-1}}$. It is not hard to verify that $H := \text{PSL}(2, q)$ has Sylow 2-subgroup $E \cong D_{2^{n+1}}$. More precisely, E can be generated by the following matrices

$$v := \begin{pmatrix} \omega & 0 \\ 0 & \omega^{-1} \end{pmatrix}, \qquad x := \begin{pmatrix} 0 & 1 \\ -1 & 0 \end{pmatrix}$$

where $\omega \in \mathbb{F}_q^\times$ has order 2^{n+1}, and the matrices are regarded modulo $Z(\text{SL}(2, q)) = \langle -1_2\rangle$. Now consider the matrix $a_1 := \begin{pmatrix} 0 & \omega \\ -1 & 0 \end{pmatrix} \in \text{GL}(2, q)/Z(\text{SL}(2, q))$. Then a_1 acts on H and a calculation shows $^{a_1}v = v^{-1}$ and $^{a_1}x = vx$. Let γ_1 be the Frobenius automorphism of $\mathbb{F}_q$ with respect to $\mathbb{F}_5$, i.e. $\gamma_1(\tau) = \tau^5$ for $\tau \in \mathbb{F}_q$. As usual we may regard γ_1 as an automorphism of H. Let $\gamma := \gamma_1^{2^{n-m-1}}$ so that $|\langle\gamma\rangle| = 2^m$. Recall that $(\mathbb{Z}/2^{n+1}\mathbb{Z})^\times = \langle 5 + 2^{n+1}\mathbb{Z}\rangle \times \langle -1 + 2^{n+1}\mathbb{Z}\rangle \cong C_{2^{n-1}} \times C_2$. It is easy to show that $\langle 5^{2^{n-m-1}} + 2^{n+1}\mathbb{Z}\rangle$ and $\langle 1 - 2^{n-m+1} + 2^{n+1}\mathbb{Z}\rangle$ are subgroups of $(\mathbb{Z}/2^{n+1}\mathbb{Z})^\times$ of order 2^m. Since

$$5^{2^{n-m-1}} \equiv 1 - 2^{n-m+1} \pmod{8},$$

it follows that

$$\langle 5^{2^{n-m-1}} + 2^{n+1}\mathbb{Z}\rangle = \langle 1 - 2^{n-m+1} + 2^{n+1}\mathbb{Z}\rangle.$$

In particular we can find an odd integer v such that $5^{2^{n-m-1}v} \equiv 1 - 2^{n-m+1} \pmod{2^{n+1}}$. Now we set

$$a := a_1\gamma^v.$$

Since γ_1 fixes x, we obtain $^a v = v^{-1+2^{n-m+1}}$ and $^a x = vx$. It remains to show that $a^{2^m} = v^{2^{n-1}} =: z$. Here we identify H with $\text{Inn}(H) \cong H$. For an element $u \in H$ we have

$$a^2(u) = (a_1 \gamma^v a_1 \gamma^v)(u) = (a_1 \gamma^v(a_1)) \gamma^{2v}(u)(a_1 \gamma^v(a_1))^{-1}$$

$$= \left(\begin{pmatrix} \omega & 0 \\ 0 & \omega^{5^{2^{n-m-1}}v} \end{pmatrix} \gamma^{2v} \right)(u).$$

After multiplying the matrix in the last equation by $\begin{pmatrix} \omega & 0 \\ 0 & \omega \end{pmatrix}^h \in Z(\text{GL}(2,q))$ for $h := -(5^{2^{n-m-1}}v + 1)/2$, we obtain

$$a^2(u) = \left(\begin{pmatrix} \omega^{2^{n-m}} & 0 \\ 0 & \omega^{-2^{n-m}} \end{pmatrix} \gamma^{2v} \right)(u),$$

since $(1 - 5^{2^{n-m-1}}v)/2 \equiv 2^{n-m} \pmod{2^n}$. Using induction and the same argument we get

$$a^{2^j} = \begin{pmatrix} \omega^{h_j} & 0 \\ 0 & \omega^{-h_j} \end{pmatrix} \gamma^{2^j v}$$

where $2^{n-m+j-1} \mid h_j$ and $2^{n-m+j} \nmid h_j$ for $j \geq 1$. In particular, $a^{2^m} = z$ as claimed. Now Theorem 15.3.1 in [102] shows that the following non-split extension exists

$$G := H \langle a \rangle \cong \text{PSL}(2, 5^{2^{n-1}}).C_{2^m}.$$

Moreover, the construction shows that G has Sylow 2-subgroup P. Since H is non-abelian simple, $\mathscr{F}_P(G)$ is non-nilpotent. Hence, $\mathscr{F} = \mathscr{F}_P(G)$.

Case (1b): $a^{2^m} = 1$.

Here $P \cong D_{2^{n+1}} \rtimes C_{2^m}$. Moreover, by Eq. (10.3) we have $n - m + 2 \leq i$. As in case (1a) we may assume that $\text{Aut}_{\mathscr{F}}(Q)$ acts on $\langle x, z \rangle$ using the following automorphism of P if necessary:

$$v \mapsto v, \qquad\qquad x \mapsto x\lambda^{2^{m-2}}, \qquad\qquad a \mapsto av^{2^{n-2}}.$$

Now assume $i < n$ (and thus $m, n \geq 3$). Here we consider the following map

$$v \mapsto v^{1+2^{n-1}} = vz =: \tilde{v}, \qquad x \mapsto x, \qquad a \mapsto a^{1+2^{n-i}} =: \tilde{a}.$$

It can be seen that $\tilde{v}$, x and $\tilde{a}$ generate P and satisfy the same relations as v, x and a. Moreover, as above we have $\lambda^{2^{n-i}} = za^{2^{n-i+1}}$. This shows

$$\lambda \mapsto \tilde{v}^{-2^{i-1}} \tilde{a}^2 = v^{-2^{i-1}} a^{2+2^{n-i+1}} = \lambda^{1+2^{n-i}} z = (\lambda z)^{1+2^{n-i}}.$$

Hence, we obtain $C_Q(\mathrm{Aut}_{\mathscr{F}}(Q)) = \langle\lambda\rangle$ after applying this automorphism if necessary. This determines $\mathscr{F}$ completely, and we will construct $\mathscr{F}$ later.

We continue by looking at the case $i = n$. Here we show that $\lambda = za^2$ is not a square in P. Assume the contrary, i.e. $za^2 = (v^j x^k a^l)^2$ for some $j, k, l \in \mathbb{Z}$. Of course, l must be odd. In case $k = 0$ we get the contradiction $(v^j a^l)^2 = a^{2l}$. Thus, $k = 1$. Then $[v, xa^l] = 1$ and $(v^j xa^l)^2 = v^{2j}(xaxa^{-1})a^{2l} = v^{2j-1}a^{2l}$. Again a contradiction. Hence, λ is in fact a non-square. However, $\lambda z = a^2$ is a square and so is every power. As a consequence, it turns out that the two possibilities $C_Q(\mathrm{Aut}_{\mathscr{F}}(Q)) = Z(\mathscr{F}) = \langle\lambda\rangle$ or $C_Q(\mathrm{Aut}_{\mathscr{F}}(Q)) = Z(\mathscr{F}) = \langle a^2\rangle$ give in fact non-isomorphic fusion systems. We denote the latter possibility by $\mathscr{F}'$, i.e. $Z(\mathscr{F}') = \langle a^2\rangle$. Now for every $i \in \{\max(2, n-m+2), \ldots, n\}$ we construct P and $\mathscr{F}$. After that we explain how to obtain $\mathscr{F}'$ for $i = n$. This works quite similar as in case (1a). Let q, H, v, x, a_1 and γ_1 as there. It is easy to see that $\langle 1 - 2^i + 2^{n+1}\mathbb{Z}\rangle$ has order 2^{n+1-i} as a subgroup of $(\mathbb{Z}/2^{n+1}\mathbb{Z})^\times$. Set $\gamma := \gamma_1^{2^{i-2}}$. Then $\gamma^{2^m} = 1$, since $m + i - 2 \geq n$. Again we can find an odd integer v such that $5^{2^{i-2}v} \equiv 1 - 2^i$ (mod 2^{n+1}). Setting $a := a_1\gamma^v \in \mathrm{Aut}(H)$ we get $^a v = v^{-1+2^i}$ and $^a x = vx$. It remains to prove $a^{2^m} = 1$. As above we obtain

$$a^2 = \begin{pmatrix} \omega^{2^{i-1}} & 0 \\ 0 & \omega^{-2^{i-1}} \end{pmatrix} \gamma^{2v}.$$

This leads to $a^{2^m} = 1$. Now we can define $G := H \rtimes \langle a\rangle$ (notice that the action of $\langle a\rangle$ on H is usually not faithful). It is easy to see that in fact $P \in \mathrm{Syl}_2(G)$ and $\mathscr{F}_P(G)$ is non-nilpotent. Hence, for $i < n$ we get $\mathscr{F} = \mathscr{F}_P(G)$ immediately. Now assume $i = n$. Since $\omega^{2^n} = -1 \in \mathbb{F}_q$, we can choose ω such that $\omega^{2^{n-1}} = 2 \in \mathbb{F}_5 \subseteq \mathbb{F}_q$. Define

$$\alpha := \begin{pmatrix} 3 & 1 \\ 2 & 1 \end{pmatrix} \in H.$$

A calculation shows that α has order 3 and acts on $\langle x, z\rangle$ non-trivially. Moreover, $\gamma^{2v} = 1$, and a^2 is the inner automorphism induced by z. In particular, a^2 does not fix α. We can view α as an element of $\mathrm{Aut}_{\mathscr{F}_P(G)}(Q)$. Then $C_Q(\mathrm{Aut}_{\mathscr{F}_P(G)}(Q)) = \langle\lambda\rangle = Z(\mathscr{F})$ is generated by a non-square in P. This shows again $\mathscr{F} = \mathscr{F}_P(G)$. It remains to construct $\mathscr{F}'$. Observe that γ acts trivially on $\langle v, x\rangle$, since $5^{2^{n-2}} \equiv 1$ (mod 2^n). Hence, we can replace the automorphism a just by $a_1 = \begin{pmatrix} 0 & \omega \\ -1 & 0 \end{pmatrix}$ without changing the isomorphism type of P. Again we define $G := H \rtimes \langle a_1\rangle$. Then it turns out that $a_1^2 = \begin{pmatrix} \omega & 0 \\ 0 & \omega \end{pmatrix} \in Z(\mathrm{GL}(2, q))$. In particular, a_1^2 is fixed by the element $\alpha \in \mathrm{Aut}_{\mathscr{F}_P(G)}(Q)$ above. So here $Z(\mathscr{F}_P(G)) = \langle a_1^2\rangle$ is generated by a square in P. Thus, we obtain $\mathscr{F}' = \mathscr{F}_P(G)$.

Case (2): $Q \cong C_{2^t} \times Q_8$ is $\mathscr{F}$-essential in P for some $t \geq 1$.

We have seen above that E cannot be dihedral. Hence, E is (generalized) quaternion, i.e. $x^2 = z$. Now $|Q : Z(Q)| = 4$ implies $Q \cap E \cong Q_8$. After conjugation of Q we may assume $Q \cap E = \langle v^{2^{n-2}}, x \rangle$. Proposition 10.8 implies $z \in Z(\mathscr{F})$. In particular, $Q/\langle z \rangle \cong C_{2^t} \times C_2^2$ is an $\mathscr{F}/\langle z \rangle$-essential subgroup of $P/\langle z \rangle$ (see Theorem 6.3 in [184]). So by the first part of the proof and Proposition 10.4 (for $n = 2$) we get $t = m-1$, and Q is the only $\mathscr{F}$-essential subgroup up to conjugation. Since $C_Q(x) \cap \Phi(P)$ is still non-cyclic, we have $C_{\Phi(P)}(x) = \langle \lambda \rangle \times \langle z \rangle \cong C_{2^{m-1}} \times C_2$ as in case (1). Moreover, a^2 fixes $v^{2^{n-2}}$, and it follows that $Q = \langle v^{2^{n-2}}, x, \lambda \rangle$.

Here we can handle the uniqueness of $\mathscr{F}$ uniformly without discussing the split and non-split case separately. Since $\text{Inn}(Q) \cong C_2^2$, $\text{Aut}_{\mathscr{F}}(Q)$ is a group of order 24 which is generated by $N_P(Q)/Z(Q)$ and an automorphism $\alpha \in \text{Aut}_{\mathscr{F}}(Q)$ of order 3. Hence, in order to describe the action of $\text{Aut}_{\mathscr{F}}(Q)$ on Q (up to automorphisms from $\text{Aut}(P)$), it suffices to know how α acts on Q. First of all, α acts on only one subgroup $Q_8 \cong R \le Q$. It is not hard to see that $Q' = \langle z \rangle \subseteq R$ and thus $R \trianglelefteq Q$. In particular, R is invariant under inner automorphisms of Q. Now let β be an automorphism of Q coming from $N_P(Q)/Q \le \text{Out}_{\mathscr{F}}(Q)$. Then $\beta\alpha \equiv \alpha^{-1}\beta$ (mod $\text{Inn}(Q)$). In particular $\beta(R) = \alpha^{-1}(\beta(R)) = R$. Looking at the action of $N_P(Q)$, we see that $R \in \{\langle v^{2^{n-2}}, x \rangle, \langle v^{2^{n-2}}, x\lambda^{2^{m-2}} \rangle\}$. Again the automorphism

$$v \mapsto v, \qquad\qquad x \mapsto x\lambda^{2^{m-2}}, \qquad\qquad a \mapsto av^{2^{n-2}}$$

leads to $R = \langle v^{2^{n-2}}, x \rangle$. The action of α on R is not quite unique. However, after inverting α if necessary, we have $\alpha(x) \in \{v^{2^{n-2}}, v^{-2^{n-2}}\}$. If we conjugate α with the inner automorphism induced by x in doubt, we end up with $\alpha(x) = v^{2^{n-2}}$. Since α has order 3, it follows that $\alpha(v^{2^{n-2}}) = xv^{2^{n-2}}$. So we know precisely how α acts on R. Since α is unique up to conjugation in $\text{Aut}(Q)$, we have $C_Q(\alpha) = Z(Q) = \langle \lambda, z \rangle$. Hence, the action of $\text{Aut}_{\mathscr{F}}(Q)$ on Q is uniquely determined. By Alperin's Fusion Theorem, $\mathscr{F}$ is unique. For the construction of $\mathscr{F}$ we split up the proof again.

Case (2a): $a^{2^m} = z$.

Again $n > m > 1$ and $i = n - m + 1$ by Eq. (10.3). So the isomorphism type of P is determined by m and n. We construct P and $\mathscr{F}$ in a similar manner as above. For this set $q := 5^{2^{n-2}}$ and $H := \text{SL}(2, q)$. Then a Sylow 2-subgroup H is given by $E := \langle v, x \rangle \cong Q_{2^{n+1}}$ where v and x are defined quite similar as in case (1a). The only difference is that $\omega \in \mathbb{F}_q^\times$ has now order 2^n and the matrices are not considered modulo $Z(\text{SL}(2, q))$ anymore. Also the element a_1 as above still satisfies $^{a_1}v = v^{-1}$ and $^{a_1}x = vx$. Now we can repeat the calculations in case (1a) word by word. Doing so, we obtain $G := H\langle a \rangle \cong \text{SL}(2, q).C_{2^m}$ and $\mathscr{F} = \mathscr{F}_P(G)$.

Case (2b): $a^{2^m} = 1$.

Here Eq. (10.3) gives $\max(n + m + 2, 2) \le i \le n$. For every i in this interval we can again construct P and $\mathscr{F}$ in the same manner as before. We omit the details.

Case (3): $Q \cong C_{2^t} * Q_8$ is $\mathscr{F}$-essential in P for some $t \geq 2$.

Again the argumentation above reveals that E is a quaternion group and $x^2 = z$. Moreover, $Q \cap E = \langle v^{2^{n-2}}, x \rangle \cong Q_8$ after conjugation if necessary. Going over to $P/\langle z \rangle$, it follows that $t = m$. Assume $n = m = i$ and $a^{2^m} = z$ for a moment. Then $(ax)^2 = vza^2$ and $F_1 := \langle v, ax \rangle \cong C_{2^n}^2$ is maximal in P. Since $P/\Phi(F_1)$ is non-abelian, we get $P \cong C_{2^n} \wr C_2$ (compare with the proof of Proposition 6.11). Thus, in case $n = m$ and $a^{2^m} = z$ we assume $i < n$ in the following. We will see later that other parameters cannot lead to a wreath product. After excluding this special case, it follows as before that Q is the only $\mathscr{F}$-essential subgroup up to conjugation. Since $C_Q(x)$ contains an element of order 2^m, we have $C_{\Phi(P)}(x) = \langle \lambda \rangle$. Hence, we have to replace Eq. (10.3) by

$$z = \lambda^{2^{m-1}} = v^{2^{m+i-2}} a^{2^m}$$

where v is an odd number. Moreover, $Q = \langle v^{2^{n-2}}, x, \lambda \rangle$. If $a^{2^m} = z$, then $\max(n - m+2, 2) \leq i \leq n$. On the other hand, if $a^{2^m} = 1$, then $n > m > 1$ and $i = n-m+1$. Hence, these cases complement exactly the case (2) above.

The uniqueness of $\mathscr{F}$ is a bit easier than for the other types of essential subgroups. Again $\mathrm{Aut}_{\mathscr{F}}(Q)$ has order 24 and is generated by $N_P(Q)/Z(Q)$ and an automorphism $\alpha \in \mathrm{Aut}_{\mathscr{F}}(Q)$ of order 3. It suffices to describe the action of α on Q up to automorphisms from $\mathrm{Aut}(P)$. By considering $Q/Q' \cong C_{2^{m-1}} \times C_2^2$ we see that $R := \langle v^{2^{n-2}}, x \rangle$ is the only subgroup of Q isomorphic to Q_8. In particular, α must act on R. Here we also can describe the action precisely by changing α slightly. Moreover, $C_Q(\alpha) = Z(Q) = \langle \lambda \rangle$, since α is unique up to conjugation in $\mathrm{Aut}(Q)$. This shows that $\mathscr{F}$ is uniquely determined (up to isomorphism). Now we distinguish the split and non-split case in order to construct P and $\mathscr{F}$.

Case (3a): $a^{2^m} = 1$.

At first glance one might think that the construction in case (2) should not work here. However, it does. We denote q, H and so on as in case (2a). Then a^{2^m} is the inner automorphism on H induced by z. But since $z \in Z(H)$, a^{2^m} is in fact the trivial automorphism. Hence, we can construct the semidirect product $G = H \rtimes \langle a \rangle$ which does the job.

Case (3b): $a^{2^m} = z$.

Here we do the opposite as in case (3a). With the notation of case (3a), a is an automorphism of H such that $a^{2^m} = 1$ and a fixes $z \in Z(H)$. Using Theorem 15.3.1 in [102] we can build a non-split extension $G := H \langle a \rangle$ such that $a^{2^m} = z$. This group fulfills our conditions.

Finally we show that different parameters in all these group presentations give non-isomorphic groups. Obviously the metacyclic groups are pairwise non-isomorphic and not isomorphic to non-metacyclic groups. Hence, it suffices to look at the groups coming from Theorem 4.4 in [140]. So let P be as in Eq. (10.2) together with additional dependence between x^2 and the choice of i as in the statement of our theorem (this restriction is important). Assume that P is isomorphic to a similar group P_1 where we attach an index 1 to all elements and parameters of P_1. Then we have $2^{n+m+1} = |P| = |P_1| = 2^{n_1+m_1+1}$ and $2^n = |P'| = |P_1'| = 2^{n_1}$. This already shows $n = n_1$ and $m = m_1$. As proved above, P admits a non-nilpotent fusion system with essential subgroup $C_{2^{m-1}} \times C_2^2$ if and only if $x^2 = 1$. Hence, $x^2 = 1$ if and only if $x_1^2 = 1$. Now we show $i = i_1$. For this we consider $\Phi(P) = \langle v, a^2 \rangle$. Since $\Phi(P)$ is metacyclic, it follows that $\Phi(P)' = \langle [v, a^2] \rangle = \langle v^{2^{i+1}} \rangle \cong C_{2^\eta}$ where $\eta := \max(n - i - 1, 0)$. Since $i, i_1 \leq n$, we may assume $i, i_1 \in \{n - 1, n\}$. In case $i = n$ the subgroup $C := \langle v, ax \rangle$ is abelian. By Theorem 4.3(f) in [140], C is a metacyclic maximal subgroup of P. However, in case $i = n - 1$ it is easy to see that the two metacyclic maximal subgroups $\langle v, a \rangle$ and $\langle v, ax \rangle$ of P are both non-abelian. This gives $i = i_1$. It remains to show: $a^{2^m} = 1 \iff a_1^{2^{m_1}} = 1$. For this we may assume $x^2 = z$ and $x_1^2 = z_1$. In case $i = n - m + 1$ (and $n > m > 1$) we have $a^{2^m} = 1$ if and only if P provides a fusion system with essential subgroup $C_{2^m} * Q_8$. A similar equivalence holds for $\max(n - m + 2, 2) \leq i \leq n$ (even in case $n = m = i$). This completes the proof. $\square$

We present an example to shed more light on the alternative in part (10) of Theorem 10.17. Let us consider the smallest case $n = m = i = 2$. The group $N := A_6 \cong PSL(2, 3^2)$ has Sylow 2-subgroup D_8. Let $H := \langle h \rangle \cong C_4$. It is well-known that $\mathrm{Aut}(N)/N \cong C_2^2$, and the three subgroups of $\mathrm{Aut}(N)$ of index 2 are isomorphic to S_6, $PGL(2, 9)$ and the Mathieu group M_{10} of degree 10. We choose two homomorphisms $\varphi_j : H \to \mathrm{Aut}(N)$ for $j = 1, 2$ such that $\varphi_1(h) \in PGL(2, 9) \setminus N$ is an involution and $\varphi_2(h) \in M_{10} \setminus N$ has order 4 (we do not define φ_j precisely). Then it turns out that the groups $G_j := N \rtimes_{\varphi_j} H$ for $j = 1, 2$ have Sylow 2-subgroup P as in part (10). Moreover, one can show that $\mathscr{F}_1 := \mathscr{F}_P(G_1) \neq \mathscr{F}_P(G_2) =: \mathscr{F}_2$. More precisely, $Z(\mathscr{F}_1) = Z(G_1) = \langle \varphi_1(h)^2 \rangle$ is generated by a square in P and $Z(\mathscr{F}_2)$ is not. The indices of G_j in the Small Groups Library are [1440, 4592] and [1440, 4595] respectively. It should be clarified that this phenomenon is not connected to the special behavior of A_6, since it occurs for all n with $PSL(2, 5^{2^{n-1}})$.

As a second remark we indicate a more abstract way to establish the non-exoticness of our fusion systems. It suffices to look at the cases (9)–(14) in Theorem 10.17. If P does not contain an abelian $\mathscr{F}$-essential subgroup, then Proposition 10.8 shows $Z(\mathscr{F}) \neq 1$. Here Theorem 2.4(b) in [211] reduces the question of exoticness to a fusion system on the smaller bicyclic group $P/\langle z \rangle$. Hence, we may assume that there is an $\mathscr{F}$-essential subgroup $Q = \langle z, x, \lambda \rangle \cong C_{2^{m-1}} \times C_2^2$. Moreover, we can assume that $Z(\mathscr{F}) = 1$. Now we construct the *reduced* fusion system of $\mathscr{F}$ (see Definition 2.1 in [17]). By Proposition 1.5 in [17] we have $O_2(\mathscr{F}) \leq Q \cap {}^a Q \subseteq \langle z, \lambda \rangle$. Since $O_2(\mathscr{F})$ is strongly closed in P, we have $z \notin O_2(\mathscr{F})$. Hence, $O_2(\mathscr{F})$ is cyclic and $\Omega(O_2(\mathscr{F})) \subseteq Z(\mathscr{F}) = 1$. This shows $O_2(\mathscr{F}) = 1$. So in the definition of the reduced fusion system we have $\mathscr{F}_0 = \mathscr{F}$. Now we determine $\mathscr{F}_1 := O^2(\mathscr{F})$. Since $E = \langle x, vx \rangle$, it turns out that the hyperfocal subgroup of $\mathscr{F}$ is $E \cong D_{2^{n+1}}$. Using Definitions 1.21 and 1.23 in [17] it is easy to see that $\mathscr{F}_1$ has two essential subgroups isomorphic to C_2^2 up to conjugation. That is $\mathscr{F}_1 = \mathscr{F}_E(\mathrm{PSL}(2, 5^{2^{n-1}}))$. Moreover, we have $\mathscr{F}_2 := O^{2'}(\mathscr{F}_1) = \mathscr{F}_1$. So it follows that $\mathscr{F}_1$ is the reduction of $\mathscr{F}$. By Proposition 4.3 in [17], $\mathscr{F}_1$ is tame in the sense of Definition 2.5 in [17]. Without using the classification of the finite simple groups, Theorem 2.10 in [17] implies that $\mathscr{F}_1$ is even strongly tame. Hence, also $\mathscr{F}$ is tame by Theorem 2.20 in [17]. In particular $\mathscr{F}$ is not exotic.

However, using this approach it is a priori not clear if these (non-nilpotent) fusion systems exist at all. But this might be handled in an abstract manner as follows. Let Q be (a candidate for) an $\mathscr{F}$-essential subgroup of P. By definition, the fusion system $N_{\mathscr{F}}(Q)$ on $N := N_P(Q)$ is constrained and thus can be realized by a finite group H. Then Theorem 1 in [236] shows that $\mathscr{F}$ is the fusion system of the (infinite!) free product $H *_N P$ with amalgamated subgroup N. However, it is not clear if this construction yields saturated fusion systems. Another problem which remains on these lines is the uniqueness of $\mathscr{F}$. The different possibilities for $\mathscr{F}$ differ by the ways one can embed N into H in the construction of $H *_N P$.

As another comment, we observe that the 2-groups in parts (11)–(14) have 2-rank 2. Hence, these are new examples in the classification of all fusion systems on 2-groups of 2-rank 2 which was started in [63]. It is natural to ask what happens if we interchange the restrictions on i in case (9) and case (10) in Theorem 10.17. We will see in the next theorem that this does not result in new groups.

Theorem 10.18 *Let P be a bicyclic, non-metacyclic 2-group. Then P admits a non-nilpotent fusion system if and only if P' is cyclic.*

Proof By Theorem 10.15 it suffices to prove only one direction. Let us assume that P' is cyclic. Since P is non-metacyclic, it follows that $P' \neq 1$. In case $|P'| = 2$, Theorem 4.1 in [140] implies that P is minimal non-abelian of type $(n, 1)$ for some $n \geq 2$. We have already shown that there is a non-nilpotent fusion system on this group. Thus, we may assume $|P'| > 2$. Then we are again in Theorem 4.4 in [140]. After adapting notation, P is given as in Eq. (10.2). In case $x^2 = z$ there is always a non-nilpotent fusion system on P by Theorem 10.17. Hence, let $x^2 = 1$. Then it remains to deal with two different pairs of parameters.

Case 1: $a^{2^m} = 1$ and $i = n - m + 1 \geq 2$.

Set $\tilde{x} := xa^{2^{m-1}}$. Then

$$\tilde{x}^2 = xa^{2^{m-1}}xa^{2^{m-1}} = (v^{-1}a)^{2^{m-1}}a^{2^{m-1}} = v^{2^{i+m-2}}v^{a^{2^m}} = z$$

for an odd integer v. Moreover, $^x v = v^{-1}$, $^a \tilde{x} = vxa^{2^{m-1}} = v\tilde{x}$. This shows that P is isomorphic to a group with parameters $x^2 = z$, $a^{2^m} = 1$ and $i = n - m + 1 \geq 2$. In particular Theorem 10.17 provides a non-nilpotent fusion system on P.

Case 2: $a^{2^m} = z$ and $\max(2, n - m + 2) \le i \le n$.

Again let $\tilde{x} := xa^{2^{m-1}}$. Then

$$\tilde{x}^2 = v^{2^{i+m-2}v}a^{2^m} = z.$$

Hence, P is isomorphic to a group with parameters $x^2 = a^{2^m} = z$ and $\max(2, n - m + 2) \le i \le n$. The claim follows as before. $\qquad\qquad\square$

Now we count how many interesting fusion systems we have found.

Proposition 10.19 *Let $f(N)$ be the number of isomorphism classes of bicyclic 2-groups of order 2^N which admit a non-nilpotent fusion system. Moreover, let $g(N)$ be the number of non-nilpotent fusion systems on all bicyclic 2-groups of order 2^N. Then*

N	1	2	3	≥ 4 even	≥ 5 odd
$f(N)$	0	1	2	$\frac{3}{4}N^2 - 3N + 5$	$(3N^2 + 1)/4 - 3N + 3$
$g(N)$	0	1	3	$\frac{3}{4}N^2 - 2N + 5$	$(3N^2 + 1)/4 - 2N + 5$

Proof Without loss of generality, $N \ge 4$. We have to distinguish between the cases N even and N odd. Assume first that N is even. Then we get the following five groups: $C_{2^{N/2}}^2$, D_{2^N}, Q_{2^N}, SD_{2^N} and the minimal non-abelian group of type $(N - 2, 1)$. From case (9) of Theorem 10.17 we obtain exactly $N/2 - 2$ groups. In case (10) the number of groups is

$$\sum_{n=2}^{N-3} (n - \max(2, 2n - N + 3) + 1) = \sum_{n=2}^{N/2-1} (n - 1) + \sum_{n=N/2}^{N-3} (N - n - 2) = 2\sum_{n=1}^{N/2-2} n$$

$$= (N/2 - 2)(N/2 - 1) = \frac{N^2}{4} - \frac{3N}{2} + 2.$$

The other cases are similar (observe that the wreath product cannot occur, since N is even). All together we get

$$5 + 3(N/2 - 2) + 3(N^2/4 - 3N/2 + 2) = \frac{3}{4}N^2 - 3N + 5$$

bicyclic 2-groups of order 2^N with non-nilpotent fusion system.

Now if N is odd we have the following four examples: D_{2^N}, Q_{2^N}, SD_{2^N} and the minimal non-abelian group of type $(N-2, 1)$. From case (9) of Theorem 10.17 we obtain exactly $(N-5)/2$ groups. In case (10) the number of groups is

$$\sum_{n=2}^{N-3} (n - \max(2, 2n - N + 3) + 1) = \sum_{n=2}^{(N-1)/2} (n-1) + \sum_{n=(N+1)/2}^{N-3} (N - n - 2)$$

$$= 2 \sum_{n=1}^{(N-5)/2} n + (N-3)/2$$

$$= \frac{(N-5)(N-3)}{4} + \frac{N-3}{2} = \frac{N^2 - 6N + 9}{4}.$$

Adding the numbers from the other cases (this time including the wreath product), we obtain

$$4 + 3\frac{N^2 - 4N - 1}{4} = \frac{3N^2 + 1}{4} - 3N + 3.$$

In order to obtain $g(N)$ from $f(N)$ we have to add one fusion system on D_{2^N}, one on Q_{2^N}, and two on SD_{2^N}. If N is odd, we get two more fusion systems on the wreath product. For all $N \geq 5$ we have to add $N - 4$ fusion systems coming from part (10) in Theorem 10.17. □

We present an application to finite simple groups. For this we introduce a general lemma.

Lemma 10.20 *Let G be a perfect group and $1 \neq P \in \mathrm{Syl}_p(G)$ such that $N_G(P) = P\, C_G(P)$. Then there are at least two conjugacy classes of $\mathscr{F}_P(G)$-essential subgroups in P.*

Proof Let $\mathscr{F} := \mathscr{F}_P(G)$. If there is no $\mathscr{F}$-essential subgroup, then $\mathscr{F}$ is nilpotent and G is p-nilpotent, since $\mathrm{Out}_{\mathscr{F}}(P) = N_G(P)/P\, C_G(P) = 1$. Then $G' \leq P' O_{p'}(G) < G$, because $P \neq 1$. Contradiction. Now suppose that there is exactly one $\mathscr{F}$-essential subgroup $Q \leq P$ up to conjugation. Then Q lies in a maximal subgroup $M < P$. Moreover, $P' \subseteq \Phi(P) \subseteq M$. Now the Focal Subgroup Theorem (see Theorem 7.3.4 in [94]) gives the following contradiction:

$$P = P \cap G = P \cap G' = \langle x^{-1}\alpha(x) : x \in P,\ \alpha \text{ morphism in } \mathscr{F} \rangle \subseteq P'Q \subseteq M.$$

□

Theorem 10.21 *Let G be a simple group with bicyclic Sylow 2-subgroup. Then G is one of the following groups: C_2, $\mathrm{PSL}(i, q)$, $\mathrm{PSU}(3, q)$, A_7 or M_{11} for $i \in \{2, 3\}$ and q odd.*

Proof By the Alperin-Brauer-Gorenstein Theorem [3] on simple groups of 2-rank 2, we may assume that G has 2-rank 3 (observe that a Sylow 2-subgroup of $PSU(3, 4)$ is not bicyclic, since it has rank 4). Now we could apply the Gorenstein-Harada result [95] on simple groups of sectional rank at most 4. However, we prefer to give a more elementary argument. Let $P \in \mathrm{Syl}_2(G)$ and $\mathcal{F} := \mathcal{F}_P(G)$. By Theorem 10.17, there is only one $\mathcal{F}$-essential subgroup Q in P up to conjugation. But this contradicts Lemma 10.20. $\square$

10.2   Blocks

Now we consider fusion systems coming from block theory.

Theorem 10.22 *Olsson's Conjecture holds for all blocks of finite groups with bicyclic defect groups.*

Proof Let B be a p-block of a finite group with bicyclic defect group D. Since all bicyclic p-groups for an odd prime p are metacyclic, we may assume $p = 2$ (see Corollary 8.11). If D is metacyclic, Olsson's Conjecture holds by Corollary 8.2. If D is minimal non-abelian, the same is true by Corollary 12.17 below. By results of Külshammer [160] we can also leave out the case where D is a wreath product. Let $\mathcal{F}$ be the fusion system of B. Without loss of generality, $\mathcal{F}$ is non-nilpotent. Hence, we may assume that D is given by

$$D \cong \langle v, x, a \mid v^{2^n} = 1, \ x^2, a^{2^m} \in \langle v^{2^{n-1}} \rangle, \ {}^x v = v^{-1}, \ {}^a v = v^{-1+2^i}, \ {}^a x = vx \rangle$$

where $\max(2, n-m+1) \leq i \leq n$ as in Theorem 10.17. Moreover, there is only one conjugacy class of $\mathcal{F}$-essential subgroups of D. We use Proposition 4.3. For this let us consider the subsection (a, b_a). Since a does not lie in the unique non-metacyclic maximal subgroup (see Theorem 4.4 in [140]), a does not lie in any $\mathcal{F}$-essential subgroup of D. In particular, $\langle a \rangle$ is fully $\mathcal{F}$-centralized. Thus, Lemma 1.34 implies that b_a has defect group $C_D(a)$. Obviously, $C_{\langle v \rangle}(a) = \langle z \rangle$. Now let $v^j x \in C_D(a)$ for some $j \in \mathbb{Z}$. Then $v^j x = {}^a(v^j x) = v^{1-j+2^i j} x$ and $v^{2j} = v^{1+2^i j}$, a contradiction. This shows $C_D(a) = \langle a, z \rangle$. Now by Proposition 4.3 we obtain $k_0(B) \leq |C_D(a)| = 2^{m+1} = |D : D'|$, i.e. Olsson's Conjecture holds. $\square$

Using Proposition 4.7, it is not hard to see that also Brauer's $k(B)$-Conjecture holds if for the fusion system of B one of the cases (1)–(10) in Theorem 10.17 occurs. The conjecture is open for the remaining cases.

A key feature of the groups in the next three theorems is that all their irreducible characters have degree 1 or 2. These groups can be seen as non-commutative versions of the groups in Chap. 9.

Theorem 10.23 *Let B be a non-nilpotent 2-block of a finite group with defect group*

$$D \cong \langle v, x, a \mid v^{2^n} = x^2 = a^{2^m} = 1, \ {}^x v = {}^a v = v^{-1}, \ {}^a x = vx \rangle \cong D_{2^{n+1}} \rtimes C_{2^m}$$

for some $n, m \geq 2$. *Then* $k(B) = 2^{m-1}(2^n+3)$, $k_0(B) = 2^{m+1}$, $k_1(B) = 2^{m-1}(2^n - 1)$ *and* $l(B) = 2$. *In particular Brauer's* $k(B)$-*Conjecture and Alperin's Weight Conjecture are satisfied.*

Proof Let $\mathscr{F}$ be the fusion system of B, and let $z := v^{2^{n-1}}$. Then by Theorem 10.17, $Q := \langle z, x, a^2 \rangle$ is the only $\mathscr{F}$-essential subgroup up to conjugation. In order to calculate $k(B)$ we use Theorem 1.35. We will see that it is not necessary to obtain a complete set of representatives for the $\mathscr{F}$-conjugacy classes. Since $\langle v, ax \rangle$ is an abelian maximal subgroup of D, all characters in $\mathrm{Irr}(D)$ have degree 1 or 2. In particular $k(D) = |\mathrm{Irr}(D)| = |D/D'| + (|D| - |D/D'|)/4 = 2^{m-1}(2^n + 3)$. Now we have to count how many conjugacy classes of D are fused under $\mathrm{Aut}_{\mathscr{F}}(Q)$. According to Theorem 10.17 there are two possibilities $C_Q(\mathrm{Aut}_{\mathscr{F}}(Q)) = Z(\mathscr{F}) \in \{\langle a^2 \rangle, \langle a^2 z \rangle\}$. In the first case the elements of the form xa^{2j} are conjugate to corresponding elements za^{2j} under $\mathrm{Aut}_{\mathscr{F}}(Q)$. In the second case a similar statement is true for a^{2j}. Observe that the elements xa^{2j} and xza^{2j} are already conjugate in D. Since $\langle a^2, z \rangle \subseteq Z(D)$, no more fusion can occur. Hence, the number of $\mathscr{F}$-conjugacy classes is $2^{m-1}(2^n + 3) - 2^{m-1} = 2^m(2^{n-1} + 1)$.

Now we have to determine at least some of the numbers $l(b_u)$ where $u \in D$. The group $\overline{D}_1 := D/\langle a^2 \rangle$ (resp. $\overline{D}_2 := D/\langle a^2 z \rangle$) has commutator subgroup $D'\langle a^2 \rangle / \langle a^2 \rangle$ (resp. $D'\langle a^2 z \rangle / \langle a^2 z \rangle$) of index 4. Hence, $\overline{D}_1$ (resp. $\overline{D}_2$) has maximal class. The block b_{a^2} (resp. $b_{a^2 z}$) dominates a block $\overline{b}_{a^2}$ (resp. $\overline{b}_{a^2 z}$) with defect group $\overline{D}_1$ (resp. $\overline{D}_2$). Let $\mathscr{F}_1$ (resp. $\mathscr{F}_2$) be the fusion system of $\overline{b}_{a^2}$ (resp. $\overline{b}_{a^2 z}$). Then in case $Z(\mathscr{F}) = \langle a^2 \rangle$ (resp. $Z(\mathscr{F}) = \langle a^2 z \rangle$) $\overline{Q}$ is the only $\mathscr{F}_1$-essential (resp. $\mathscr{F}_2$-essential) subgroup of $\overline{D}_1$ (resp. $\overline{D}_2$) up to conjugation. Thus, Theorem 8.1 implies $l(b_{a^2}) = l(\overline{b}_{a^2}) = 2$ (resp. $l(b_{a^2 z}) = l(\overline{b}_{a^2 z}) = 2$). The same holds for all odd powers of a^2 (resp. $a^2 z$). Next we consider the elements $u := a^{2j}$ for $2 \leq j \leq m-1$. It can be seen that the isomorphism type of $D/\langle u \rangle$ is the same as for D except that we have to replace m by j. Also the essential subgroup Q carries over to the block $\overline{b}_u$. Hence, induction on m gives $l(b_u) = 2$ as well. For all other non-trivial subsections (u, b_u) we only know $l(b_u) \geq 1$. Finally, $l(B) \geq 2$, since B is centrally controlled (Theorem 1.38). Applying Theorem 1.35 gives

$$k(B) \geq 2^m + 2^m(2^{n-1} + 1) - 2^{m-1} = 2^{m-1}(2^n + 3) = k(D).$$

We already know from Theorem 10.22 that Olsson's Conjecture holds for B, i.e. $k_0(B) \leq |D : D'| = 2^{m+1}$. Now we apply Proposition 4.7 to the subsection (z, b_z) which gives

$$|D| = 2^{m+1} + 2^{m+1}(2^n - 1) \leq k_0(B) + 4(k(B) - k_0(B)) \leq \sum_{i=0}^{\infty} 2^{2i} k_i(B) \leq |D|.$$

This implies $k(B) = k(D) = 2^{m-1}(2^n + 3)$, $k_0(B) = 2^{m+1}$, $k_1(B) = 2^{m-1}(2^n - 1)$ and $l(B) = 2$. Brauer's $k(B)$-Conjecture follows immediately. In order to prove Alerin's Weight Conjecture, it suffices to show that Q and D are the only

$\mathscr{F}$-radical, $\mathscr{F}$-centric subgroups of D (up to conjugation). Thus, assume by way of contradiction that Q_1 is another $\mathscr{F}$-radical, $\mathscr{F}$-centric subgroup. Since Q_1 is $\mathscr{F}$-centric it cannot lie inside Q. Moreover, $\mathrm{Out}_{\mathscr{F}}(Q_1)$ must provide an isomorphism of odd order, because $Q_1 < D$. However, by Alperin's Fusion Theorem $\mathscr{F}$ is generated by $\mathrm{Aut}_{\mathscr{F}}(Q)$ and $\mathrm{Aut}_{\mathscr{F}}(D)$. This gives the desired contradiction.   □

We add some remarks. If $n = 1$ we obtain the minimal non-abelian group $C_2^2 \rtimes C_{2^m}$ for which the block invariants are also known by results from the author's dissertation [244] (see Chap. 12). Moreover, it is an easy exercise to check that various other conjectures are also true in the situation of Theorem 10.23. We will not go into the details here.

The next theorem concerns defect groups which have a similar structure as the central products $Q_{2^{n+1}} * C_{2^m}$ discussed in Sect. 9.2. Also, this result is needed for the induction step in the theorem after that.

Theorem 10.24 *Let B be a non-nilpotent 2-block of a finite group with defect group*

$$D \cong \langle v, x, a \mid v^{2^n} = 1,\ a^{2^m} = x^2 = v^{2^{n-1}},\ {}^x v = {}^a v = v^{-1},\ {}^a x = vx \rangle$$

$$\cong Q_{2^{n+1}}.C_{2^m} \cong D_{2^{n+1}}.C_{2^m}$$

for some $n, m \geq 2$ and $m \neq n$. Then $k(B) = 2^{m+1}(2^{n-2} + 1)$, $k_0(B) = 2^{m+1}$, $k_1(B) = 2^{m-1}(2^n - 1)$, $k_n(B) = 2^{m-1}$ and $l(B) = 2$. In particular Brauer's $k(B)$-Conjecture and Alperin's Weight Conjecture are satisfied.

Proof First observe that the proof of Theorem 10.18 shows that in fact

$$D \cong \langle v, x, a \mid v^{2^n} = x^2 = 1,\ a^{2^m} = v^{2^{n-1}},\ {}^x v = {}^a v = v^{-1},\ {}^a x = vx \rangle \cong D_{2^{n+1}}.C_{2^m}.$$

Let $\mathscr{F}$ be the fusion system of B, and let $y := v^{2^{n-2}}$ and $z := x^2$. Then by Theorem 10.17, $Q := \langle x, y, a^2 \rangle \cong Q_8 * C_{2^m}$ is the only $\mathscr{F}$-essential subgroup up to conjugation (since $n \neq m$, D is not a wreath product). Again we use Theorem 1.35 to get a lower bound for $k(B)$. The same argumentation as in Theorem 10.23 shows that D has $2^{m-1}(2^n + 3)$ conjugacy classes and we need to know which of them are fused in Q. It is easy to see that xa^{2j} is conjugate to ya^{2j} under $\mathrm{Aut}_{\mathscr{F}}(Q)$ for $j \in \mathbb{Z}$. Observe that xa^{2j} is already conjugate to xya^{2j} and $x^{-1}a^{2j} = xa^{2j+2^m}$ in D. Since $Z(\mathscr{F}) = \langle a^2 \rangle$, this is the only fusion which occurs. Hence, the number of $\mathscr{F}$-conjugacy classes is again $2^m(2^{n-1} + 1)$.

Again $D/\langle a^2 \rangle$ has maximal class and $l(b_{a^2}) = 2$ by Theorem 8.1. The same is true for the odd powers of a^2. Now let $u := a^{2j}$ for some $2 \leq j \leq m$. Then it turns out that $D/\langle u \rangle$ is isomorphic to the group $D_{2^n} \rtimes C_{2^j}$ as in Theorem 10.23. So we obtain $l(b_u) = 2$ as well. For the other non-trivial subsections (u, b_u) we have at least $l(b_u) \geq 1$. Finally $l(B) \geq 2$, since B is centrally controlled (Theorem 1.38).

Therefore,

$$k(B) \geq 2^{m+1} + 2^m(2^{n-1} + 1) - 2^m = 2^{m+1}(2^{n-2} + 1). \tag{10.4}$$

Also, $k_0(B) \leq 2^{m+1}$ by Theorem 10.22. However, in this situation we cannot apply Proposition 4.7. So we use Theorem 4.2 for the major subsection (a^2, b_{a^2}). Let us determine the isomorphism type of $\overline{D} := D/\langle a^2 \rangle$ precisely. Since $(ax)^2 = axax = vx^2a^2 \equiv v \pmod{\langle a^2 \rangle}$, ax generates a cyclic maximal subgroup $\overline{D}$. Since ${}^a(ax) = avx = axv^{-1} \equiv (ax)^{-1} \pmod{\langle a^2 \rangle}$, $\overline{D} \cong D_{2^{n+1}}$. Hence, the Cartan matrix of b_{a^2} is given by

$$2^m \begin{pmatrix} 2^{n-1} + 1 & 2 \\ 2 & 4 \end{pmatrix}$$

up to basic sets (see Theorem 8.1). This gives $k(B) \leq 2^m(2^{n-1} + 3)$ which is not quite what we wanted. However, the restriction on $k_0(B)$ will show that this maximal value for $k(B)$ cannot be reached. For this we use the same method as in Theorem 9.18, i.e. we analyze the generalized decomposition numbers $d^u_{\chi\varphi_i}$ for $u := a^2$ and $\mathrm{IBr}(b_u) = \{\varphi_1, \varphi_2\}$. Since the argument is quite similar except that n has a slightly different meaning, we only present some key observations here. As in Sect. 9.2 we write

$$d^u_{\chi\varphi_i} = \sum_{j=0}^{2^{m-1}-1} a^i_j(\chi)\zeta^j$$

where $\zeta := e^{2\pi i/2^m}$. It follows that

$$(a^1_i, a^1_j) = (2^n + 2)\delta_{ij}, \qquad (a^1_i, a^2_j) = 4\delta_{ij}, \qquad (a^2_i, a^2_j) = 8\delta_{ij}.$$

Moreover, $h(\chi) = 0$ if and only if $\sum_{j=0}^{2^{m-1}-1} a^2_j(\chi) \equiv 1 \pmod 2$. This gives three essentially different possibilities for a^1_j and a^2_j as in Sect. 9.2. Let the numbers α, β, γ and δ be defined as there. Then

$$\gamma = 2^{m-1} - \alpha - \beta,$$

$$k(B) \leq (2^n + 6)\alpha + (2^n + 4)\beta + (2^n + 2)\gamma - \delta/2$$

$$= 2^{m+n-1} + 6\alpha + 4\beta + 2\gamma - \delta/2$$

$$= 2^{m+n-1} + 2^m + 4\alpha + 2\beta - \delta/2,$$

$$8\alpha + 4\beta - \delta \leq k_0(B) \leq 2^{m+1}.$$

This shows $k(B) \leq 2^{m+n-1} + 2^{m+1} = 2^{m+1}(2^{n-2} + 1)$. Together with (10.4) we have $k(B) = 2^{m+1}(2^{n-2} + 1)$ and $l(B) = 2$. The inequalities above also show

$k_0(B) = 2^{m+1}$. Now we can carry over the further discussion in Sect. 9.2 word by word. In particular we get $\delta = 0$,

$$k_1(B) = (2^n - 2)\alpha + (2^n - 1)\beta + 2^n\gamma = 2^{n+m-1} - 2\alpha - \beta$$
$$= 2^{n+m-1} - 2^{m-1} = 2^{m-1}(2^n - 1)$$

and finally $k_n(B) = 2^{m-1}$. The conjectures follow as usual.   □

Now we can also handle defect groups of type $Q_{2^{n+1}} \rtimes C_{2^m}$. It is interesting to see that we get the same number of characters, although the groups are non-isomorphic as it was shown in Sect. 10.1.

Theorem 10.25 *Let B be a non-nilpotent 2-block of a finite group with defect group*

$$D \cong \langle v, x, a \mid v^{2^n} = a^{2^m} = 1, x^2 = v^{2^{n-1}}, {}^x v = {}^a v = v^{-1}, {}^a x = vx \rangle \cong Q_{2^{n+1}} \rtimes C_{2^m}$$

for some $n, m \geq 2$. Then $k(B) = 2^{m+1}(2^{n-2} + 1)$, $k_0(B) = 2^{m+1}$, $k_1(B) = 2^{m-1}(2^n - 1)$, $k_n(B) = 2^{m-1}$ and $l(B) = 2$. In particular Brauer's $k(B)$-Conjecture and Alperin's Weight Conjecture are satisfied.

Proof Let $\mathscr{F}$ be the fusion system of B, and let $y := v^{2^{n-2}}$ and $z := x^2$. Then by Theorem 10.17, $Q := \langle x, y, a^2 \rangle \cong Q_{2^{n+1}} \times C_{2^{m-1}}$ is the only $\mathscr{F}$-essential subgroup up to conjugation. Again we use Theorem 1.35 to get a lower bound for $k(B)$.

The same argument as in Theorem 10.23 shows that D has $2^{m-1}(2^n + 3)$ conjugacy classes and we need to know which of them are fused in Q. It is easy to see that xa^{2j} is conjugate to ya^{2j} under $\mathrm{Aut}_{\mathscr{F}}(Q)$ for $j \in \mathbb{Z}$. Since $\mathrm{Z}(\mathscr{F}) = \langle z, a^2 \rangle$, this is the only fusion which occurs. Hence, the number of $\mathscr{F}$-conjugacy classes is again $2^m(2^{n-1} + 1)$. In case $n = 2$ the group $D/\langle z \rangle \cong C_2^2 \rtimes C_{2^m}$ is minimal non-abelian, and we get $l(b_z) = 2$ from Theorem 12.4 below. Otherwise, $D/\langle z \rangle$ is isomorphic to one of the groups in Theorem 10.23. Hence, again $l(b_z) = 2$. As usual the groups $D/\langle a^2 \rangle$ and $D/\langle a^2 z \rangle$ have maximal class and it follows that $l(b_{a^2}) = l(b_{a^2 z}) = 2$. The same holds for all odd powers of a^2 and $a^2 z$. For $2 \leq j \leq m - 1$ the group $D/\langle u \rangle$ with $u := a^{2^j}$ has the same isomorphism type as D where m has to be replaced by j. So induction on m shows $l(b_u) = 2$. It remains to deal with $u := a^{2^j} z$. Here $D/\langle u \rangle \cong Q_{2^{n+1}}.C_{2^j}$ is exactly the group from Theorem 10.24. Thus, for $j \neq n$ we have again $l(b_u) = 2$. In case $j = n$, $D/\langle u \rangle \cong C_{2^n} \wr C_2$. Then (7.G) in [160] gives $l(b_u) = 2$ as well. Now Theorem 1.35 reveals

$$k(B) \geq 2^{m+1} + 2^m(2^{n-1} + 1) - 2^m = 2^{m+1}(2^{n-2} + 1).$$

For the opposite inequality we apply Theorem 4.2 to the major subsection (u, b_u) where $u := a^2 z$. A similar calculation as in Theorem 10.24 shows that $D/\langle u \rangle \cong Q_{2^{n+2}}$. Hence, the Cartan matrix of b_u is given by

$$2^m \begin{pmatrix} 2^{n-1} + 1 & 2 \\ 2 & 4 \end{pmatrix}$$

up to basic sets (see Theorem 8.1). This is the same matrix as in Theorem 10.24, but the following discussion is slightly different, because a^2 has only order 2^{m-1} here. So we copy the proof of Theorem 9.28. In fact we just have to replace m by $m + 1$ and n by $n - 2$ in order to use this proof word by word. The claim follows. □

We describe the structure of these group extensions in a more generic way.

Proposition 10.26 *Let D be an extension of the cyclic group $\langle a \rangle \cong C_{2^n}$ by a group M which has maximal class or is the four-group. Suppose that the corresponding coupling $\omega : \langle a \rangle \to \mathrm{Out}(M)$ satisfies the following: If $\omega \neq 0$, then the coset $\omega(a)$ of $\mathrm{Inn}(M)$ contains an involution which acts non-trivially on $M/\Phi(M)$. Moreover, assume that $D \ncong C_{2^m} \wr C_2$ for all $m \geq 3$. Then the invariants for every block of a finite group with defect group D are known.*

Proof Assume first that $M \cong C_2^2$. Then in case $\omega = 0$ we get the groups $C_{2^n} \times C_2^2$ and $C_{2^{n+1}} \times C_2$ for which the block invariants can be calculated by Usami [270] and Kessar et al. [148]. So let $\omega \neq 0$. If D is non-split, it must contain a cyclic maximal subgroup. In particular, D is metacyclic and the block invariants are known. If the extension splits, then we obtain the minimal non-abelian group $C_2^2 \rtimes C_{2^n}$. Here the block invariants are known by results from the author's dissertation [244] (see Chap. 12).

Hence, let M be a 2-group of maximal class. Then $|Z(M)| = 2$. Thus, for $\omega = 0$ we obtain precisely two extensions for every group M. All these cases were handled in Chap. 9. Let us now consider the case $\omega \neq 0$. Since the three maximal subgroups of a semidihedral group are pairwise non-isomorphic, M must be a dihedral or quaternion group. Write $M = \langle v, x \mid v^{2^m} = 1, x^2 \in \langle v^{2^{m-1}} \rangle, {}^x v = v^{-1} \rangle$. Let $\alpha \in \mathrm{Aut}(M)$ be an involution which acts non-trivially on $M/\Phi(M)$. Then there is an odd integer i such that ${}^\alpha x = v^i x$. Since $\alpha^2 = 1$, it follows that ${}^\alpha v = v^{-1}$. Hence, the coset $\alpha \mathrm{Inn}(M) \in \mathrm{Out}(M)$ is determined uniquely. Hence, ω is unique. So we get four group extensions for every pair (n, m). Two of them are isomorphic and all cases are covered in Theorems 10.23–10.25 (and [160] for $C_4 \wr C_2$). □

Chapter 11
Defect Groups of p-Rank 2

In this chapter which is taken from [114,252] we will use Theorem 4.12 to show that Olsson's Conjecture is satisfied for controlled blocks B with certain defect groups D. Recall that in this situation all subgroups of D are fully $\mathscr{F}$-normalized where $\mathscr{F}$ is the fusion system of B. In particular for a subsection (u, b_u) the block b_u has defect group $C_D(u)$ (cf. Lemma 1.34). Our strategy will be to find a subsection (u, b_u) such that $l(b_u) = 1$ and $|C_D(u)| = |D : D'|$. Then Olsson's Conjecture follows from Theorem 4.12. Observe that the inequality $|D : C_D(u)| \le |D'|$ always holds by elementary group theory. The next proposition gives a general criterion for this situation.

Proposition 11.1 *Let B be a controlled p-block of a finite group with defect group D. Suppose that there exists an element $u \in D$ such that $|D : C_D(u)| = |D'|$ and $C_{\mathrm{Aut}_{\mathscr{F}}(D)}(u)$ is a p-group. Then Olsson's Conjecture holds for B.*

Proof By Proposition 2.1 in [7], b_u is a controlled block. Thus, it suffices to show $e(b_u) = 1$ (see Theorem 4.12). Let $\mathscr{F}$ be the fusion system of B. Since $\mathscr{F}$ is controlled, b_u has defect group $C_D(u)$ and fusion system $\mathscr{C} := C_{\mathscr{F}}(\langle u \rangle)$ (see Lemma 1.34). Hence, every automorphism in $\mathrm{Aut}_{\mathscr{C}}(C_D(u))$ extends to an element of $\mathrm{Aut}_{\mathscr{F}}(D)$. By hypothesis, $\mathrm{Aut}_{\mathscr{C}}(C_D(u))$ is a p-group, and the claim follows. $\square$

Since the inertial quotient $\mathrm{Out}_{\mathscr{F}}(D)$ is always a p'-group, we can formulate Proposition 11.1 in the following abstract setting. Let P be a finite p-group and let A be a p'-group of automorphisms on P. Then we can form the semidirect product $G := P \rtimes A$. The conclusion of Proposition 11.1 applies if we find an element $u \in P$ such that $|P : C_P(u)| = |P'|$ and $C_G(u) \le P$. Observe that the requirement $C_A(u) = 1$ alone is not sufficient, since for a P-conjugate v of u we might have $C_A(v) \ne 1$. In the following results we verify this condition for several families of 2-generator p-groups. Most ideas here are due to Héthelyi and Külshammer. We start with a useful lemma.

© Springer International Publishing Switzerland 2014 159
B. Sambale, *Blocks of Finite Groups and Their Invariants*, Lecture Notes in Mathematics 2127, DOI 10.1007/978-3-319-12006-5_11

Lemma 11.2 *Let P be a p-group such that $|P : \Phi(P)| \leq p^2$. Let $A \leq \operatorname{Aut}(P)$ be a p'-group and $G = P \rtimes A$. If P contains an A-invariant maximal subgroup C, then there is an element $u \in P \setminus C$ such that $\mathrm{C}_G(u) \leq P$.*

Proof In case $|P : \Phi(P)| = p$ the claim is trivial. Hence, assume $|P : \Phi(P)| = p^2$. By Maschke's Theorem there is another A-invariant maximal subgroup C_1 of P. Let $u \in P \setminus (C \cup C_1)$. Then $\mathrm{C}_A(u)$ acts trivially on $\langle u \rangle \Phi(P)/\Phi(P)$. Since $P/\Phi(P) = C/\Phi(P) \times C_1/\Phi(P)$, it follows that $\mathrm{C}_A(u)$ acts trivially on $C/\Phi(P)$ and on P/C. This shows $\mathrm{C}_A(u) = 1$, because A is a p'-group. By way of contradiction assume that $\mathrm{C}_G(u)$ is not a p-group. Let $\alpha \in \mathrm{C}_G(u)$ be a non-trivial p'-element. By Schur-Zassenhaus α is P-conjugate to an element of A. In particular $\mathrm{C}_A(v) \neq 1$ for some P-conjugate v of u. However, this contradicts the first part of the proof, since $v \in P \setminus (C \cup C_1)$. $\qquad\square$

Proposition 11.3 *Let p be an odd prime, and let P be a p-group of maximal class with $|P| \geq p^4$. If $A \leq \operatorname{Aut}(P)$ is a p'-group and $G = P \rtimes A$, then there exists an element $u \in P$ such that $|P : \mathrm{C}_P(u)| = |P'|$ and $\mathrm{C}_G(u) \leq P$.*

Proof Let $|P| = p^n$. By Hilfssatz III.14.4 in [128], $P_1 := \mathrm{C}_P(\mathrm{K}_2(P)/\mathrm{K}_4(P))$ is a characteristic maximal subgroup of P. Moreover, Hauptsatz III.14.6(a) tells us that the set $\{\mathrm{C}_P(\mathrm{K}_i(P)/\mathrm{K}_{i+2}(P)) : 2 \leq i \leq n - 2\}$ contains at most one subgroup $C := \mathrm{C}_P(\mathrm{K}_{n-2}(P)) < P$ different from P_1. By (the proof of) Lemma 11.2 there exists an element $u \in P \setminus (P_1 \cup C)$ such that $\mathrm{C}_G(u) \leq P$. By Hilfssatz III.14.13 in [128] we also have $|P : \mathrm{C}_P(u)| = |P'|$. $\qquad\square$

Proposition 11.4 *Let p be an odd prime, and let P be a p-group such that P' is cyclic, $|P : \Phi(P)| = p^2$ and $|P| \geq p^4$. If $A \leq \operatorname{Aut}(P)$ is a p'-group and $G = P \rtimes A$, then there exists an element $u \in P$ such that $|P : \mathrm{C}_P(u)| = |P'|$ and $\mathrm{C}_G(u) \leq P$.*

Proof Assume first that P is abelian. By Lemma 11.2 we may assume $P \cong C_{p^s} \times C_{p^s}$ for some $s \geq 2$. Since $\mathrm{C}_G(u) = P\,\mathrm{C}_A(u)$ for all $u \in P$, it suffices to show $\mathrm{C}_A(u) = 1$ for some $u \in P$. After replacing P by $\Omega_2(P)$, we may also assume that $s = 2$. Let $x \in P \setminus \Phi(P)$. Suppose that $A_1 := \mathrm{C}_A(x) \neq 1$. Since A_1 acts faithfully on $\Omega_1(P)$, we have $\mathrm{C}_P(A_1) = \langle x \rangle$. The group $A_2 := \mathrm{C}_A(x^p)$ must be cyclic, since it acts faithfully on $\Omega_1(P)/\langle x^p \rangle$. Thus, it follows from $A_1 \leq A_2$ that A_2 acts on $\langle x \rangle = \mathrm{C}_P(A_1)$. But since A_2 fixes $x^p \in \Omega_1(\langle x \rangle)$, we derive $A_1 = A_2$. Now choose an element $u \in P$ such that $\Omega_1(P) \subseteq \langle x, u \rangle$ and $\langle u^p \rangle = \langle x^p \rangle$. Then $\mathrm{C}_A(u) = \mathrm{C}_A(u) \cap \mathrm{C}_A(u^p) = \mathrm{C}_{A_2}(u) = \mathrm{C}_{A_1}(u) \subseteq \mathrm{C}_A(\Omega_1(P)) = 1$.

Now suppose that P has class 2. Then for $P = \langle a, b \rangle$ we have $P' = \langle [a, b] \rangle = \{[a, b^n] : n \in \mathbb{Z}\} = \{[a, x] : x \in P\}$. In particular, $|P : \mathrm{C}_P(u)| = |P'|$ for all $u \in P \setminus \Phi(P)$. Hence, it suffices to show $\mathrm{C}_A(u) = 1$ for all u in a certain P-conjugacy class lying in $P \setminus \Phi(P)$ (compare with proof of Lemma 11.2). For this we may replace P by P/P'. In case $|P : P'| > p^2$ the claim follows from the arguments above. Thus, assume $|P : P'| = p^2$. Then $P' = \mathrm{Z}(P)$ and $|P'| = p$. This contradicts $|P| \geq p^4$. $\qquad\square$

Finally let P be a group of class at least 3. Then $P' \not\subseteq Z(P)$ and $1 \neq P/C_P(P') \leq \mathrm{Aut}(P')$ is cyclic. Hence, $C := C_P(P')\Phi(P)$ is a characteristic maximal subgroup of P. By Lemma 11.2 there is an element $u \in P \setminus C$ such that $C_G(u) \leq P$. Choose $x \in C_P(P')$ such that $P = \langle u, x \rangle$. Now $N := \langle x \rangle P'$ is an abelian normal subgroup of P, and $P/N = \langle uN \rangle$ is cyclic. Thus, Aufgabe 2 on page 259 of [128] implies that $P' = \{[y, u] : y \in N\} = \{[y, u] : y \in P\}$; in particular, we have $|P'| = |P : C_P(u)|$. $\qquad\square$

We observe that $\mathrm{GL}(2, p)$ contains a p'-subgroup A of order $2(p - 1)^2$ which is bigger than p^2 for $p > 3$. Thus, when P is elementary abelian of order p^2, then there is no regular orbit of A on P.

Next we turn to p-groups of p-rank 2. For the convenience of the reader we recall Blackburn's classification of these groups from Theorems A.1 and A.2 in [67].

Theorem 11.5 (Blackburn) *Let P be p-group of p-rank at most 2 for an odd prime p. Then one of the following holds:*

(i) *P is metacyclic.*
(ii) *$P \cong C(p, n) := \langle a, b, c \mid a^p = b^p = c^{p^{n-2}} = [a, c] = [b, c] = 1, [a, b] = c^{p^{n-3}} \rangle$ for some $n \geq 3$.*
(iii) *$P \cong G(p, n, \epsilon) := \langle a, b, c \mid a^p = b^p = c^{p^{n-2}} = [b, c] = 1, [a, b^{-1}] = c^{\epsilon p^{n-3}}, [a, c] = b \rangle$ where $n \geq 4$ and ϵ is 1 or a quadratic non-residue modulo p.*
(iv) *P is a 3-group of maximal class, but not $C_3 \wr C_3$. More precisely, $P \cong B(3, n; \beta, \gamma, \delta)$ is defined by generators $s, s_1, \ldots, s_{n-1}$ and relations*

- *$s_i = [s_{i-1}, s]$ for $i = 2, 3, \ldots, n - 1$,*
- *$[s_1, s_2] = s_{n-1}^{\beta}$,*
- *$[s_1, s_i] = 1$ for $i = 3, 4, \ldots, n - 1$,*
- *$s^3 = s_{n-1}^{\delta}$,*
- *$s_1^3 s_2^3 s_3 = s_{n-1}^{\gamma}$,*
- *$s_i^3 s_{i+1}^3 s_{i+2} = 1$ for $i = 2, 3, \ldots, n - 1$ where $s_n := s_{n+1} := 1$.*

Moreover, $|P| = 3^n$ and one of the following holds

- *$n \geq 4$ and $(\beta, \gamma, \delta) = (0, 0, 1)$,*
- *$n \geq 5$ and $(\beta, \gamma, \delta) \in \{(0, 0, 0), (0, 1, 0), (1, 0, 0), (1, 0, 1), (1, 0, 2)\}$,*
- *$n \geq 6$ is even and $(\beta, \gamma, \delta) = (0, 2, 0)$.*

Proposition 11.6 *Let p be an odd prime, and let P be a p-group of p-rank 2 with $|P| \geq p^4$. If $A \leq \mathrm{Aut}(P)$ is a p'-group and $G = P \rtimes A$, then there exists an element $u \in P$ such that $|P : C_P(u)| = |P'|$ and $C_G(u) \leq P$.*

Proof By Theorem 11.5, there are four cases to consider. The metacyclic case follows from Proposition 11.4. If P is a 3-group of maximal class, then the result holds by Proposition 11.3.

Now suppose that $P \cong C(p, n)$ for some $n \geq 4$. Then it is easy to see that $P = \Omega_1(P) * Z(P)$, where $\Omega_1(P) = \langle a, b \rangle$ is a non-abelian group of order p^3 and exponent p, and $Z(P) = \langle c \rangle$ is cyclic of order p^{n-2}. Thus, $|P'| = p$. Then

$$\mathcal{U} := \{x \in P \setminus Z(P) : |\langle x \rangle| = p^{n-2}\} \neq \varnothing.$$

For $u \in \mathcal{U}$ we have $C_A(u) \leq C_A(u^p) = C_A(c^p)$. Hence, $C_A(u)$ acts trivially on $Z(P) = \langle c \rangle$ and on $\langle u, c \rangle$. Now Problem 4D.1 in [132] implies $C_A(u) = 1$ for all $u \in \mathcal{U}$. Since $\mathcal{U}$ is closed under conjugation in P, we obtain $C_G(u) \leq P$ easily (compare with proof of Lemma 11.2). Obviously, we also have $|P : C_P(u)| = p = |P'|$ for all $u \in \mathcal{U}$.

Finally, it remains to handle the case $P \cong G(p, n, \epsilon)$. Obviously, $P = \langle a, c \rangle$ and $P' = \langle b, c^{p^{n-3}} \rangle \cong C_p \times C_p$. Moreover, $C_P(P') = \langle b, c \rangle$ is abelian and maximal in P. Hence, by Lemma 11.2 we find an element $u \in P \setminus C_P(P')$ such that $C_G(u) \leq P$. It remains to show $|P : C_P(u)| = |P'|$. By way of contradiction suppose that $C_P(u)$ is maximal in P. Then $\Phi(P) = C_P(P') \cap C_P(u) \subseteq C_P(\langle C_P(P'), u \rangle) = Z(P)$. Thus, P is minimal non-abelian and we get the contradiction $|P'| = p$. This completes the proof. $\square$

Theorem 11.7 *Let D be a finite p-group, where p is an odd prime, and suppose that one of the following holds:*

 (i) *D has p-rank 2,*
 (ii) *D has maximal class,*
(iii) *D' is cyclic and $|D : \Phi(D)| = p^2$.*

Then Olsson's Conjecture holds for all controlled blocks with defect group D.

Proof In case $|D| \leq p^3$ the claim follows easily from Proposition 4.3 (observe that D is not elementary abelian of order p^3). The other cases are consequences of the previous propositions. $\square$

In connection with (11.7) in Theorem 11.7 we mention that by a result of Burnside, D' is already cyclic if $Z(D')$ is (see Satz III.7.8 in [128]).

If u is an element of D such that $|D : C_D(u)| = |D'|$, then $D' = \{[u, v] : v \in D\}$; in particular, every element in D' is a commutator. Thus, one cannot expect to prove Olsson's Conjecture for all possible defect groups in this way (see for example [100]).

Now we discuss Olsson's Conjecture for blocks which are not necessarily controlled. We begin with a special case for which the previous method does not suffice. For this reason we use the classification of finite simple groups.

Proposition 11.8 *Let B be a block of a finite group G with a non-abelian defect group D of order 5^3 and exponent 5. Suppose that the fusion system $\mathcal{F}$ of B is the same as the fusion system of the sporadic simple Thompson group Th for the prime 5. Then B is Morita equivalent to the principal 5-block of Th. In particular, Olsson's Conjecture holds for B.*

Proof By the Second Fong Reduction, we may assume that $O_{5'}(G)$ is central and cyclic. The ATLAS [59] shows that Th has a unique conjugacy class of elements of order 5. Thus, by our hypothesis, all non-trivial B-subsections are conjugate in G. In particular, all B-subsections are major. Since $O_5(G) \leq D$, this implies that $O_5(G) = 1$. Thus, $F(G) = Z(G) = O_{5'}(G)$.

Let $N/Z(G)$ be a minimal normal subgroup of $G/Z(G)$. By the First Fong Reduction, we may assume that B covers a unique block b of N. Then $D \cap N$ is a defect group of b. We may also assume that $D \cap N \neq 1$. Since all non-trivial B-subsections are conjugate in G this implies that $D \cap N = D$, i.e. $D \subseteq N$. In particular, $N/Z(G)$ is the only minimal normal subgroup of $G/Z(G)$. Hence, $N = F^*(G)$, and $E(G)$ is a central product of the components $L_1, \ldots, L_n$ of G.

For $i = 1, \ldots, n$, b covers a unique block b_i of L_i. Let D_i be a defect group of b_i. Then $D \cong D_1 \times \ldots \times D_n$ by Lemma 7.5. This shows that we must have $n = 1$. Hence, $E(G)$ is quasisimple, and $S := E(G)/Z(E(G))$ is simple. Since $F^*(G) = E(G) F(G) = E(G) Z(G)$, we conclude that $C_G(E(G)) = C_G(F^*(G)) = Z(F(G)) = Z(G)$, so that $G/Z(G)$ is isomorphic to a subgroup of $\mathrm{Aut}(E(G))$.

Now we discuss the various possibilities for S, by making use of the classification of finite simple groups. In each case we apply [12].

If S is an alternating group then, by Sect. 2 in [12], the block b cannot exist. Similarly, if S is exceptional group of Lie type then, by Theorem 5.1 in [12], the block b cannot exist.

Now suppose that S is a classical group. Then, by Theorem 4.5 in [12], $p = 5$ must be the defining characteristic of S. Moreover, S has to be isomorphic to $\mathrm{PSL}(3, 5)$ or $\mathrm{PSU}(3, 5)$. Also, D is a Sylow 5-subgroup of $E(G)$. But now the ATLAS shows that S contains non-conjugate elements $\overline{x}$ and $\overline{y}$ of order 5 such that $|C_S(\overline{x})| \neq |C_S(\overline{y})|$. Thus, there are elements x and y of order 5 in $E(G)$ which are not conjugate in G. This contradicts the fact that all non-trivial B-subsections are conjugate in G.

The only remaining possibility is that S is a sporadic simple group. Then Table 1 in [12] implies that $S \in \{HS, McL, Ru, Co_2, Co_3, Th\}$. In all cases D is a Sylow 5-subgroup of S. In the first five cases we derive a contradiction as above, using the ATLAS. So we may assume that $S = Th$. Since Th has trivial Schur multiplier and trivial outer automorphism group, we must have $G = S \times Z(G)$. Thus, $B \cong b \otimes_{\mathcal{O}} \mathcal{O} \cong b$, and b is the principal 5-block of Th, by Uno [269]. Moreover, we have $k_0(B) = k_0(b) = 20 \leq |D : D'|$. This completes the proof. $\square$

Theorem 11.9 *Let $p > 3$. Then Olsson's Conjecture holds for all p-blocks with defect groups of p-rank 2.*

Proof Let B be a p-block with defect group D of p-rank 2 for $p > 3$. Then, by the Theorems 4.1–4.3 in [67], B is controlled unless D is non-abelian of order p^3 and exponent p (see also [265]). Hence, by Theorem 11.7 we may assume that D is non-abelian of order p^3 and exponent p.

If in addition $p > 7$, Hendren [108] has shown that there is at least one non-major B-subsection. In this case the result follows easily from Proposition 4.3. Now let $p = 7$. Then the fusion system $\mathcal{F}$ of B is one of the systems given in [241]. Kessar and Stancu showed using the classification of finite simple groups that three of them cannot occur for blocks (see [153]). In the remaining cases the number of $\mathcal{F}$-radical and $\mathcal{F}$-centric subgroups of D is always less than $p + 1 = 8$. In particular, there is an element $u \in D \setminus Z(D)$ such that $\langle u \rangle Z(D)$ is not $\mathcal{F}$-radical, $\mathcal{F}$-centric. Then by Alperin's Fusion Theorem, $\langle u \rangle$ is not $\mathcal{F}$-conjugate to $Z(D)$. Hence, the subsection (u, b_u) is non-major, and Olsson's Conjecture follows from Proposition 4.3.

In case $p = 5$ the same argument shows that we can assume that $\mathcal{F}$ is the fusion system of the principal 5-block of Th. However, in this case Olsson's Conjecture holds by Proposition 11.8. $\square$

As usual we denote the non-abelian (extraspecial) group of order p^3 and exponent p by p_+^{1+2}. For $p = 3$, there are two fusion systems on p_+^{1+2} in [241] such that all subsections are major. These correspond to the simple groups $^2F_4(2)'$ and J_4. It appears to be very difficult to prove Olsson's Conjecture for these fusion systems. Using the Cartan method (plus additional arguments) I was able to show $k(B) \leq 15$ for the fusion system of $^2F_4(2)'$. However, Olsson's Conjecture holds for the 3-blocks of $^2F_4(2)'$, $^2F_4(2)$, J_4, Ru and $2.Ru$ (see [5,6,15,16]; cf. Remark 1.3 in [241]). More generally, Olsson's Conjecture is known to hold for all principal blocks with defect group 3_+^{1+2} by Remark 64 in [198]. In addition to 3-blocks of defect 3, there are infinitely many non-controlled 3-blocks whose defect groups have 3-rank 2. In the following we consider these cases in detail. The results come from [252].

Proposition 11.10 *Let B be a 3-block of a finite group with defect group D. Assume that D has 3-rank 2, but not maximal class. Then Olsson's Conjecture holds for B.*

Proof By Theorem 11.7 we may assume that the fusion system $\mathcal{F}$ of B is not controlled. Then $|D| \geq 3^4$, since D does not have maximal class. By Theorems 4.1 and 4.2 in [67] it remains to handle the groups $D = G(3, r, \epsilon)$ of order 3^r where $r \geq 5$ and $\epsilon \in \{\pm 1\}$ as in Theorem 4.7 in [67] (by Remark A.3 in [67], $G(3, 4, \epsilon)$ has maximal class). Assume the notation of Theorem 11.5. Consider the element $x := ac$. By Lemma A.8 in [67], x is not contained in the unique $\mathcal{F}$-essential ($\mathcal{F}$-Alperin) subgroup $C(3, r - 1) = \langle a, b, c^3 \rangle$. In particular, $\langle x \rangle$ is fully $\mathcal{F}$-centralized, and the block b_x of the subsection (x, b_x) has defect group $C_D(x)$. It is easy to see that $D' = \langle b, c^{3^{r-3}} \rangle \cong C_p \times C_p$. It follows that $x^{3^{r-4}} \equiv c^{3^{r-4}} \not\equiv 1$ (mod D') and $|\langle x \rangle| \geq 3^{r-3}$. As usual we have $|C_D(x)| \geq |D : D'| = 3^{r-2}$. In case $|C_D(x)| \geq 3^{r-1}$ we get the contradiction $b \in D' \subseteq C_D(x)$. Hence, $|C_D(x)| = |D : D'|$ and $C_D(x)/\langle x \rangle$ is cyclic. Now Olsson's Conjecture for B follows from Proposition 4.3. $\square$

The next theorem says that for a given defect group order, we can prove Olsson's Conjecture for all but one defect group.

Theorem 11.11 *Let B be a 3-block of a finite group with defect group D of 3-rank 2. Assume that D is neither isomorphic to 3_+^{1+2} nor to $B(3, n; 0, 0, 0)$ for some $n \geq 4$. Then Olsson's Conjecture holds for B.*

Proof By Proposition 11.10 we may assume that D has maximal class of order at least 3^4. By Theorem 11.7 we may assume that the fusion system $\mathcal{F}$ of B is not controlled. Then $\mathcal{F}$ is given as in Theorem 5.10 in [67]. In particular $D = B(3, r; 0, \gamma, 0)$ where $\gamma \in \{1, 2\}$. Let $D_1 := C_D(K_2(D)/ K_4(D))$. Observe that in the notation of [28, 67] we have $D_1 = \gamma_1(D)$. Proposition A.9 in [67] shows $x := ss_1 \notin D_1$. Moreover, we have $x^3 \neq 1$ also by Proposition A.9 in [67]. Then by Lemma A.15 in [67], x does not lie in one of the centric subgroups D_1, E_i or V_i for $i \in \{-1, 0, 1\}$. This shows that x is not $\mathscr{F}$-conjugate to an element in D_1. By Satz III.14.17 in [128], D is not an exceptional group. In particular, Hilfssatz III.14.13 in [128] implies $|C_D(y)| = 9 = |D : D'|$ for all $y \in D \setminus D_1$. Hence, $\langle x \rangle$ is fully $\mathscr{F}$-centralized. Thus, the block b_x of the subsection (x, b_x) has defect group $C_D(x)$. Now Olsson's Conjecture follows from Proposition 4.3. $\square$

We remark that the method in Theorem 11.11 does not work for the groups $B(3, r; 0, 0, 0)$. For example, every block of a subsection of the principal 3-block of $^3D_4(2)$ has defect at least 3 (here $r = 4$). However, $|D : D'| = 3^2$ for every 3-group of maximal class.

Chapter 12
Minimal Non-abelian Defect Groups

A non-abelian group G is *minimal non-abelian* if all its proper subgroups are abelian.

Lemma 12.1 *A finite p-group P is minimal non-abelian if and only if P has rank 2 and $|P'| = p$.*

Proof Assume first that P is minimal non-abelian. Choose two non-commuting elements $x, y \in P$. Then $\langle x, y \rangle$ is non-abelian and $P = \langle x, y \rangle$ has rank 2. Every element $x \in P$ lies in a maximal subgroup $M \le P$. Since M is abelian, $M \subseteq C_P(x)$. In particular, all conjugacy classes of P have length at most p. By a result of Knoche (see Aufgabe III.24b) in [128]) we obtain $|P'| = p$.

Next, suppose that P has rank 2 and $|P'| = p$. Then $P' \le Z(P)$. For $x, y \in P$ we have $[x^p, y] = [x, y]^p = 1$ (see Hilfssatz III.1.3 in [128]). Hence, $\Phi(P) = P'\langle x^p : x \in P \rangle \le Z(P)$. Since P is non-abelian, we obtain $\Phi(P) = Z(P)$. For any maximal subgroup $M \le P$ it follows that $|M : Z(P)| = |M : \Phi(P)| = p$. Therefore, M is abelian and P is minimal non-abelian. □

Rédei [242] classified all minimal non-abelian p-groups as follows.

Theorem 12.2 (Rédei) *Every minimal non-abelian p-group is isomorphic to one of the following groups*

1. $\langle x, y \mid x^{p^r} = y^{p^s} = 1, \ xyx^{-1} = y^{1+p^{r-1}} \rangle$ *for* $r \ge 2$ *and* $s \ge 1$,
2. $MNA(r, s) := \langle x, y \mid x^{p^r} = y^{p^s} = [x, y]^p = [x, x, y] = [y, x, y] = 1 \rangle$ *for* $r \ge s \ge 1$,
3. Q_8.

It can be seen that the groups in Theorem 12.2 are metacyclic except in case (ii). For the metacyclic, minimal non-abelian groups with have complete information by Theorems 8.1 and 8.13. In case (ii) we say that P is of *type* (r, s). The group structure is clarified by the following result.

B. Sambale, *Blocks of Finite Groups and Their Invariants*, Lecture Notes in Mathematics 2127, DOI 10.1007/978-3-319-12006-5_12

Lemma 12.3 *Let P be a minimal non-abelian group of type (r, s). Then the following holds:*

(i) $|P| = p^{r+s+1}$.
(ii) $\Phi(P) = Z(P) = \langle x^2, y^2, [x, y] \rangle \cong C_{p^{r-1}} \times C_{p^{s-1}} \times C_p$.
(iii) $P' = \langle [x, y] \rangle \cong C_p$.

Proof It is easy to see that $|P| \leq p^{r+s+1}$. Conversely, Rédei constructed groups of order p^{r+s+1} with the given generators and relations. Hence, $|P| = p^{r+s+1}$. The other properties can be easily verified.   □

It seems natural to compute the invariants of blocks with minimal non-abelian defect groups. For $p = 2$ this project was started in the author's dissertation [244] (see also [243]) and later completed in [75]. Preliminary work was done by Olsson [214]. For primes $p > 2$ we present a minor result from [114].

12.1   The Case $p = 2$

First, we state the main result of this section.

Theorem 12.4 *Let B be a 2-block of a finite group G with a minimal non-abelian defect group D. Then one of the following holds:*

1. *B is nilpotent. Then $k(B) = \frac{5}{8}|D|$, $k_0(B) = \frac{1}{2}|D|$, $k_1(B) = \frac{1}{8}|D|$ and $l(B) = 1$. Moreover, $|\{\chi(1) : \chi \in \mathrm{Irr}(B)\}| = 2$.*
2. *$|D| = 8$. Then Theorem 8.1 applies.*
3. *$D \cong MNA(r, 1)$ for some $r \geq 2$. Then $k(B) = 5 \cdot 2^{r-1}$, $k_0(B) = 2^{r+1}$, $k_1(B) = 2^{r-1}$ and $l(B) = 2$. The decomposition and Cartan matrices of B are given by*

$$\begin{pmatrix} 1 \cdots 1 \ . \ \cdots \ . \ 1 \cdots 1 \\ . \ \cdots \ . \ 1 \cdots 1 \ 1 \cdots 1 \end{pmatrix}^T \qquad and \qquad 2^{r-1}\begin{pmatrix} 3 & 1 \\ 1 & 3 \end{pmatrix}$$

   up to basic sets (here the characters are ordered with respect to their heights). Moreover, $\mathrm{Irr}_0(B)$ contains four 2-rational characters and two families of 2-conjugate characters of size 2^i for $i = 1, \ldots, r - 1$. The characters of height 1 split into two 2-rational characters and one family of 2-conjugate characters of size 2^i for $i = 2, \ldots, r - 2$. Also, the characters of height 1 have the same degree and $|\{\chi(1) : \chi \in \mathrm{Irr}_0(B)\}| \leq 2$.
4. *$D \cong MNA(r, r)$ for some $r \geq 2$. Then B is Morita equivalent to $\mathcal{O}[D \rtimes C_3]$. In particular, $k(B) = (5 \cdot 2^{2r-2} + 16)/3$, $k_0(B) = (2^{2r} + 8)/3$, $k_1(B) = (2^{2r-2} + 8)/3$ and $l(B) = 3$. Moreover, the Cartan matrix of B is*

$$\frac{2}{3}\begin{pmatrix} 2^{2r} + 2 & 2^{2r} - 1 & 2^{2r} - 1 \\ 2^{2r} - 1 & 2^{2r} + 2 & 2^{2r} - 1 \\ 2^{2r} - 1 & 2^{2r} - 1 & 2^{2r} + 2 \end{pmatrix}$$

   up to basic sets.

Let B be as in Theorem 12.4. If D is metacyclic, we see from Theorems 12.2 and 8.1 that B is nilpotent or $|D| = 8$. Hence, assume that B is nilpotent. Since $|D'| = 2$, we get $k_0(B) = |D : D'| = \frac{1}{2}|D|$. Since $|D : Z(D)| = 4$, the number of conjugacy classes of D is

$$k(B) = k(D) = |Z(D)| + \frac{|D| - |Z(D)|}{2} = \frac{5}{8}|D|.$$

Now $|D|$ is the square sum of $k(D)$ character degrees. This shows $k_1(B) = k_1(D) = k(D) - k_0(D) = \frac{1}{8}|D|$. The claim about the character degrees of $\mathrm{Irr}(B)$ follows from the action of $D/\mathfrak{foc}(B) = D/D'$ (see Proposition 1.31). This proves the first two parts of Theorem 12.4. Thus for the remainder of the section, we may assume $D \cong MNA(r, s)$ with $r \geq 2$. The following results are extracted from [243].

Lemma 12.5 *The automorphism group* $\mathrm{Aut}(D)$ *is a 2-group, if and only if $r \neq s$.*

Proof It is easy to see using Lemma 12.3 that the maximal subgroups of D are isomorphic to $C_{2^{r-1}} \times C_{2^s} \times C_2$, $C_{2^r} \times C_{2^{s-1}} \times C_2$ and $C_{2^r} \times C_{2^{s-1}} \times C_2$ respectively. If $r \neq s$, then an automorphism of order 3 on D cannot permutes these maximal subgroups. Hence, in these cases $\mathrm{Aut}(D)$ must be a 2-group. Thus, we may assume $r = s \geq 2$. Then one can show that the map $x \mapsto y$, $y \mapsto x^{-1}y^{-1}$ is an automorphism of order 3. $\qquad\square$

Lemma 12.6 *Let* $P \cong C_{2^{n_1}} \times \ldots \times C_{2^{n_k}}$ *with* $n_1, \ldots, n_k, k \in \mathbb{N}$. *Then* $\mathrm{Aut}(P)$ *is a 2-group, if and only if the n_i are pairwise distinct.*

Proof See for example Lemma 2.7 in [214]. $\qquad\square$

Proposition 12.7 *Let* $\mathscr{F}$ *be a fusion system on* D. *Then one of the following holds*

   (i) $\mathscr{F}$ *is nilpotent,*
   (ii) $r = s \geq 2$ *and* $\mathscr{F} = \mathscr{F}_D(D \rtimes C_3)$ *is controlled,*
   (iii) $s = 1$ *and* $\mathscr{F} = \mathscr{F}_D(A_4 \rtimes C_{2^r})$ *is constrained.*

Proof Let $Q < D$ be an $\mathscr{F}$-essential subgroup. Since Q is also $\mathscr{F}$-centric, we get $C_P(Q) = Q$. This shows that Q is a maximal subgroup of D. By Lemma 12.6, one of the following holds:

   (i) $r = s = 2$ and $Q \in \{\langle x^2, y, z \rangle, \langle x, y^2, z \rangle, \langle xy, x^2, z \rangle\}$,
   (ii) $r > s = 2$ and $Q \in \{\langle x, y^2, z \rangle, \langle xy, x^2, z \rangle\}$,
   (iii) $r = s + 1$ and $Q = \langle x^2, y, z \rangle$,
   (iv) $s = 1$ and $Q = \langle x^2, y, z \rangle$.

We show that the first three cases cannot occur. In all these cases $\Omega(Q) \subseteq Z(P)$. Let us consider the action of $\mathrm{Aut}_{\mathscr{F}}(Q)$ on $\Omega(Q)$. The subgroup $1 \neq P/Q = N_P(Q)/C_P(Q) \cong \mathrm{Aut}_P(Q) \leq \mathrm{Aut}_{\mathscr{F}}(Q)$ acts trivially on $\Omega(Q)$. On the other hand every non-trivial automorphism of odd order acts non-trivially on $\Omega(Q)$ (see for example 8.4.3 in [159]). Hence, the kernel of this action is a non-trivial normal 2-

subgroup of $\text{Aut}_{\mathscr{F}}(Q)$. In particular $O_2(\text{Aut}_{\mathscr{F}}(Q)) \neq 1$. But then $\text{Aut}_{\mathscr{F}}(Q)$ cannot contain a strongly 2-embedded subgroup. This shows that $\mathscr{F}$ is controlled unless $s = 1$. Hence, if $r \neq s \neq 1$, Lemma 12.5 shows that $\mathscr{F}$ is nilpotent.

Now let $s = 1$. There is only one candidate $Q = \langle x^2, y, z \rangle$ for an $\mathscr{F}$-essential subgroup and $\text{Aut}_{\mathscr{F}}(Q) \cong S_3$ by Proposition 6.12. Since $\text{Aut}(Q)$ has only one element of order 3 up to conjugation, the action of $\text{Aut}_{\mathscr{F}}(Q)$ on Q is essentially unique. Moreover, P/Q acts non-trivially on $\langle y, z \rangle$ and on $\langle x^{2^{r-1}} y, z \rangle$. After replacing y by $x^{2^{r-1}} y$ if necessary, we may assume that $\text{Aut}_{\mathscr{F}}(Q)$ acts non-trivially on $\langle y, z \rangle$. Similarly, we may assume that x^2 is fixed by $\text{Aut}_{\mathscr{F}}(Q)$ after replacing x by xy if necessary. In particular, $\mathscr{F}$ is unique up to isomorphism and it suffices to construct a non-trivial constrained fusion system. Let A_4 be the alternating group of degree 4, and let $H := \langle \tilde{x} \rangle \cong C_{2^r}$. Moreover, let $\varphi : H \to \text{Aut}(A_4) \cong S_4$ such that $\varphi_{\tilde{x}} \in \text{Aut}(A_4)$ has order 4. Write $\tilde{y} := (12)(34) \in A_4$ and choose φ such that $\varphi_{\tilde{x}}(\tilde{y}) := (13)(24)$. Finally, let $G := A_4 \rtimes_{\varphi} H$. Since all 4-cycles in S_4 are conjugate, G is uniquely determined up to isomorphism. Because $[\tilde{x}, \tilde{y}] = (13)(24)(12)(34) = (14)(23)$, we get $\langle \tilde{x}, \tilde{y} \rangle \cong D$. The fusion system $\mathscr{F}_D(G)$ is nonnilpotent, since A_4 (and therefore G) is not 2-nilpotent. $\square$

Now we are in a position to give a new proof of case (iii) of Theorem 12.4 which is much shorter than the one in [243]. Since we know the fusion system $\mathscr{F}$ (on $A_4 \times C_{2^r}$ for example), it is easy to see that $|D : \mathfrak{foc}(B)| = 2^r$. In particular, $2^r \mid k_0(B)$ by Proposition 1.31. Moreover, Proposition 1.32 implies $2^{r+1} \leq k_0(B)$. Since $|Z(D) : Z(D) \cap \mathfrak{foc}(B)| = 2^{r-1}$, we also have $2^{r-1} \mid k_1(B)$ by Proposition 1.31. Finally, there is an element $z \in Z(D)$ such that $C_{\mathscr{F}}(\langle z \rangle)$ is trivial. Hence, Proposition 4.7 yields

$$2^{r+2} \leq k_0(B) + 4k_1(B) \leq \sum_{i=0}^{\infty} k_i(B) 2^{2i} \leq |D| = 2^{r+2}.$$

This gives $k_0(B) = 2^{r+2}$, $k_1(B) = 2^{r-1}$ and $k(B) = k_0(B) + k_1(B) = 5 \cdot 2^{r-1}$. It is easy to see that D has 2^{r+1} $\mathscr{F}$-conjugacy classes where 2^{r-1} of them lie in $Z(\mathscr{F}) = \langle x^2 \rangle$. Using induction on r and Theorem 1.35 we obtain $l(B) = 2$. Now we consider the decomposition and Cartan matrices of B.

Proposition 12.8 *Let B be a non-nilpotent 2-block with defect group $MNA(r, 1)$ for some $r \geq 2$. Then the decomposition and Cartan matrices of B are given by*

$$\begin{pmatrix} 1 \cdots 1 \ . \ \cdots \ . \ 1 \cdots 1 \\ . \ \cdots \ . \ 1 \cdots 1 \ 1 \cdots 1 \end{pmatrix}^T \qquad and \qquad 2^{r-1} \begin{pmatrix} 3 & 1 \\ 1 & 3 \end{pmatrix}$$

up to basic sets.

Proof Let C be the Cartan matrix of B. First we show that C has elementary divisor 2^{r-1}. Let $z := x^2 \in Z(\mathscr{F})$. Then $l(b_z) = 2$. Moreover, b_z covers a block $\overline{b}_z$ of $C_G(z)$ with defect group $D/\langle z \rangle \cong D_8$. By Theorem 8.1, b_z has Cartan matrix $2^{r-1} \begin{pmatrix} 3 & 2 \\ 2 & 4 \end{pmatrix}$. By Lemma 1.44, $\langle z \rangle$ is a lower defect group of b_z. It is easy to see that $N_G(\langle z \rangle, b_z) =$

$C_G(z)$. Therefore, Lemma 1.42 implies $m_B^{(1)}(\langle z \rangle, b_z) = m_{b_z}^{(1)}(\langle z \rangle) > 0$. This shows that C has elementary divisors 2^{r-1} and $|D| = 2^{r+2}$. Hence, $\tilde{C} := 2^{1-r}C$ is an integral matrix with elementary divisors 1 and 8. The reduction theory of quadratic forms shows that $\tilde{C}$ is

$$\begin{pmatrix} 1 & 0 \\ 0 & 8 \end{pmatrix} \qquad \text{or} \qquad \begin{pmatrix} 3 & 1 \\ 1 & 3 \end{pmatrix}$$

up to basic sets (see Eq. (3.1)). By way of contradiction, suppose that $\tilde{C} = \begin{pmatrix} 1 & 0 \\ 0 & 8 \end{pmatrix}$. As we have seen above, $\mathrm{Irr}(B)$ splits under the action of $D/\mathfrak{foc}(B)$ into three orbits. Two orbits have length 2^r and consists of characters of height 0 and one orbit has length 2^{r-1} and consists of characters of height 1. For two characters in the same orbit the corresponding decomposition numbers coincide. Hence, there are non-negative integers $\alpha, \beta, \gamma, \delta, \epsilon, \xi$ such that

$$\tilde{C} = \begin{pmatrix} \alpha & \alpha & \gamma & \gamma & \epsilon \\ \beta & \beta & \delta & \delta & \xi \end{pmatrix} \begin{pmatrix} \alpha & \beta \\ \alpha & \beta \\ \gamma & \delta \\ \gamma & \delta \\ \epsilon & \xi \end{pmatrix} = \begin{pmatrix} 2\alpha^2 + 2\gamma^2 + \epsilon^2 & 2\alpha\beta + 2\gamma\delta + \epsilon\xi \\ 2\alpha\beta + 2\gamma\delta + \epsilon\xi & 2\beta^2 + 2\delta^2 + \xi^2 \end{pmatrix}.$$

It follows that $\alpha = \gamma = 0$ and $\epsilon = 1$. Moreover, $\xi = 0$ and $\beta^2 + \delta^2 = 4$. However, this means that $\beta = 0$ or $\delta = 0$ and Proposition 1.36 gives a contradiction. Therefore, $\tilde{C} = \begin{pmatrix} 3 & 1 \\ 1 & 3 \end{pmatrix}$ and we get $\epsilon\xi = 1$, $\alpha\beta = \gamma\delta = 0$ and $\alpha^2 + \gamma^2 = \beta^2 + \delta^2 = 1$. This gives the decomposition matrix. $\square$

The proof of Proposition 12.8 gives evidence for Question A. It remains to determine the distribution into 2-conjugate and 2-rational characters.

Proposition 12.9 *Let B be a non-nilpotent 2-block with defect group $MNA(r, 1)$ for some $r \geq 2$. Then $\mathrm{Irr}_0(B)$ contains four 2-rational characters and two families of 2-conjugate characters of size 2^i for $i = 1, \ldots, r - 1$. The characters of height 1 split into two 2-rational characters and one family of 2-conjugate characters of size 2^i for $i = 2, \ldots, r - 2$. Also, the characters of height 1 have the same degree and $|\{\chi(1) : \chi \in \mathrm{Irr}_0(B)\}| \leq 2$.*

Proof Since $\mathrm{Irr}_1(B)$ is just one orbit under $D/\mathfrak{foc}(B)$, we see that all characters of height 1 have the same degree. The same argument gives $|\{\chi(1) : \chi \in \mathrm{Irr}_0(B)\}| \leq 2$. We note that it is conjectured that $|\{\chi(1) : \chi \in \mathrm{Irr}_0(B)\}| = 2$ (see [187]).

As usual, we study the action of the Galois group $\mathcal{G}$ first (see Sect. 1.2). Let $\mathcal{R}$ be a set of representatives for the $\mathcal{F}$-conjugacy classes of D. As we have already seen, $|\mathcal{R}| = 2^{r+1}$. The columns $\{d_{\chi\varphi_u}^u : \chi \in \mathrm{Irr}(B)\}$ with $u \in \mathcal{R} \setminus Z(D)$ split in two orbits of length 2^{r-1}. For $i = 1, 2$, the columns $\{d_{\chi\varphi_i}^u : \chi \in \mathrm{Irr}(B)\}$ with $u \in \langle x^2 \rangle$ and $\mathrm{IBr}(b_u) = \{\varphi_1, \varphi_2\}$ split in r orbits of lengths $1, 1, 2, 4, \ldots, 2^{r-2}$ respectively. Finally, the columns $\{d_{\chi\varphi_u}^u : \chi \in \mathrm{Irr}(B)\}$ with $u \in Z(D) \setminus \langle x^2 \rangle$

consist of r orbits of lengths $1, 1, 2, 4, \ldots, 2^{r-2}$ respectively. This gives $3r + 2$ orbits altogether. By Lemma IV.6.10 in [81] there also exist exactly $3r + 2$ families of 2-conjugate characters. (Since $\mathcal{G}$ is noncyclic, one cannot conclude a priori that also the lengths of the orbits of these two actions coincide.)

Now consider the generalized decomposition numbers $d^x := (d^x_{\chi\varphi_x} : \chi \in \mathrm{Irr}(B))$ where $\mathrm{IBr}(b_x) = \{\varphi_x\}$. As usual we can write $d^x = \sum_{i=0}^{2^{r-1}-1} a_i \zeta^i$ for a primitive 2^r-th root of unity ζ. By Proposition 5.1 we obtain $(a_i, a_i) = 4$ for $i = 0, \ldots, 2^{r-1} - 1$. On the other hand all 2^{r+1} entries of d^x for characters of height 0 must be non-zero by Proposition 1.36. This shows that for every $\chi \in \mathrm{Irr}_0(B)$ there is a i such that $d^x_{\chi\varphi_x} = a_i(\chi)$. It follows that the irreducible characters of height 0 split in at most $2(r + 1)$ orbits of lengths $1, 1, 1, 1, 2, 2, 4, 4, \ldots, 2^{r-1}, 2^{r-1}$ respectively. Finally let $u := x^2 z \in Z(D)$ and $d^u = \sum_{i=0}^{2^{r-2}-1} a_i \zeta^i$ for a primitive 2^{r-1}-th root of unity ζ. Again by Proposition 5.1 we have $(a_i, a_i) = 16$ for $i = 0, \ldots, 2^{r-2} - 1$. Also $2 \mid a_i(\chi)$ provided $h(\chi) = 1$ by Lemma 1.37. Since all entries of d^u are non-zero, it follows that there is a $\chi \in \mathrm{Irr}_1(B)$ such that $d^u_{\chi\varphi_u} = 2\zeta^i$ for some $i \in \mathbb{Z}$. Since $Z(D)/Z(D) \cap \mathfrak{foc}(B) = \langle u(Z(D) \cap \mathfrak{foc}(B)) \rangle$, it follows that $\{d^u_{\chi\varphi_u} : \chi \in \mathrm{Irr}_1(B)\} = \{2\zeta^i : i = 0, \ldots, 2^{r-1} - 1\}$. Thus, there are at most r orbits of lengths $1, 1, 2, 4, \ldots, 2^{r-2}$ of characters of height 1. Since $2(r+1) + r = 3r + 2$, these orbits do not merge further, and the claim is proved. $\qquad\qquad\square$

If G is solvable, one has also information on the Brauer character degrees and the precise Cartan matrix of B (see [131, 206]).

In [243], I did not prove the Ordinary Weight Conjecture. This was done later in [171], and we will provide the result here with a simpler proof.

Proposition 12.10 *Let B be a 2-block of a finite group with minimal non-abelian defect group $MNA(r, 1)$ for some $r \geq 2$. Then the Ordinary Weight Conjecture holds for B.*

Proof Let $D \cong MNA(r, 1)$ be a defect group of B, and let $Q \leq D$ be an $\mathcal{F}$-centric, $\mathcal{F}$-radical subgroup where $\mathcal{F}$ is the fusion system of B. Since $C_D(Q) \subseteq Q$, $|D : Q| \leq 2$. It follows from Proposition 12.7 that $Q = \langle x^2, y, z \rangle$ or $Q = D$. In both cases we have $\mathrm{H}^2(\mathrm{Out}_\mathcal{F}(Q), F^\times) = 1$. Hence, all 2-cocycles appearing in the OWC are trivial. Therefore the conjecture asserts that $k_i(B)$ only depends on $\mathcal{F}$. Since the conjecture is known to hold for the principal block of the solvable group $G = A_4 \rtimes C_{2^r}$, the claim follows. $\qquad\qquad\square$

The rest of this section is devoted to the proof of part (iv) in Theorem 12.4. Many arguments here are due to Eaton, one of the coauthors of [75]. The classification of the finite simple groups is needed.

We gather together some useful facts about blocks with defect groups as in (iv).

Lemma 12.11 *Let B be a 2-block of a finite group G with defect group $D \cong MNA(r, r)$ $(r \geq 2)$ and fusion system $\mathcal{F}$. Then*

  (i) $\mathscr{F}$ is controlled;

  (ii) either B is nilpotent or $e(B) = 3$, and in the latter case $z := [x, y]$ is the only non-trivial fixed point of $Z(D)$ under the action of $I(B)$;

  (iii) if B is not nilpotent, then $O_2(Z(G)) \leq \langle z \rangle$;

  (iv) if $Q \leq Z(D)$ and $Q \not\leq D'$, then there is a B-subpair (Q, b_Q) with b_Q nilpotent;

  (v) if $D \in \mathrm{Syl}_2(G)$, then G is solvable.

Proof The first two parts follow from Proposition 12.7 and Lemma 12.5 (and its proof). Now we turn to the third part of the lemma. As usual $O_2(Z(G)) \leq O_2(G) \leq D$. Hence, the second part shows the claim. Let $Q \leq Z(D)$ and $Q \not\leq D'$. Since $\mathscr{F}$ is controlled, Q is fully $\mathscr{F}$-normalized, and b_Q has defect group $C_D(Q) = D$ and fusion system $C_{\mathscr{F}}(Q) = \mathscr{F}_D(D)$.

It remains to prove the last part. By Feit-Thompson, we may assume $O_{2'}(G) = 1$. Now the Z^*-Theorem (see Theorem XII.8.1 in [81]) implies $z \in Z(G)$, and it suffices to show that $\overline{G} := G/\langle z \rangle$ is solvable. Obviously, $\overline{G}$ has Sylow 2-subgroup $D/\langle z \rangle \cong C_{2^r}^2$. A result of Brauer (see Theorem XII.5.1 in [81]) shows the claim. $\square$

In our proof of Theorem 12.4, the following result will be very useful.

Lemma 12.12 *Let G, B, D be as in Theorem 12.4(iv). Moreover, let b be a 2-block of a normal subgroup H of G which is covered by B. If a defect group d of b satisfies $|d| < |D|$, then b is nilpotent.*

Proof It is well-known that d is conjugate to $D \cap H$. Replacing D by a conjugate if necessary, we may assume that $d = D \cap H$. If $d < D$ then also $d\Phi(D) < D$. By Lemma 12.11, B has inertial index $e(B) = 3$. Since $|D : \Phi(D)| = 4$, this implies that $N_G(D)$ permutes the three maximal subgroups of D transitively. Since $d\Phi(D)$ is normal in $N_G(D)$, we must have $|D : d\Phi(D)| \geq 4$. But then $d \subseteq \Phi(D)$, and $[N_H(D), D] \subseteq D \cap H = d \subseteq \Phi(D)$. Thus, $N_H(D)$ acts trivially on $D/\Phi(D)$. Hence, $N_H(D)/C_H(D)$ is a 2-group. Let β be the unique 2-block of DH covering b. Then D is a defect group of β, by Theorem E in [161]. Let β_D be a 2-block of $D\,C_{DH}(D)$ such that $(\beta_D)^{DH} = \beta$. Then $N_H(D, \beta_D)/C_H(D)$ and $N_{DH}(D, \beta_D)/C_{DH}(D)$ are also 2-groups, i.e. β has inertial index $e(\beta) = 1$. Since β is a controlled block, by Lemma 12.11 this implies that β is a nilpotent block. But now Theorem 7.3 shows that b is also nilpotent. $\square$

Corollary 12.13 *Let G, B, D be as in Theorem 12.4(iv). If $H \lhd G$ has index a power of 2, then $D \leq H$.*

Proof There is a block b of H covered by B with defect group $D \cap H$. If $D \not\leq H$, then by Lemma 12.12, b is nilpotent. But then by Theorem 7.3, B is nilpotent, a contradiction. $\square$

Proof (Proof (of Theorem 12.4(iv)).) We assume that Theorem 12.4(iv) fails, and choose a counterexample G, B, D such that $|G : Z(G)|$ is as small as possible. Moreover, among all such counterexamples, we choose one where $|G|$ is minimal. Then, by the First Fong Reduction, the block B is quasiprimitive, i.e. for every normal subgroup N of G, there is a unique block of N covered by B; in particular, this block of N is G-stable. Moreover, by the Second Fong Reduction $O_{2'}(G)$ is cyclic and central.

We claim that $Q := O_2(G) \subseteq D'$. Since $Q \trianglelefteq G$ we certainly have $Q \subseteq D$. If $Q = D$, then B has a normal defect group, and B is Morita equivalent to $\mathcal{O}[D \rtimes C_3]$, by Theorem 1.19. Thus, we may assume that $1 < Q < D$; in particular, Q is abelian. Let B_Q be a block of $Q C_G(Q) = C_G(Q)$ such that $(B_Q)^G = B$. Since $C_G(Q) \trianglelefteq G$, the block B_Q has defect group $C_D(Q)$, and either $C_D(Q) = D$ or $|D : C_D(Q)| = 2$. Since B has inertial index $e(B) = 3$, $N_G(D)$ permutes the maximal subgroups of D transitively. Since $C_D(Q) \trianglelefteq N_G(D)$, we must have $C_D(Q) = D$, i.e. $Q \subseteq Z(D)$.

Thus, B_Q is a 2-block of $C_G(Q)$ with defect group D. If $Q \nsubseteq D'$ then B_Q is nilpotent, by Lemma 12.11. Then, by Theorem 7.3, B is Morita equivalent to a block of $N_G(D)$ with defect group D, and we are done by Theorem 1.19.

This shows that we have indeed $O_2(G) \subseteq D'$; in particular, $|O_2(G)| \le 2$ and thus $O_2(G) \subseteq Z(G)$. Hence, also $F(G) = Z(G)$.

Let b be a block of $E(G)$ covered by B. If b is nilpotent, then, by Theorem 7.3, B is Morita equivalent to a 2-block $\tilde{B}$ of a finite group $\tilde{G}$ having a nilpotent normal subgroup $\tilde{N}$ such that $\tilde{G}/\tilde{N} \cong G/E(G)$, and the defect groups of $\tilde{B}$ are isomorphic to D. Thus by minimality, we must have $E(G) = 1$. Then $F^*(G) = F(G) = Z(G)$, and $G = C_G(Z(G)) = C_G(F^*(G)) = Z(F^*(G)) = Z(G)$, a contradiction.

Thus, b is not nilpotent. By Lemma 12.12, b has defect group D. Let $L_1, \ldots, L_n$ be the components of G and, for $i = 1, \ldots, n$, let b_i be a block of L_i covered by b. If $b_1, \ldots, b_n$ were nilpotent, then b would also be nilpotent by Lemma 7.5, a contradiction. Thus, we may assume that b_1 is a non-nilpotent 2-block (of the quasisimple group L_1). By Lemma 12.12, D is a defect group of b_1. But now the following proposition gives a contradiction. $\square$

Proposition 12.14 *Let $D \cong MNA(r, r)$ for some $r \ge 2$, and let G be a quasisimple group. Then G does not have a 2-block B with defect group D.*

Note that the proposition holds for classical groups by An and Eaton [12], where blocks whose defect groups have derived subgroup of prime order are classified. However, since our situation is less general we are able to give new and more direct arguments here.

Proof We assume the contrary. Then we may also assume that B is faithful. Note that by An and Eaton [14], B cannot be nilpotent since D is non-abelian. By Lemma 12.11, D is not a Sylow 2-subgroup of G; in particular, $64 = 2^6$ divides $|G|$.

Suppose first that $\overline{G} := G/Z(G) \cong A_n$ for some $n \ge 5$. If $|Z(G)| > 2$, then $n \in \{6, 7\}$ and $|Z(G)| \mid 6$, by Gorenstein et al. [97]. But then $|G|$ is not divisible by 64, a contradiction. Thus, we must have $|Z(G)| \le 2$. Then $Z(G) \subseteq D$, and B dominates a unique 2-block $\overline{B}$ of $\overline{G}$ with defect group $\overline{D} := D/Z(G) \ne 1$. Let $\mathscr{B}$ be a 2-block of S_n covering $\overline{B}$. Then $\mathscr{B}$ has a defect group $\mathscr{D}$ such that $\overline{D} \subseteq \mathscr{D}$ and $|\mathscr{D} : \overline{D}| = 2$, by Theorem 7.10. Let w denote the weight of $\mathscr{B}$. Then, by

Theorem 7.8, $\mathscr{D}$ is conjugate to a Sylow 2-subgroup of S_{2w}. We may assume that $\mathscr{D}$ is a Sylow 2-subgroup of S_{2w}. Then $\overline{D} = \mathscr{D} \cap A_n = \mathscr{D} \cap S_{2w} \cap A_n = \mathscr{D} \cap A_{2w}$ is a Sylow 2-subgroup of A_{2w}, and D is a Sylow 2-subgroup of A_{2w} or $C_2.A_{2w}$. Thus, A_{2w} is solvable by Lemma 12.11, so that $w \leq 2$ and $|\overline{D}| \leq 4$, $|D| \leq 8$. Since $|D| \geq 32$, this is a contradiction.

Suppose next that $\overline{G}$ is a sporadic simple group. Then, using Table 1 in [14], we get a contradiction immediately unless $G = \text{Ly}$ and $|D| = 2^7$. In this remaining case, we get a contradiction since, by Landrock [175], D is a Sylow 2-subgroup of $C_2.A_8$, and A_8 is non-solvable.

Now suppose that G is a group of Lie type in characteristic 2. Then, by Theorem 7.11, the 2-blocks of G have either defect zero or full defect. Thus, again Lemma 12.11 leads to a contradiction.

It remains to deal with the groups of Lie type in odd characteristic. We use three strategies to deal with the various subcases.

Suppose first that $\overline{G} \cong \text{PSL}(n, q)$ or $\text{PSU}(n, q)$ where $1 < n \in \mathbb{N}$ and q is odd. Except in the cases $\text{PSL}(2, 9)$ and $\text{PSU}(4, 3)$, there is $E \cong \text{SL}(n, q)$ or $\text{SU}(n, q)$ such that G is a homomorphic image of E with kernel W say. We may rule out the cases $G/Z(G) \cong \text{PSL}(2, 9)$ or $\text{PSU}(4, 3)$ using [266]. Let $H \cong \text{GL}(n, q)$ or $\text{GU}(n, q)$ with $E \lhd H$. There is a block B_E of E with defect group D_E such that $D_E W/W \cong D$. Let B_H be a block of H covering B_E with defect group D_H such that $D_H \cap E = D_E$. Now B_H is labeled by a semisimple element $s \in H$ of odd order such that $D_H \in \text{Syl}_2(C_H(s))$ (see, for example, [45, 3.6]). It follows that $D \in \text{Syl}_2(C_E(s)/W)$ and so $C_E(s)/W$ is solvable by Lemma 12.11. Now W and H/E are solvable, so $C_H(s)$ is also solvable. By [85, 1A], $C_H(s)$ is a direct product of groups of the form $\text{GL}(n_i, q^{m_i})$ and $\text{GU}(n_i, q^{m_i})$. Write

$$C_H(s) \cong \prod_{i=1}^{t_1} \text{GL}(n_i, q^{m_i}) \times \prod_{i=t_1+1}^{t_2} \text{GU}(n_i, q^{m_i})$$

where $t_1, t_2 \in \mathbb{N}$, $n_1, \ldots, n_{t_2} \in \mathbb{N}$, and $m_1, \ldots, m_{t_2} \in \mathbb{N}$, with $n_i \geq 3$ for $i > t_1$. Solvability implies that $t_2 = t_1$ and that for $i = 1, \ldots, t_1$ we have either $n_i = 1$ or $n_i = 2$, where in the latter case $m_i = 1$ and $q = 3$. Since D, D_E, and D_H are non-abelian, we cannot have $n_i = 1$ for all $i = 1, \ldots, t_1$. Thus, we must have $q = 3$ and, without loss of generality, $n_1 = 2$, $m_1 = 1$. Then D_H is a direct product of factors which are either cyclic or isomorphic to SD_{16}. Moreover, we have $|D_H : D_E| \leq 2$ and $|W| \leq 2$. Since $|D : \Phi(D)| = 4$, we also have $|D_E : \Phi(D_E)| \leq 8$ and $|D_H : \Phi(D_H)| \leq 16$.

Suppose first that $|D_H : \Phi(D_H)| = 16$. Then $|D_E : \Phi(D_E)| = 8$, $|D_H : D_E| = 2$, and $|W| = 2$. Since $W \nsubseteq \Phi(D_E)$, $D_E \cong D \times W$. If $D_H \cong SD_{16} \times SD_{16}$, then $|D_H| = 2^8$ and $|D| = 2^6$ which is impossible.

Thus, we must have $D_H \cong SD_{16} \times C_k \times C_l$ where k and l are powers of 2. Observe that $\Phi(D_E) \subseteq \Phi(D_H)$ and $|D_H : \Phi(D_H)| = 16 = |D_H : \Phi(D_E)|$. So we must have $\Phi(D_E) = \Phi(D_H)$. Since $\Phi(D_E) \cong \Phi(D) \cong C_{2^{r-1}} \times C_{2^{r-1}} \times C_2$ and $\Phi(D_H) \cong C_4 \times C_{k/2} \times C_{l/2}$, this implies that $4 = 2^{r-1}$, i.e. $r = 3$ and

$\Phi(D) \cong \Phi(D_E) \cong C_4 \times C_4 \times C_2$. So we may assume that $k = 8$, $l = 4$. Thus, $D_E \cong D \times C_2$ and $D_H \cong SD_{16} \times C_8 \times C_4$. Hence, $D'_E = D' \times 1$, $|D'_E| = 2$ and $D'_E \subseteq D'_H \cap Z(D_H) \cong Z(SD_{16}) \times 1 \times 1$, so that $D'_E = Z(SD_{16}) \times 1 \times 1$. Moreover, $D_E/D'_E \cong C_8 \times C_8 \times C_2$ is a subgroup of $D_H/D'_E \cong D_8 \times C_8 \times C_4$. Hence, $\mho_2(C_8 \times C_8 \times C_2) \cong C_2 \times C_2$ is isomorphic to a subgroup of $\mho_2(D_8 \times C_8 \times C_4) \cong C_2$ which is impossible.

Next we consider the case $|D_H : \Phi(D_H)| = 8$. In this case we have $D_H \cong SD_{16} \times C_k$ where k is a power of 2. Then $\Phi(D_E) \subseteq \Phi(D_H) \cong C_4 \times C_{k/2}$ and $\Phi(D) \cong \Phi(D_E W/W) = \Phi(D_E)W/W$. However, this contradicts $\Phi(D) \cong C_{2^{r-1}} \times C_{2^{r-1}} \times C_2$.

The case $|D_H : \Phi(D_H)| \leq 4$ is certainly impossible.

A similar argument applies to the other classical groups, at least when they are defined over fields with $q > 3$ elements, and we give this now. Suppose that G is a classical quasisimple group of type $B_n(q)$, $C_n(q)$, $D_n(q)$ or $^2D_n(q)$, where $q > 3$ is a power of an odd prime. Note that in these cases there is no exceptional cover.

Let E be the Schur cover of $G/Z(G)$, so that G is a homomorphic image of E with kernel W say. Note that $Z(E)$, and so W, is a 2-group. There is a block B_E of E with defect group D_E such that $D \cong D_E/W$. Details of the following may be found in [55] and [51]. We may realize E as $\mathbf{E}^F$, where $\mathbf{E}$ is a simple, simply-connected group of Lie type defined over the algebraic closure of a finite field, $F : \mathbf{E} \to \mathbf{E}$ is a Frobenius map (in this setting F is not a field!) and $\mathbf{E}^F$ is the group of fixed points under F. Write $\mathbf{E}^*$ for the group dual to $\mathbf{E}$, with corresponding Frobenius map F^*. Note that if $\mathbf{H}$ is an F-stable connected reductive subgroup of $\mathbf{E}$, then $\mathbf{H}$ has dual $\mathbf{H}^*$ satisfying $|\mathbf{H}^F| = |(\mathbf{H}^*)^{F^*}|$.

By [78, 1.5] there is a semisimple element $s \in \mathbf{E}^*$ of odd order such that D_E is a Sylow 2-subgroup of $\mathbf{L}^F$, where $\mathbf{L} \leq \mathbf{E}$ is dual to $C_{\mathbf{E}^*}^0(s)$, the connected component of $C_{\mathbf{E}^*}(s)$ containing the identity element. Now $W \leq Z(E) \leq D_E \leq \mathbf{L}^F$. Hence, $D_E/W \in \mathrm{Syl}_2(\mathbf{L}^F/W)$. By Lemma 12.11, $\mathbf{L}^F/W$, and so $\mathbf{L}^F$, is solvable. Now by [54] $C_{\mathbf{E}^*}(s)$ factorizes as $\mathbf{MT}$, where $\mathbf{T}$ is a torus and $\mathbf{M}$ is semisimple, $C_{(\mathbf{E}^*)^{F^*}}(s) = C_{\mathbf{E}^*}(s)^{F^*} = \mathbf{M}^{F^*}\mathbf{T}^{F^*}$ and the components of $\mathbf{M}^{F^*}$ are classical groups defined over fields of order a power of q. Hence, $C_{(\mathbf{E}^*)^{F^*}}(s)$ is either abelian or non-solvable. It follows that $\mathbf{L}^F$ is either abelian or non-solvable, in either case a contradiction.

Let G be a quasisimple finite group of Lie type with $|G|$ minimized such that there is a block B of G with defect group D as in Theorem 12.4(iv). We have shown that G cannot be defined over a field of characteristic two, of type $A_n(q)$ or $^2A_n(q)$ or of classical type for $q > 3$.

We group the remaining cases into two.

Case 1. Suppose that G is a quasisimple finite group of Lie type with center of odd order, and further that $q = 3$ if G is classical. We analyze $C_G(z)$, where we recall that $D' = \langle z \rangle$. There is a non-nilpotent block b_z of $C_G(z)$ with defect group D. As z is semisimple, $C_G(z)$ may be described in detail. By [97, 4.2.2] $C_G(z)$ has a normal subgroup C^0 such that $C_G(z)/C^0$ is an elementary abelian 2-group and $C^0 = LT$, where $L = L_1 * \cdots * L_m \lhd C^0$ is a central product of quasisimple groups of Lie type and T is an abelian group acting on each L_i by inner-diagonal automorphisms.

If G is a classical group or any exceptional group of Lie type except $E_6(q)$, $^2E_6(q)$ or $E_7(q)$, then by [97, 4.5.1] and [97, 4.5.2], T is a 2-group. In particular $C_G(z)/L$ is a 2-group, so by Corollary 12.13, $D \leq L$. Let b_L be a block of L covered by b_z with defect group D. If b_L is nilpotent, then by Theorem 7.3 b_z is also nilpotent since $C_G(z)/L$ is a 2-group, a contradiction. Hence, b_L is not nilpotent. By Lemma 12.12, for each i we have that b_L either covers a nilpotent block of L_i, or $D \leq L_i$. It follows that either $D \leq L_i$ for some i or b_L covers a nilpotent block of each L_i. In the latter case by Lemma 7.5, b_L would be nilpotent, a contradiction. Hence, $D \leq L_i$ for some i and there is a non-nilpotent block of L_i with defect group D. But $|L_i| < |G|$ and L_i is quasisimple, contradicting minimality.

If G is of type $E_6(q)$ or $^2E_6(q)$, then in the notation of [97, 4.5.1] G has (up to isomorphism of centralizers) two conjugacy classes of involutions, with representatives t_1 and t_2. Suppose first of all that z is of type t_1. In this case $C_G(z)$ has a normal subgroup X of index a power of 2 such that X is a central product of $L = L_1$ and a cyclic group A. Arguing as above, b_z either covers a nilpotent block of X, and so is itself nilpotent (a contradiction) or $D \leq X$. So b_z covers a non-nilpotent block b_X of X with defect group D. Applying the argument again, either b_X covers nilpotent blocks of L and A, in which case b_X would be nilpotent by Lemma 7.5 (a contradiction), or $D \leq L$. We have $|L| < |G|$ and L is quasisimple, so by minimality we obtain a contradiction. Consider now the case that z has type t_2. Then $C_G(z)$ has a normal subgroup of index 2 which is a central product of quasisimple groups, and we can argue as above to again get a contradiction.

If G is of type $E_7(q)$, then in the notation of [97, 4.5.1] G has (up to isomorphism of centralizers) five conjugacy classes of involutions, with representatives t_1, t_4, t_4', t_7 and t_7'. In the first three of these cases T is a 2-group and we may argue exactly as above. In case t_7 and t_7', we have $|C_G(z) : C^0| = 2$ and by a now familiar argument $D \leq C^0$ and b_z covers a non-nilpotent block of C^0 with defect group D. There is $X \lhd C^0$ of index 3 such that $X = LA$, where $L = L_1$ and A is cyclic of order $q \pm 1$. Now by Lemma 12.11, $O_2(Z(A)) = \langle z \rangle$, so $|A|_2 = 2$ and $D \leq L$. By minimality this situation cannot arise since L is quasisimple, and we are done in this case.

Case 2. Suppose that G is a quasisimple group of Lie type with center of even order, and further that $q = 3$ if G is classical. Note that G cannot be of type $A_n(q)$ or $^2A_n(q)$. Here we must use a different strategy since we may have $C_G(z) = G$. Let $u \in Z(D)$ be an involution with $u \neq z$. By Lemma 12.11 there is a nilpotent block b_u of $C_G(u)$ with $b_u^G = B$. As before we refer to [97, 4.5.2] for the structure of $C_G(u)$, and $C_G(u) \cong LT$, where L is a central product of either one or two quasisimple groups and T is an abelian group acting on L by inner-diagonal automorphisms. We take a moment to discuss types $D_n(3)$ for $n \geq 4$ even and $^2D_n(3)$. In these two cases the universal version of the group has center of order 4, and the information given in [97, 4.5.2] applies only to the full universal version. In order to extract the required information when

$|Z(G)| = 2$ it is necessary to use [97, 4.5.1], taking advantage of the fact that if Y is a finite group, $X \leq Z(Y)$ with $|X| = 2$ and $y \in Y$ is an involution, then $|C_{Y/X}(yX) : C_Y(y)/X|$ divides 2. Note also that [97, 4.5.2] gives the fixed point group of an *automorphism* of order 2 acting on G, and that not every such automorphism is realized by an involution in G (this information is contained in the column headed $|\hat{t}|$). We will make no further reference to this fact.

Now $Z(C_G(u))$ and T are both 2-groups, and in each case there is a direct product E of quasisimple groups of Lie type and abelian 2-groups, with $W \leq Z(E)$ such that $L \cong E/W$ and W is a 2-group, and there is a direct product H of finite groups of Lie type such that $E \leq H$ has index a power of 2 and H/W has a subgroup isomorphic to $C_G(u)$ of index a power of 2. Since W and H/E are 2-groups, by [169, 6.5] there are nilpotent blocks B_E of E and B_H of H with defect groups D_E and D_H such that $D_E \leq D_H$ and D_E/W has a subgroup isomorphic to D. By Lemma 7.5, B_E is a product of nilpotent blocks of finite groups of Lie type, and so by An and Eaton [14], D_E is abelian. But then D is abelian, a contradiction.   □

Proposition 12.15 *Let B be as in Theorem 12.4(iv). Then D is the vertex of the simple B-modules.*

Proof First we consider the situation in the group $D \rtimes C_3$. Here the three irreducible Brauer characters are linear and can be extended to irreducible ordinary characters. By Theorem 12.4 there is a Morita equivalence between $\mathcal{O}[D \rtimes C_3]$ and B. Under this equivalence the three ordinary linear characters map to irreducible characters of height 0 in B. These characters are again extensions of three distinct Brauer characters, since the decomposition matrix is also preserved under Morita equivalence. Now the claim follows from Theorem 19.26 in [64].   □

Corollary 12.16 *Let $D \cong MNA(r, r)$ for some $r \geq 2$. Then Donovan's Conjecture holds for 2-blocks of finite groups with defect group D.*

Corollary 12.17 *Every 2-block B with minimal non-abelian defect groups satisfies the following conjectures:*

- *Alperin's Weight Conjecture*
- *Brauer's $k(B)$-Conjecture*
- *Brauer's Height-Zero Conjecture*
- *Olsson's Conjecture*
- *Alperin-McKay Conjecture*
- *Ordinary Weight Conjecture*
- *Gluck's Conjecture*
- *Eaton's Conjecture*
- *Eaton-Moretó Conjecture*
- *Malle-Navarro Conjecture*
- *Robinson's Conjecture*

Moreover, the Gluing Problem for B has a unique solution.

Proof Most conjectures are obviously true by Theorem 12.4 and Proposition 12.10. Gluck's Conjecture only applies if B is nilpotent or $|D| = 8$. We have already seen in Corollary 8.2 that the conjecture holds here. It remains to deal with the Gluing Problem (which was done in [243]). The nilpotent case and case (iv) are controlled and thus uninteresting (see Example 5.3 in [183]). Now let B be a 2-block with defect group $MNA(r, 1)$ for some $r \geq 2$. Let $\mathscr{F}$ be the fusion system of B. Then the $\mathscr{F}$-centric subgroups of D are given by $M_1 := \langle x^2, y, z \rangle$, $M_2 := \langle x, z \rangle$, $M_3 := \langle xy, z \rangle$ and D. Moreover, $\text{Aut}_{\mathscr{F}}(M_1) \cong \text{Out}_{\mathscr{F}}(M_1) \cong S_3$, $\text{Aut}_{\mathscr{F}}(M_i) \cong D/M_i \cong C_2$ for $i = 2, 3$ and $\text{Aut}_{\mathscr{F}}(D) \cong D/Z(D) \cong C_2^2$. Using this, we get $\text{H}^i(\text{Aut}_{\mathscr{F}}(\sigma), F^\times) = 0$ for $i = 1, 2$ and every chain σ of $\mathscr{F}$-centric subgroups (see proof of Corollary 2.2 in [219]). Hence, $\text{H}^0([S(\mathscr{F}^c)], \mathscr{A}_{\mathscr{F}}^2) = \text{H}^1([S(\mathscr{F}^c)], \mathscr{A}_{\mathscr{F}}^1) = 0$. Now the claim follows from Theorem 1.1 in [219]. □

12.2 The Case $p > 2$

For odd primes p, Gao, Yang and Zeng already obtained some incomplete results about minimal non-abelian defect groups (see [90, 289]). Here we settle Olsson's Conjecture in almost all cases. The result was obtained in [114].

Theorem 12.18 *Let B be a block of a finite group with minimal non-abelian defect group $D \not\cong 3_+^{1+2}$. Then Olsson's Conjecture holds for B.*

Proof By Theorem 12.4 we may assume $p \geq 3$. Then by Rédei's classification Theorem 12.2, we may assume that $D \cong MNA(r, s)$ for $r \geq s \geq 1$. We set $z := [x, y] \in Z(D)$. Observe that $\Phi(D) = Z(D) = \langle x^p, y^p, z \rangle$ and $D' = \langle z \rangle$. Let $\mathscr{F}$ be the fusion system of B.

First assume $s \geq 2$. Then we show that B is controlled. By Alperin's Fusion Theorem it suffices to show that D does not contain $\mathscr{F}$-essential subgroups. By way of contradiction, assume that $Q < D$ is $\mathscr{F}$-essential. Since $C_D(Q) \subseteq Q$, Q is a maximal subgroup of D. Let $a \in D$ be an element of order p. Then also $aD' \in D/D' \cong C_{p^r} \times C_{p^s}$ has order p. Since $r \geq s \geq 2$, we see that $a \in Z(D)$ and $\Omega_1(D) \subseteq Z(D)$. This shows that $1 \neq D/Q = \text{Aut}_D(Q) \leq \text{Aut}_{\mathscr{F}}(Q)$ acts trivially on $\Omega_1(Q)$. On the other hand every p'-automorphism of $\text{Aut}_{\mathscr{F}}(Q)$ acts non-trivially on $\Omega_1(Q)$ (see Theorem 5.2.4 in [94]). Hence, $O_p(\text{Aut}_{\mathscr{F}}(Q)) \neq 1$ which contradicts the choice of Q. Thus, we have proved that B is a controlled block. Now the claim follows from Theorem 11.7(iii).

Now assume that $s = 1$. If also $r = 1$, then D is non-abelian of order p^3 and exponent p. By hypothesis, $p > 3$ here. In this case we have seen in the proof of Theorem 11.9 that Olsson's Conjecture holds for B. Thus, let $r \geq 2$. Since $Z(D)$ has exponent p^{r-1}, we see that x is not $\mathscr{F}$-conjugate to an element in $Z(D)$. In particular (x, b_x) is a non-major B-subsection. Moreover, $\langle x \rangle$ is fully $\mathscr{F}$-centralized, since $C_D(x)$ is a maximal subgroup of D. Hence, $C_D(x)$ is a defect group of b_x by Lemma 1.34. Now the claim follows from Proposition 4.3. □

Chapter 13
Small Defect Groups

13.1 Results on the $k(B)$-Conjecture

After we have computed the block invariants for many specific defect groups, it is interesting to see what is the smallest open case. Obviously, the smallest non-metacyclic group C_2^3 comes to mind. Landrock [176] gave partial results here, and later the case was settled by Kessar, Koshitani and Linckelmann [148] using the classification of the finite simple groups. The result also follows easily from Theorem 7.14 (still using the classification). We add some information about Cartan matrices.

Theorem 13.1 (Kessar-Koshitani-Linckelmann) *Let B be a block of a finite group with elementary abelian defect group of order 8. Then $k(B) = k_0(B) = 8$, and one of the following holds:*

(i) $e(B) = l(B) = 1$ *and B is nilpotent.*
(ii) $e(B) = l(B) = 3$ *and the Cartan matrix of B is*

$$2 \begin{pmatrix} 2 & 1 & 1 \\ 1 & 2 & 1 \\ 1 & 1 & 2 \end{pmatrix}$$

up to basic sets.
(iii) $e(B) = l(B) = 7$ *and the Cartan matrix of B is $(1 + \delta_{ij})_{1 \le i,j \le 7}$ up to basic sets.*

© Springer International Publishing Switzerland 2014
B. Sambale, *Blocks of Finite Groups and Their Invariants*, Lecture Notes in Mathematics 2127, DOI 10.1007/978-3-319-12006-5_13

(iv) $e(B) = 21$, $l(B) = 5$ and the Cartan matrix of B is

$$\begin{pmatrix} 2 & . & . & . & 1 \\ . & 2 & . & . & 1 \\ . & . & 2 & . & 1 \\ . & . & . & 2 & 1 \\ 1 & 1 & 1 & 1 & 4 \end{pmatrix}$$

up to basic sets.

Proof Let D be a defect group of B. As usual, $I(B) \le \mathrm{Aut}(D) \cong \mathrm{GL}(3,2)$ has odd order. Hence, $e(B) \in \{1,3,7,21\}$. By [148] there is a so-called isotypy between B and its Brauer correspondent b in $\mathrm{N}_G(D)$. Since b has normal defect group D, we may compute the invariants in the group algebra $D \rtimes I(B)$ by Theorem 1.19.   □

The Cartan matrices in Theorem 13.1 can also be determined by the Cartan method (Sect. 4.2) without using the isotypy (after one knows $k(B) = k_0(B)$). The next interesting case of a small defect group is the elementary abelian group of order 9. Here we have already mentioned the incomplete results by Kiyota [154] (see also Watanabe [282]). For example, it is still open whether Alperin's Conjecture holds in case $D \cong C_3^2$ and $I(B) \cong C_8$.

This shifts the focus to 2-blocks of defect 4. It turns out that we have already handled the non-abelian defect groups of order 16. Next we settle the elementary abelian case which is taken from [74, 171].

Theorem 13.2 *Let B be a block of a finite group with elementary abelian defect group D of order* 16. *Then one of the following holds:*

  (i) B is nilpotent. Then $e(B) = l(B) = 1$ and $k(B) = k_0(B) = 16$.
 (ii) $e(B) = l(B) = 3$, $\mathrm{C}_D(I(B)) = 1$ and $k(B) = k_0(B) = 8$.
(iii) $e(B) = l(B) = 3$, $|\mathrm{C}_D(I(B))| = 4$ and $k(B) = k_0(B) = 16$.
 (iv) $e(B) = l(B) = 5$ and $k(B) = k_0(B) = 8$.
  (v) $e(B) = l(B) = 7$ and $k(B) = k_0(B) = 16$.
 (vi) $e(B) = l(B) = 9$ and $k(B) = k_0(B) = 16$.
(vii) $e(B) = 9$, $l(B) = 1$ and $k(B) = k_0(B) = 8$.
(viii) $e(B) = l(B) = 15$ and $k(B) = k_0(B) = 16$.
 (ix) $e(B) = 21$, $l(B) = 5$ and $k(B) = k_0(B) = 16$.

Moreover, all cases actually occur.

Proof First of all by Theorem 7.14 we have $k(B) = k_0(B)$. The inertial quotient $I(B)$ is a subgroup of $\mathrm{Aut}(D) \cong \mathrm{GL}(4,2)$ of odd order. It follows that $e(B) \in \{1,3,5,7,9,15,21\}$ (this can be shown with GAP [266]). If $e(B) \ne 21$, the inertial quotient is necessarily abelian. Then by Corollary 1.2(ii) in [232] there is a nontrivial subsection (u,b) such that $l(b) = 1$. Hence, Lemma 1.37 implies that $|D| = 16$ is a sum of $k(B)$ odd squares. This shows $k(B) \in \{8,16\}$ for these cases. In order to determine $l(B)$ we calculate the numbers $l(b)$ for all non-trivial subsections

(u, b). Here it suffices to consider a set of representatives of the orbits of D under $I(B)$, since B is a controlled block. If $e(B) = 1$, the block is nilpotent and the result is clear. We discuss the remaining cases separately:

Case 1: $e(B) = 3$

Here by results of Usami and Puig (see [227, 270]) there is a perfect isometry between B and its Brauer correspondent in $N_G(D)$. According to two different actions of $I(B)$ on D, we get $k(B) = 8$ if $C_D(I(B)) = 1$ or $k(B) = 16$ if $|C_D(I(B))| = 4$. In both cases we have $l(B) = 3$.

Case 2: $e(B) = 5$

Then there are four subsections $(1, B)$, (u_1, b_1), (u_2, b_2) and (u_3, b_3) with $l(b_1) = l(b_2) = l(b_3) = 1$ up to conjugation. By way of contradiction, suppose $k(B) = 16$. We derive a contradiction using the Cartan method. It is easy to see that the three columns of the generalized decomposition matrix corresponding to b_1, b_2 and b_3 can be arranged in the form

$$
\begin{pmatrix}
1\ 1\ 1\ 1\ \ 1 & 1 & 1 & 1 & 1 & 1 & 1 & 1 & 1 & 1 & 1 \\
1\ 1\ 1\ 1\ \ 1 & 1 & 1 & 1 & -1 & -1 & -1 & -1 & -1 & -1 & -1 & -1 \\
1\ 1\ 1\ 1\ -1 & -1 & -1 & -1 & 1 & 1 & 1 & 1 & -1 & -1 & -1 & -1
\end{pmatrix}^{\mathrm{T}} .
$$

Hence, the Cartan matrix C of B is given by

$$
C =
\begin{pmatrix}
4 & 3 & 3 & 3 & 1\ 1\ 1\ 1\ 1\ 1 & -1 & -1 & -1 \\
3 & 4 & 3 & 3 & 1\ 1\ 1\ 1\ 1\ 1 & -1 & -1 & -1 \\
3 & 3 & 4 & 3 & 1\ 1\ 1\ 1\ 1\ 1 & -1 & -1 & -1 \\
3 & 3 & 3 & 4 & 1\ 1\ 1\ 1\ 1\ 1 & -1 & -1 & -1 \\
1 & 1 & 1 & 1 & 2\ 1\ 1\ .\ .\ . & .\ & .\ & . \\
1 & 1 & 1 & 1 & 1\ 2\ 1\ .\ .\ . & .\ & .\ & . \\
1 & 1 & 1 & 1 & 1\ 1\ 2\ .\ .\ . & .\ & .\ & . \\
1 & 1 & 1 & 1 & .\ .\ .\ 2\ 1\ 1 & .\ & .\ & . \\
1 & 1 & 1 & 1 & .\ .\ .\ 1\ 2\ 1 & .\ & .\ & . \\
1 & 1 & 1 & 1 & .\ .\ .\ 1\ 1\ 2 & .\ & .\ & . \\
-1 & -1 & -1 & -1 & .\ .\ .\ .\ .\ . & 2 & 1 & 1 \\
-1 & -1 & -1 & -1 & .\ .\ .\ .\ .\ . & 1 & 2 & 1 \\
-1 & -1 & -1 & -1 & .\ .\ .\ .\ .\ . & 1 & 1 & 2
\end{pmatrix}
$$

up to basic sets. In particular $\det C = 256$. However, this contradicts Proposition 1.46. Therefore, $k(B) = 8$ and $l(B) = 5$.

Case 3: $e(B) = 7$

There are again four subsections $(1, B)$, (u_1, b_1), (u_2, b_2) and (u_3, b_3) up to conjugation. But in this case $l(b_1) = l(b_2) = 1$ and $l(b_3) = 7$ by Theorem 13.1. Thus, $k(B) = 16$ and $l(B) = 7$.

Case 4: $e(B) = 9$

There are four subsections $(1, B)$, (u_1, b_1), (u_2, b_2) and (u_3, b_3) such that $l(b_1) = 1$ and $l(b_2) = l(b_3) = 3$ up to conjugation. This gives the possibilities (13.2) and (13.2).

Case 5: $e(B) = 15$

This case was handled in [74] as a byproduct. We will not give the proof which is very complicated. It turns out that (13.2) occurs.

Case 6: $e(B) = 21$

Here $I(B)$ is non-abelian. Hence, we get four subsections $(1, B)$, (u_1, b_1), (u_2, b_2) and (u_3, b_3) up to conjugation. We have $l(b_1) = l(b_2) = 3$ and $l(b_3) = 5$ by Theorem 13.1. Since $I(B)$ has a fixed point on D, it follows that $l(B) = 5$ and $k(B) = 16$ by Theorem 1.39.

For all cases except (13.2) examples are given by the principal block of $D \rtimes I(B)$. In case (13.2) we can take a non-principal block of the group

$$\texttt{SmallGroup(432,526)} \cong D \rtimes E$$

where E is the extraspecial group of order 27 and exponent 3 (see Small Groups Library and Proposition 1.20). $\square$

In order to prove Alperin's Weight Conjecture, we investigate the differences between the cases (13.2) and (13.2).

Lemma 13.3 *Let B be a block of a finite group G with elementary abelian defect group D of order 16. If $e(B) = l(B) = 9$, then the elementary divisors of the Cartan matrix of B are $1, 1, 1, 1, 4, 4, 4, 4, 16$. Moreover, the two $I(B)$-stable subgroups of D of order 4 are lower defect groups of B. Both occur with 1-multiplicity 2.*

Proof Let C be the Cartan matrix of B. As in the proof of Theorem 13.2 there are four subsections $(1, B)$, (u_1, b_1), (u_2, b_2) and (u_3, b_3) such that $l(b_1) = 1$ and $l(b_2) = l(b_3) = 3$ up to conjugation. In order to determine C up to basic sets, we need to investigate the generalized decomposition numbers $d_{rs}^{u_i}$ for $i = 1, 2, 3$. The block b_2 dominates a block $\overline{b_2}$ of $C_G(u_2)/\langle u_2 \rangle$ with defect group $D/\langle u_2 \rangle$ and inertial index 3. Thus, by Theorem 13.1 the Cartan matrix of b_2 has the form

$$4 \begin{pmatrix} 2 & 1 & 1 \\ 1 & 2 & 1 \\ 1 & 1 & 2 \end{pmatrix}$$

up to basic sets. Since $k(B) = 16$, we may assume that the numbers $d_{rs}^{u_2}$ take the form

$$
\begin{pmatrix}
1\,1\,1\,1\,1\,1\,1\,1\,.\,.\,.\,.\,.\,.\,.\,. \\
1\,1\,1\,1\,.\,.\,.\,.\,1\,1\,1\,1\,.\,.\,.\,. \\
1\,1\,1\,1\,.\,.\,.\,.\,.\,.\,.\,.\,1\,1\,1\,1
\end{pmatrix}^{\mathrm{T}} .
$$

For the column of decomposition numbers $d_{rs}^{u_1}$ we have essentially the following possibilities:

$(i) : (1, 1, 1, 1, -1, -1, -1, -1, -1, -1, -1, -1, -1, -1, -1, -1)^{\mathrm{T}},$

$(ii) : (1, 1, 1, -1, 1, -1, -1, -1, 1, -1, -1, -1, 1, -1, -1, -1)^{\mathrm{T}},$

$(iii) : (1, 1, -1, -1, 1, 1, -1, -1, 1, 1, -1, -1, 1, 1, -1, -1)^{\mathrm{T}}.$

Now we use a GAP program to enumerate the possible decomposition numbers $d_{rs}^{u_3}$. After that the ordinary decomposition matrix M can be calculated as the orthogonal space. Then $C = M^{\mathrm{T}}M$ up to basic sets. It turns out that in some cases C has 2 as an elementary divisor. Using the notion of lower defect groups we show that these cases cannot occur. If 2 is an elementary divisor of C, then there exists a lower defect group $Q \le D$ of order 2 such that $m_B^{(1)}(Q, b_Q) > 0$ by Proposition 1.41. Since $N_G(Q) = C_G(Q)$, it follows from Lemma 1.42 that also $m_{b_Q}^{(1)}(Q) > 0$. Hence, 2 is also an elementary divisors of the Cartan matrix of b_Q. Since (Q, b_Q) is a B-subsection, we see that b_Q is conjugate to b_2 or b_3. But we have seen above that all elementary divisors of the Cartan matrix of b_2 (and also b_3) must be divisible by 4. This contradiction shows that 2 does not occur as elementary divisor of C. After excluding these cases the GAP program reveals the following two possibilities for the elementary divisors of C: $1, 1, 1, 1, 4, 4, 4, 4, 16$ or $1, 1, 4, 4, 4, 4, 4, 4, 16$.

Now Proposition 1.41 implies

$$
4 \le m(4) = \sum_{R \in \mathscr{R}} m_B^{(1)}(R, b_R)
$$

where $\mathscr{R}$ is a set of representatives for the $I(B)$-conjugacy classes of subgroups of D of order 4. Let $Q \le D$ be of order 4 such that $m_B^{(1)}(Q, b_Q) > 0$. Then by Lemma 1.42 we have $m_{B_Q}^{(1)}(Q) > 0$ where $B_Q := b_Q^{N_G(Q, b_Q)}$. If Q is not fixed under $I(B)$, then we would have the contradiction $e(B_Q) = l(B_Q) = 1$. Thus, we have shown that Q is stable under $I(B)$. Hence,

$$
4 \le m_B^{(1)}(Q, b_Q) + m_B^{(1)}(P, b_P) \tag{13.1}
$$

where $P \ne Q$ is the other $I(B)$-stable subgroup of D of order 4. Since 16 is always an elementary divisor of C, we have $m_{B_Q}^{(1)}(D) = 1$. Observe that b_Q has defect group D and inertial index 3, so that $l(b_Q) = 3$ by Theorem 13.2. Thus, Lemma 1.43 shows

$$3 = l(b_Q) \geq m_{B_Q}^{(1)}(Q) + m_{B_Q}^{(1)}(D).$$

Therefore, $m_{B_Q}^{(1)}(Q) \leq 2$ and similarly $m_{B_P}^{(1)}(P) \leq 2$. Equation (13.1) yields

$$m_B^{(1)}(Q, b_Q) = m_B^{(1)}(P, b_P) = 2.$$

In particular, 4 occurs as elementary divisor of C with multiplicity 4. It is easy to see that we also have $m_B^{(1)}(Q) = m_B^{(1)}(P) = 2$ which proves the last claim. □

Proposition 13.4 *Let B be a block of a finite group G with elementary abelian defect group D of order 16. If $e(B) = 9$, then Alperin's Weight Conjecture holds for B.*

Proof Let b_D be a Brauer correspondent of B in $C_G(D)$, and let B_D be the Brauer correspondent of B in $N_G(D, b_D)$. Then it suffices to show that $l(B) = l(B_D)$. By Theorem 13.2 we have to consider two cases $l(B) \in \{1, 9\}$. We start with the assumption $l(B) = 9$. Then by Lemma 13.3 there is an $I(B)$-stable subgroup $Q \leq D$ of order 4 such that $m_{B_Q}^{(1)}(Q) = m_B^{(1)}(Q, b_Q) > 0$ where $B_Q := b_Q^{N_G(Q, b_Q)}$. In particular $l(B_Q) = 9$. Let $P \leq D$ be the other $I(B)$-stable subgroup of order 4. Moreover, let $b_P' := b_D^{N_G(Q, b_Q) \cap C_G(P)}$ such that (P, b_P') is a B_Q-subpair. Then by the same argument we get

$$m_\beta^{(1)}(P) = m_{B_Q}^{(1)}(P, b_P') > 0$$

where $\beta := (b_P')^{N_G(Q, b_Q) \cap N_G(P, b_P')}$ is a block with defect group D and $l(\beta) = 9$. Now $D = QP$ implies

$$N_G(D, b_D) \leq N_G(Q, b_Q) \cap N_G(P, b_P') \leq N_G(D).$$

Since $B_D^{N_G(Q, b_Q) \cap N_G(P, b_P')} = \beta$, it follows that $l(B_D) = 9$ as desired.

Now let us consider the case $l(B) = 1$. Here we can just follow the same lines except that we have $m_{B_Q}^{(1)}(Q) = 0$ and $m_\beta^{(1)}(P) = 0$. □

We want to point out that Usami showed in [272] that in case $2 \neq p \neq 7$ there is a perfect isometry between p-blocks with abelian defect group D and inertial quotient C_3^2 and their Brauer correspondents in $N_G(D)$.

Now we present a result on Gluck's Conjecture whose proof is new.

Proposition 13.5 *Gluck's Conjecture holds for the 2-blocks of defect at most 4.*

Proof By Corollary 3.2 and Theorem B in [92] the claim holds for 2-blocks of defect at most 3. Thus, let B be a 2-block with defect group D of order 16.

We may assume that D has exponent 4 and nilpotency class 2 by Lemma 2.1 in [92]. Moreover, by Lemma 3.1 in [92] it suffices to show that the generalized decomposition numbers $d_{\chi\varphi}^u$ of B are (rational) integers. This is trivial if $|\langle u \rangle| \leq 2$. Hence, assume $|\langle u \rangle| = 4$. Let $\mathscr{F}$ be the fusion system of B. Since D is rational, u is not $\mathscr{F}$-conjugate to an element of $Z(D)$. In particular, b_u has defect group $C_D(u)$ of order at most 8. As usual, b_u dominates a block $\overline{b_u}$ with defect at most 1. This shows $l(b_u) = 1$. Now Lemma 3.3 in [92] implies that $d_{\chi\varphi}^u$ is integral. $\qquad \square$

We collect the state of the conjectures for the 2-block of defect at most 4.

Theorem 13.6 *Let B be a 2-block of a finite group with defect at most 4. Then the following conjectures are satisfied for B:*

- *Alperin's Weight Conjecture*
- *Brauer's $k(B)$-Conjecture*
- *Brauer's Height-Zero Conjecture*
- *Olsson's Conjecture*
- *Alperin-McKay Conjecture*
- *Ordinary Weight Conjecture*
- *Gluck's Conjecture*
- *Eaton's Conjecture*
- *Eaton-Moretó Conjecture*
- *Malle-Navarro Conjecture*
- *Robinson's Conjecture*

Moreover, the Gluing Problem for B has a unique solution.

Proof We may assume that B has defect group D of order 16. Then the situation splits into the following possibilities:

(a) D is metacyclic
(b) D is minimal non-abelian
(c) D is abelian, but non-metacyclic
(d) $D \cong D_8 \times C_2$
(e) $D \cong Q_8 \times C_2$
(f) $D \cong D_8 * C_4$

The metacyclic case was done in Corollary 8.2 and the minimal non-abelian case follows from Corollary 12.17. In the last three cases we refer to Theorem 9.1. It remains to consider the abelian case. Here it is known that the Gluing Problem has a unique solution (see [183]). We have two possibilities: $D \cong C_4 \times C_2 \times C_2$ or D is elementary abelian. We may assume that B is non-nilpotent.

In case $D \cong C_4 \times C_2 \times C_2$, 3 is the only odd prime divisor of $|\mathrm{Aut}(D)|$. Thus, by Usami and Puig (see [227, 270]) there is a perfect isometry between B and its Brauer correspondent in $N_G(D)$. Then it is easy to see that the conjectures are true.

Now we consider the elementary abelian case. By Theorem 13.2, Brauer's $k(B)$-Conjecture, Brauer's Height-Zero Conjecture, Olsson's Conjecture, Eaton's Conjecture, the Eaton-Moretó Conjecture, the Malle-Navarro Conjecture and Robinson's

Conjecture are satisfied. Alperin's Weight Conjecture is equivalent to $l(B) = k(I(B))$ unless $e(B) = 9$. However, for $e(B) = 9$, AWC holds by Proposition 13.4. Since $k(B) - l(B) = k_0(B) - l(B)$ is determined locally, the Alperin-McKay Conjecture follows from Alperin's Weight Conjecture. Now consider the Ordinary Weight Conjecture. In case $9 \neq e(B) \neq 21$, the OWC reduces to

$$k(B) = \sum_{\chi \in \mathrm{Irr}(D)/I(B)} |I(\chi)| \tag{13.2}$$

which is true. Now assume $e(B) = 21$. Here the number of 2-blocks of defect 0 in $F[I(B)]$ is 5. We have to insert this number for $|I(\chi)|$ in Eq. (13.2) if χ is invariant under $I(B)$. Finally, let $e(B) = 9$. Here the Brauer correspondent b of B in $N_G(D)$ is Morita equivalent to a twisted group algebra of $D \rtimes I(B)$ (see Theorem 1.19). If the corresponding 2-cocycle α is trivial, we have $l(B) = 9$ and $l(B) = 1$ otherwise. In turn we have $z(F_\alpha I(B)) = 9$ or $z(F_\alpha I(B)) = 1$ respectively. Now the OWC follows as before.   □

Even more information about 2-blocks of defect 4 can be found in [246]. For example in most cases Cartan matrices and the distribution into 2-rational and 2-conjugate characters are known. We omit these information here, since they are of no further use.

We use the previous results to obtain a major theorem about Brauer's $k(B)$-Conjecture.

Theorem 13.7 *Brauer's $k(B)$-Conjecture holds for defect groups with a central cyclic subgroup of index at most 16. In particular, the $k(B)$-Conjecture holds for the 2-blocks of defect at most 5 and 3-blocks of defect at most 3.*

Proof Let B be a p-block of a finite group G with defect group D. By hypothesis, there exists an element $u \in Z(D)$ such that $|D/\langle u \rangle| \leq 16$. Let (u, b_u) be a corresponding (major) subsection. Then b_u dominates a block $\overline{b}_u$ with defect group $D/\langle u \rangle$. Hence, we can apply the previous results. If $D/\langle u \rangle$ is cyclic, then D is abelian of rank at most 2. In this case Brauer's $k(B)$-Conjecture has been known for a long time (see (7D) in [39]). By Theorem 4.10 we may assume that $l(\overline{b}_u) \geq 4$ for $p = 2$. It follows that $D/\langle u \rangle$ is elementary abelian of order 8, 9 or 16. Assume first that $|D/\langle u \rangle| = 8$. Then by Theorem 13.1 we have $l(\overline{b}_u) \in \{5, 7\}$. In case $l(\overline{b}_u) = 7$, Brauer's $k(B)$-Conjecture follows from Theorem 4.2. This also works for $l(\overline{b}_u) = 5$, but here we need to take the quadratic form q corresponding to the positive definite matrix

$$\frac{1}{2} \begin{pmatrix} 2 & 1 & . & . & -1 \\ 1 & 2 & . & . & -1 \\ . & . & 2 & . & -1 \\ . & . & . & 2 & -1 \\ -1 & -1 & -1 & -1 & 2 \end{pmatrix}.$$

Now let $|D/\langle u \rangle| = 9$ (and $p = 3$). Again we use the Cartan method. For sake of simplicity, we assume that B itself has defect group C_3^2. By Theorem 4.9 we may assume $l(B) \geq 3$. Let C be the Cartan matrix of B. By Kiyota's result [154], we need to handle the following cases.

Case 1: $e(B) = 4$.

If the inertial group $I(B)$ is cyclic, we obtain C up to basic sets as follows

$$\begin{pmatrix} 3 & 2 & 2 & 2 \\ 2 & 3 & 2 & 2 \\ 2 & 2 & 3 & 2 \\ 2 & 2 & 2 & 3 \end{pmatrix}$$

from Puig-Usami [227]. If $I(B)$ is non-cyclic, we may also assume that $l(B) = 4$. Here it follows from [226] that C is given by

$$\begin{pmatrix} 4 & 1 & 2 & 2 \\ 1 & 4 & 2 & 2 \\ 2 & 2 & 4 & 2 \\ 2 & 2 & 1 & 4 \end{pmatrix}$$

up to basic sets. In both cases Theorem 4.2 applies. (Later we will handle these situations in a generic way, see Lemma 14.4.)

Case 2: $I(B) \cong C_8$.

Then $I(B)$ acts regularly on $D \setminus \{1\}$. Thus, there are just two B-subsections $(1, B)$ and (u, b) with $l(b) = 1$ up to conjugation. Kiyota did not obtain the block invariants in this case. Hence, we have to consider some possibilities. By Lemma (1D) in [154] we may assume $k(B) \in \{6, 9\}$. Since u is conjugate to u^{-1} in $I(B)$, the generalized decomposition numbers d_{ij}^u are integers. In case $k(B) = 6$ (which contradicts Alperin's Weight Conjecture) the column corresponding to (u, b) in the generalized decomposition matrix is given by $(\pm 2, \pm 1, \pm 1, \pm 1, \pm 1, \pm 1)^{\mathrm{T}}$, and C is

$$\begin{pmatrix} 2 & 1 & 1 & 1 & . \\ 1 & 2 & 1 & 1 & 1 \\ 1 & 1 & 2 & 1 & 1 \\ 1 & 1 & 1 & 2 & 1 \\ . & 1 & 1 & 1 & 3 \end{pmatrix}$$

up to basic sets. In case $k(B) = 9$ we get $C = (1 + \delta_{ij})_{1 \leq i,j \leq 8}$ up to basic sets. In both cases Theorem 4.2 works.

Case 3: $I(B) \cong D_8$.

By Proposition (2F) in [154] we may assume $k(B) = 9$ and $l(B) = 5$. There are three subsections $(1, B)$, (u_1, b_1) and (u_2, b_2) with $l(b_1) = l(b_2) = 2$ up to conjugation. The Cartan matrix of b_1 and b_2 is given by $\left(\begin{smallmatrix} 6 & 3 \\ 3 & 6 \end{smallmatrix}\right)$. The numbers $d_{ij}^{u_1}$ and $d_{ij}^{u_2}$ are integers (see Subcase (a) on page 39 in [154]). Thus, we may assume that the numbers $d_{ij}^{u_1}$ form the two columns

$$\begin{pmatrix} 1\,1\,1\,1\,1\,1 \; . \; . \; . \\ . \; . \; . \,1\,1\,1\,1\,1\,1 \end{pmatrix}^{\mathrm{T}}.$$

Now we use a GAP program to enumerate the possibilities for the columns $(d_{1j}^{u_2}, d_{2j}^{u_2}, \ldots, d_{9j}^{u_2})$ $(j = 1, 2)$. It turns out that C is

$$\begin{pmatrix} 3 & . & 1 & . & 1 \\ . & 3 & 1 & . & 1 \\ 1 & 1 & 3 & 1 & . \\ . & . & 1 & 3 & 1 \\ 1 & 1 & . & 1 & 3 \end{pmatrix}$$

up to basic sets. Here we can take the positive definite quadratic form q corresponding to the matrix

$$\frac{1}{2}\begin{pmatrix} 2 & . & -1 & . & -1 \\ . & 2 & -1 & 1 & -1 \\ -1 & -1 & 2 & -1 & 1 \\ . & 1 & -1 & 2 & -1 \\ -1 & -1 & 1 & -1 & 2 \end{pmatrix}$$

in Theorem 4.2.

Case 4: $I(B) \cong Q_8$.

Then $I(B)$ acts regularly on $D \setminus \{1\}$. Hence, the result follows as in the case $I(B) \cong C_8$.

Case 5: $e(B) = 16$.

Then there are two B-subsections $(1, B)$ and (u, b) up to conjugation. This time we have $l(b) = 2$. By Watanabe [282] we have $k(B) = 9$ and $l(B) = 7$. The Cartan matrix of b is given by $\left(\begin{smallmatrix} 6 & 3 \\ 3 & 6 \end{smallmatrix}\right)$. By way of contradiction, suppose that the columns $d_1 := (d_{11}^u, d_{21}^u, \ldots, d_{91}^u)$ and $d_2 := (d_{12}^u, d_{22}^u, \ldots, d_{92}^u)$ are algebraic conjugate. We write $d_1 = a + b\zeta$ with $a, b \in \mathbb{Z}^9$ and $\zeta := e^{2\pi i/3}$. Then $d_2 = a + b\bar{\zeta}$. The orthogonality relations show that $(a, a) = 5$, $(b, b) = 2$ and $(a, b) = 1$ (cf.

Sect. 5.2). This gives the contradiction $k(B) \leq 6$. Hence, the columns d_1 and d_2 have the form

$$\begin{pmatrix} 1\ 1\ 1\ 1\ 1\ 1\ .\ .\ . \\ .\ .\ .\ 1\ 1\ 1\ 1\ 1\ 1 \end{pmatrix}^{\mathrm{T}}.$$

Thus, we obtain C up to basic sets as follows:

$$\begin{pmatrix} 2\ 1\ .\ .\ .\ .\ 1 \\ 1\ 2\ .\ .\ .\ .\ 1 \\ .\ .\ 2\ 1\ .\ .\ 1 \\ .\ .\ 1\ 2\ .\ .\ 1 \\ .\ .\ .\ .\ 2\ 1\ 1 \\ .\ .\ .\ .\ 1\ 2\ 1 \\ 1\ 1\ 1\ 1\ 1\ 1\ 3 \end{pmatrix}.$$

In this case we can take the positive definite quadratic form q corresponding to the matrix

$$\frac{1}{2}\begin{pmatrix} 2\ -1\ .\ .\ .\ .\ -1 \\ -1\ 2\ .\ .\ .\ .\ . \\ .\ .\ 2\ -1\ .\ .\ -1 \\ .\ .\ -1\ 2\ .\ 1\ . \\ .\ .\ .\ .\ 2\ -1\ -1 \\ .\ .\ .\ 1\ -1\ 2\ . \\ -1\ .\ -1\ .\ -1\ .\ 2 \end{pmatrix}$$

in Theorem 4.2.

Finally, it remains to deal with the case $|D/\langle u \rangle| = C_2^4$. Again we replace $\overline{b}_u$ by B. By Theorem 4.10 we may suppose that $l(B) \geq 4$. We have to settle the following cases according to Theorem 13.2.

Case 1: $e(B) = 5$.

By Theorem 13.2 we have $e(B) = l(B)$. Hence, by the main theorem of [281] we may assume that $D \trianglelefteq G$. Using Proposition 15.2 we obtain

$$C = \begin{pmatrix} 4\ 3\ 3\ 3\ 3 \\ 3\ 4\ 3\ 3\ 3 \\ 3\ 3\ 4\ 3\ 3 \\ 3\ 3\ 3\ 4\ 3 \\ 3\ 3\ 3\ 3\ 4 \end{pmatrix}$$

up to basic sets. The claim follows with Theorem 4.2.

Case 2: $e(B) = 7$.

Again by Theorem 13.2, the main theorem of [281], and Proposition 15.2, we obtain $C = 2(1 + \delta_{ij})_{1 \le i,j \le 7}$ up to basic sets. The claim follows again by Theorem 4.2.

Case 3: $e(B) = l(B) = 9$.

Here we use the inverse Cartan method (see Sect. 4.3). As in Lemma 13.3 we obtain a list of possible Cartan matrices of B. However, since we are considering 9×9 matrices it is very hard to see if two of these candidates only differ by basic sets. In order to reduce the set of possible Cartan matrices further we apply various ad hoc matrix manipulations as permutations of rows and columns and elementary row/column operations. After this procedure we end up with a list of only ten possible Cartan matrices of B which might be all equal up to basic sets. For the purpose of illustration, we display one of these matrices:

$$
\begin{pmatrix}
4 & -1 & 1 & . & 1 & 1 & 2 & . & . \\
1 & 4 & . & 1 & -1 & 1 & . & 1 & 1 \\
1 & . & 4 & 1 & -1 & 1 & 2 & -1 & -1 \\
. & 1 & 1 & 4 & . & . & . & 2 & . \\
1 & -1 & -1 & . & 4 & . & 1 & 1 & 1 \\
1 & 1 & 1 & . & . & 4 & 1 & 1 & 1 \\
2 & . & 2 & . & 1 & 1 & 4 & . & -2 \\
. & 1 & -1 & 2 & 1 & 1 & . & 4 & . \\
. & 1 & -1 & . & 1 & 1 & -2 & . & 4
\end{pmatrix}
$$

(the full list can be found in [256]). It can be seen that all diagonal entries are 4 (for every one of these ten matrices). In order to apply Theorem 4.4, let C be one of these ten matrices. Then we have a positive definite integral quadratic form q corresponding to the matrix $16C^{-1}$. We need to find the minimal non-zero value of q among all integral vectors. More precisely, we have to check if a value strictly smaller than 9 is assumed by q. By Lemma 4.5 it suffices to consider only vectors with entries in $\{0, \pm1\}$. Hence, there are only 3^9 values to consider. An easy computer computation shows that in fact the minimum of q is at least 9. So Brauer's $k(B)$-Conjecture follows from Theorem 4.4.

Case 4: $e(B) = 15$.

There are just two subsections $(1, B)$ and (u, b) with $l(b) = 1$ up to conjugation. The usual argument gives $C = (1+\delta_{ij})_{1 \le i,j \le 15}$ up to basic sets. Hence, Theorem 4.2 applies.

Case 5: $e(B) = 21$.

There are four subsections $(1, B)$, (u_1, b_1), (u_2, b_2) and (u_3, b_3) up to conjugation. We have $l(b_1) = l(b_2) = 3$ and $l(b_3) = 5$ by Theorem 13.1. The Cartan matrix of b_3 is given by

$$\begin{pmatrix} 2 & . & . & . & 1 \\ . & 2 & . & . & 1 \\ . & . & 2 & . & 1 \\ . & . & . & 2 & 1 \\ 1 & 1 & 1 & 1 & 4 \end{pmatrix}.$$

Using this, it is easy to deduce that the generalized decomposition numbers corresponding to (u_3, b_3) can be arranged in the form

$$\begin{pmatrix} 1 & 1 & 1 & 1 & . & . & . & . & . & . & . & . & . & . & . & . \\ . & . & . & . & 1 & 1 & 1 & 1 & . & . & . & . & . & . & . & . \\ . & . & . & . & . & . & . & . & 1 & 1 & 1 & 1 & . & . & . & . \\ . & . & . & . & . & . & . & . & . & . & . & . & 1 & 1 & 1 & 1 \\ . & . & 1 & 1 & . & . & 1 & 1 & . & . & 1 & 1 & . & . & 1 & 1 \end{pmatrix}^{\mathrm{T}}.$$

It is also easy to see that the columns of generalized decomposition numbers corresponding to b_1 and b_2 consist of eight entries ± 1 and eight entries 0. The theory of lower defect groups shows that $m_B^{(1)}(\langle u_1 \rangle, b_1) = m_{b_1}^{(1)}(\langle u_1 \rangle) > 0$. In particular, 2 occurs as elementary divisor of C. Now we use GAP to enumerate all possible arrangements of these columns. It turns out that C is equivalent to the Cartan matrix of b_3. The claim follows.   □

It seems reasonable that one can avoid the use of the classification of the finite simple groups in the proof of Theorem 13.7 just by considering more cases. For example, the original proof of Brauer's $k(B)$-Conjecture for 2-blocks of defect 4 does not rely on the classification (see [245]).

The $k(B)$-Conjecture for defect groups of order 27 extends results of Hendren (see Sect. 6.1 in [108]). Now we prove a similar result.

Theorem 13.8 *Let D be a 2-group and let $u \in Z(D)$ such that $D/\langle u \rangle$ is isomorphic to one of the following groups*

  (i) *a metacyclic group,*
 (ii) *a minimal non-abelian group,*
(iii) $\prod_{i=1}^{n} C_{2^{m_i}}$ *where* $|\{m_i : i = 1, \ldots, n\}| \geq n - 1$,
 (iv) $M \times C$ *where M has maximal class and C is cyclic,*
  (v) $M * C$ *where M has maximal class and C is cyclic,*
 (vi) $D_{2^n} \rtimes C_{2^m}$, $Q_{2^n} \rtimes C_{2^m}$ *and* $D_{2^n}.C_{2^m}$ *as in Theorems 10.23, 10.25 and 10.24.*

Then Brauer's $k(B)$-Conjecture holds for every block with defect group D.

Proof Let B be a block with defect group D. As usual we consider the subsection (u, b_u) and the dominated block $\overline{b_u}$. One of the groups in the theorem appears as defect group of $\overline{b_u}$. By Theorem 4.10, it suffices to show $l(\overline{b_u}) \leq 3$. By our previous results this is true except possibly in case (13.8). Thus, assume that $\overline{b_u}$ has defect group $\prod_{i=1}^{n} C_{2^{m_i}}$ where $|\{m_i : i = 1, \ldots, n\}| \geq n - 1$. Then it is easy to see that $e(\overline{b_u}) \leq 3$ (cf. Lemma 14.10). Hence, results by Usami and Puig [227, 270] imply $l(\overline{b_u}) \leq 3$.   □

It is straightforward to give similar results on $k_0(B)$ by dropping the condition $u \in Z(D)$ in the last two theorems. Here Theorem 4.13 is relevant. We leave the details to the reader.

By means of defect group orders, the next interesting case consists of 5-blocks of defect 2. It is hard to obtain strong results here, but I computed a few Cartan matrices in the unpublished note [256].

13.2   2-Blocks of Defect 5

Since our methods for the prime $p = 2$ are stronger, it is worthwhile to take a look at the defect groups of order 32. One of our aims here is to give a proof of Olsson's Conjecture (for this special case).

For the abelian defect group $C_4 \times C_2^3$ the invariants are not known so far. We handle more general abelian defect groups in the next theorem. This result relies on the classification of the finite simple groups.

Theorem 13.9 *Let B be a block of a finite group G with defect group $C_{2^n} \times C_2^3$ for some $n \geq 2$. Then we have $k(B) = k_0(B) = |D| = 2^{n+3}$ and one of the following holds:*

  (i) $e(B) = l(B) = 1$.
  (ii) $e(B) = l(B) = 3$.
  (iii) $e(B) = l(B) = 7$.
  (iv) $e(B) = 21, l(B) = 5$.

Proof Let $D = C_{2^n} \times C_2^3$. Since $\operatorname{Aut}(D)$ acts faithfully on $\Omega(D)/\Phi(D) \cong C_2^3$, we have $e(B) \in \{1, 3, 7, 21\}$. In case $e(B) = 1$, the block is nilpotent and the result is clear. Now we consider the remaining cases.

Case 1: $e(B) = 3$.

Then there are 2^{n+2} subsections (u, b_u) up to conjugation and 2^{n+1} of them satisfy $l(b_u) = 1$. For the other 2^{n+1} subsections Theorem 1.39 implies $l(b_u) = 3$. This gives $k(B) = 2^{n+3} = |D|$. Moreover, $k(B) = k_0(B)$ by Theorem 7.14.

Case 2: $e(B) = 7$.

Here we have 2^{n+1} subsections (u, b_u) up to conjugation where 2^n of them satisfy $l(b_u) = 1$. For the other 2^n subsections we use Theorem 1.39 in connection with

Theorem 13.1. This gives $l(b_u) = 7$ for these subsections. It follows that $k(B) = |D|$ and $k(B) = k_0(B)$ by Theorem 7.14.

Case 3: $e(B) = 21$.

Here we have again 2^{n+1} subsections (u, b_u) up to conjugation. But this time 2^n subsections satisfy $l(b_u) = 3$ and the other 2^n subsections satisfy $l(b_u) = 5$. The result follows as before. $\square$

Next we study another group of order 32 with an easy structure.

Proposition 13.10 *Let B be a non-nilpotent block of a finite group with defect group $D \cong MNA(2, 1) \times C_2$. Then $k(B) = 20$, $k_0(B) = 16$, $k_1(B) = 4$ and $l(B) = 2$. In particular Olsson's Conjecture and Alperin's Weight Conjecture hold for B.*

Proof Let $\mathscr{F}$ be the fusion system of B. Since $|D : Z(D)| = 4$, every $\mathscr{F}$-essential subgroup is maximal, and there are three candidates for these groups. Let $Z(D) < M < D$ such that $M \cong C_4 \times C_2^2$. Then $\mathrm{Aut}_{\mathscr{F}}(M)$ must act non-trivially on $\Omega(M)/\Phi(M)$. However, it can be seen that $N_D(M)$ acts trivially on $\Omega(M)/\Phi(M)$. In particular M is not $\mathscr{F}$-radical. Hence, there is only one $\mathscr{F}$-essential subgroup $Q \cong C_2^4$ (up to conjugation). Since $Q \trianglelefteq D$, $\mathscr{F}$ is constrained and thus uniquely determined by $\mathrm{Out}_{\mathscr{F}}(Q)$. By Lemma 6.13 we have some possibilities for $\mathrm{Out}_{\mathscr{F}}(Q)$. However, a GAP calculation shows that only $\mathrm{Out}_{\mathscr{F}}(Q) \cong S_3$ is realizable. Then $\mathscr{F}$ is the fusion system on the group $\mathtt{SmallGroup}(96, 194) \cong (A_4 \rtimes C_4) \times C_2$. In particular there are exactly 16 $\mathscr{F}$-conjugacy classes on D. Moreover, $Z(\mathscr{F}) \cong C_2^2$, and for $1 \neq z \in Z(\mathscr{F})$ we have $D/\langle z \rangle \in \{MNA(2, 1), D_8 \times C_2\}$. Hence, we get $l(b_z) = 2$ as usual. For all other non-trivial subsections (u, b_u) we have $l(b_u) \geq 1$. Since B is centrally controlled, Theorem 1.38 implies $l(B) \geq 2$. Theorem 1.35 gives $k(B) \geq 20$. If $x \in D$ has order 4, then $C_D(x)/\langle x \rangle$ has order 4. Hence, Olsson's Conjecture follows from Theorem 4.13, i.e. $k_0(B) \leq |D : D'| = 16$. For an element $z \in Z(D) \setminus Z(\mathscr{F})$ the block b_z is nilpotent. Thus, Proposition 4.7 implies

$$|D| = 32 \leq k_0(B) + 4(k(B) - k_0(B)) \leq \sum_{i=0}^{\infty} 2^{2i} k_i(B) \leq |D|.$$

The claim follows as usual. $\square$

Our next result handles rather unknown groups of order 32. The key observation here is that the fusion system is constrained and thus quite easy to understand.

Proposition 13.11 *Let B be a non-nilpotent block of a finite group G with defect group $D \cong \mathtt{SmallGroup}(32, q)$ for $q \in \{28, 29\}$. Then $k(B) = 14$, $k_0(B) = 8$, $k_1(B) = 6$ and $l(B) = 2$.*

Proof Let $\mathscr{F}$ be the fusion system of B. Using GAP one can show that $\mathrm{Aut}(D)$ is a 2-group. In particular $e(B) = 1$. Moreover, one can show using results in Chap. 6 that D contains only one $\mathscr{F}$-essential subgroup Q. Here $C_2^2 \times C_4 \cong Q \trianglelefteq D$. In particular $\mathscr{F}$ is constrained. Another GAP calculation shows that $\mathscr{F}$ is the fusion system of the group SmallGroup(96, 187) or SmallGroup(96, 185) for $q \in \{28, 29\}$ respectively. We have ten B-subsections up to conjugation. The center of D is a four-group and $\Phi(Q) \subseteq Z(D)$. Hence, an odd order automorphism of Q cannot act on $Z(D)$. It follows that we have four major subsections $(1, B)$, (z, b_z), (v, b_v) and (w, b_w) up to conjugation. Here we may assume that $l(b_v) = l(b_w) = 1$. On the other hand b_z dominates a non-nilpotent block with defect group $D/\langle z \rangle \cong D_8 \times C_2$. Thus, by Theorem 9.7 we have $l(b_z) = 2$. Also we find an element $u \in Q$ such that b_u is non-nilpotent with defect group Q. Here $l(b_u) = 3$ by Usami and Puig [227, 270]. The remaining non-major subsections split into one subsection (u_1, b_1) of defect 16 and four subsections (u_i, b_i) ($i = 2, 3, 4, 5$) of defect 8. Here $l(b_i) = 1$ for $i = 1, \dots, 5$. In particular Olsson's Conjecture $k_0(B) \le 8 = |D : D'|$ follows at once. Since B is centrally controlled, we also obtain $l(B) \ge 2$ and $k(B) \ge 14$. So the generalized decomposition numbers d_{ij}^v consist of eight entries ± 1 and six entries ± 2. Hence, $k(B) = 14$, $k_0(B) = 8$, $k_1(B) = 6$ and $l(B) = 2$. $\square$

Also in the next proposition the corresponding fusion system is easy to understand, since it is controlled. Another advantage here is that $k(B)$ is relatively small so that the computational effort is small as well.

Proposition 13.12 *Let D be a central cyclic extension of* SmallGroup(32, q) *for $q \in \{33, 34\}$. Then Brauer's $k(B)$-Conjecture holds for all blocks with defect group D.*

Proof As in the previous proofs, it suffices to consider a block B with defect group $D \cong$ SmallGroup(32, q) for $q \in \{33, 34\}$. GAP shows that B is a controlled block with inertial index 3. Hence, the fusion system of B is the same as the fusion system of the group $D \rtimes C_3$. It follows that there are only six B-subsections up to conjugation; two of them are major. For $1 \ne z \in Z(D)$ we have $l(b_z) = 1$. Let us denote the four non-major subsections by (u_i, b_i) for $i = 1, \dots, 4$. We may assume that b_1 has defect group C_2^3. It is easy to see that $\operatorname{Aut}_{\mathscr{F}}(D)$ restricts to the inertial group of b_1. In particular $l(b_1) = e(b_1) = 3$. The Cartan matrix of b_1 is given by $2(1 + \delta_{ij})_{1 \le i, j \le 3}$ up to basic sets (see Theorem 13.1). Moreover, b_2 has defect 3 and b_3 and b_4 have defect 4. Here, $l(b_2) = l(b_3) = l(b_4) = 1$. In particular Olsson's Conjecture $k_0(B) \le 8 = |D : D'|$ follows. Looking at d_{ij}^z we get $k(B) \le 14$. The numbers $d_{ij}^{u_1}$ can certainly be arranged in the form

$$
\begin{pmatrix}
1\,1\,1\,1 & . & . & . & . & \cdots & . \\
1\,1\,. & : & 1\,1 & . & . & \cdots & . \\
1\,1\,. & . & . & 1\,1 & . & \cdots & .
\end{pmatrix}^{\mathrm{T}}.
$$

Using the contributions it follows that $k_0(B) = 8$. We can easily add the column for (u_2, b_2) as

$$(1, 1, -1, \dots, -1, 0, \dots, 0)^{\mathrm{T}} \quad \text{or} \quad (1, -1, 1, -1, 1, -1, 1, -1, 0, \dots, 0)^{\mathrm{T}}.$$

We investigate next the elementary divisors of the Cartan matrix of B. For this we consider the multiplicity of $\langle u_1 \rangle$ as a lower defect group. The multiplicity of 2 as an elementary divisor of the Cartan matrix of b_1 is certainly 2. Since $\langle u_1 \rangle$ is the only lower defect group of order 2 of b_1, we have $m^{(1)}_B(\langle u_1 \rangle, b_1) = m^{(1)}_{b_1}(\langle u_1 \rangle) = 2$. This shows $l(B) \geq 3$ and $k(B) \geq 10$. Now we show $m(d) = 0$ for $2 < d < 32$. By way of contradiction suppose that $m^{(1)}_B(Q, b_Q) > 0$ for $Q \leq D$ such that $|Q| = d$. As usual, $m^{(1)}_{B_Q}(Q) > 0$ where $B_Q := b_Q^{N_G(Q, b_Q)}$. We conclude that B_Q is not nilpotent. Since $\mathscr{F}$ is controlled, Q is fully $\mathscr{F}$-normalized. In particular, B_Q has fusion system $N_{\mathscr{F}}(Q)$ (Lemma 1.42). By definition every morphism in $N_{\mathscr{F}}(Q)$ is a restriction of a morphism in $\mathscr{F}$ and thus a restriction from $\mathrm{Aut}_{\mathscr{F}}(D)$. Since B_Q is non-nilpotent, an automorphism $\alpha \in \mathrm{Aut}_{\mathscr{F}}(D)$ of order 3 must act on Q. A GAP calculation shows that Q is abelian and normal in D. In particular, b_Q has fusion system $C_{\mathscr{F}}(Q)$ by Theorem IV.3.19 in [19]. Since α fixes only two elements of D, we derive that b_Q is nilpotent. Now Lemma 1.43 gives the contradiction

$$1 = l(b_Q) \geq m^{(1)}_{B_Q}(Q) + m^{(1)}_{B_Q}(D) \geq 2.$$

Therefore, $m(d) = 0$ for $2 < d < 32$.

We have essentially four possibilities for the numbers d_{ij}^z:

- eight entries ± 1 and six entries ± 2,
- eight entries ± 1, two entries ± 2 and one entry ± 4,
- seven entries ± 1, four entries ± 2 and one entry ± 3,
- six entries ± 1, two entries ± 2 and two entries ± 3.

In particular $k(B)$ determines $k_i(B)$ for $i \geq 1$ uniquely. It remains to add the generalized decomposition numbers corresponding to (u_3, b_3) and (u_4, b_4). Here the situation is distinguished by $q \in \{33, 34\}$. Assume first that $q = 34$. Then u_3^{-1} (resp. u_4^{-1}) is conjugate to u_3 (resp. u_4). Hence, the numbers $d_{ij}^{u_3}$ and $d_{ij}^{u_4}$ are integers. It is easy to see that such a column must consist of the following (non-zero) entries:

- eight entries ± 1 and two entries ± 2,
- seven entries ± 1 and one entry ± 3.

In contrast, for $q = 33$ the elements u_3^{-1} and u_4 are conjugate. So we may assume $u_4 := u_3^{-1}$, and it suffices to consider the column $d_{ij}^{u_3}$ whose entries are Gaussian integers. Let us write $d_{\chi \varphi_3}^{u_3} := a(\chi) + b(\chi)i$ where $\mathrm{IBr}(b_3) = \{\varphi_3\}$, $a, b \in \mathbb{Z}^{k(B)}$ and $i := \sqrt{-1}$. Then $(a, a) = (b, b) = 8$ and $(a, b) = 0$. Since we have only one pair of algebraically conjugate subsections, there is only one pair of 2-conjugate characters (see Lemma IV.6.10 in [81]). This shows that b consists of two entries ± 2. Now $k_0(B) = 8$ implies that a has eight entries ± 1.

As usual we enumerate all these configurations of the generalized decomposition matrix and obtain the Cartan matrix of B as orthogonal space. However, we get two possibilities $l(B) \in \{3, 4\}$. We are not able to exclude the case $l(B) = 4$ despite it contradicts Alperin's Weight Conjecture. Anyway in both cases $l(B) \in \{3, 4\}$ all candidates for the Cartan matrix satisfy Theorem 4.2. The claim follows.   □

We add a short discussion about the defect group

$$D := \mathtt{SmallGroup}(32, 27)$$

$$\cong \langle a, b, c \mid a^2 = b^2 = c^2 = [a, b] = [a, {}^c a] = [{}^c a, b] = [b, {}^c b] = 1 \rangle \cong C_2^4 \rtimes C_2.$$

Let $\mathscr{F}$ be a non-nilpotent fusion system on D. It can be shown that $Q := \langle a, b, {}^c a, {}^c b \rangle \cong C_2^4$ is the only possible $\mathscr{F}$-essential subgroup. In particular, $\mathscr{F}$ is constrained or even controlled. In the controlled case we have

$$\mathscr{F} = \mathscr{F}_D(D \rtimes C_3) = \mathscr{F}_D(\mathtt{SmallGroup}(96, 70)).$$

In the non-controlled case we have various possibilities for $\mathscr{F}$ according to $\mathrm{Out}_{\mathscr{F}}(Q) \in \{S_3, D_{10}, S_3 \times C_3, \mathtt{SmallGroup}(18, 4), D_{10} \times C_3\}$ (see Lemma 6.13). These possibilities are represented by the following groups:

- $\mathtt{SmallGroup}(96, 195)$,
- $\mathtt{SmallGroup}(96, 227)$,
- $\mathtt{SmallGroup}(160, 234)$,
- $\mathtt{SmallGroup}(288, 1025)$,
- $\mathtt{SmallGroup}(288, 1026)$,
- $\mathtt{SmallGroup}(480, 1188)$.

Here observe that in case $\mathrm{Out}_{\mathscr{F}}(Q) = S_3$ there are essentially two different actions of $\mathrm{Out}_{\mathscr{F}}(Q)$ on Q. The cases $\mathrm{Out}_{\mathscr{F}}(Q) \in \{S_3 \times C_3, \mathtt{SmallGroup}(18, 4)\}$ also differ by $\mathrm{Out}_{\mathscr{F}}(D) \in \{C_3, 1\}$ respectively. Additionally, in case $\mathrm{Out}_{\mathscr{F}}(Q) = \mathtt{SmallGroup}(18, 4)$ there exists a non-trivial 2-cocycle on $\mathrm{Out}_{\mathscr{F}}(Q)$ (on the other hand the Künneth formula implies $\mathrm{H}^2(S_3 \times C_3, F^\times) = 0$). This gives even more examples for blocks with defect group D. For example a non-principal 2-block of $\mathtt{SmallGroup}(864, 3996)$ has defect group D and only one irreducible Brauer character. In all these examples $l(B)$ assumes the values $1, 2, 3, 5, 6, 9$. We will not consider the block invariants in full generality although it might be possible. We also end the discussion about the remaining groups of order 32. In most cases (especially when 9×9 Cartan matrices show up) the computational effort to compute the corresponding block invariants is too big. We also do not state the partial results on the extraspecial defect groups $D_8 * D_8$ and $D_8 * Q_8$ which were obtained in [252].

In Table 13.1 we enumerate all groups of order 32 by using the Small Groups Library and give information about blocks with corresponding defect groups. In many cases it can be shown with GAP that there are no non-trivial fusion systems. These cases were also determined in [275]; however with the Hall-Senior enumeration [103]. Using a conversion between both enumerations provided by Eamonn O'Brien (see [197, 209]), we confirm the results in [275]. We denote the modular group of order $2^n \geq 16$ by M_{2^n}, i.e. the unique group of class 2 with a cyclic maximal subgroup.

We prove some consequences.

Table 13.1 Defect groups of order 32

Small group id	Structure	Invariants	Comments	Reference
1	C_{32}	Known	Nilpotent	
2	$MNA(2,2)$	Known	Controlled	Theorem 12.4
3	$C_8 \times C_4$	Known	Nilpotent	
4	$C_8 \rtimes C_4$	Known	Nilpotent	Theorem 8.1
5	$MNA(3,1)$	Known		Theorem 12.4
6	$MNA(2,1) \rtimes C_2$	Known	Nilpotent	GAP
7	$M_{16} \rtimes C_2$	Known	Nilpotent	GAP
8	$C_2.MNA(2,1)$	Known	Nilpotent	GAP
9	$D_8 \rtimes C_4$	Known	Bicyclic	Theorem 10.23
10	$Q_8 \rtimes C_4$	Known	Bicyclic	Theorem 10.25
11	$C_4 \wr C_2$	Known		[160]
12	$C_4 \rtimes C_8$	Known	Nilpotent	Theorem 8.1
13	$C_8 \rtimes C_4$	Known	Nilpotent	Theorem 8.1
14	$C_8 \rtimes C_4$	Known	Nilpotent	Theorem 8.1
15	$C_8.C_4$	Known	Nilpotent	Theorem 8.1
16	$C_{16} \times C_2$	Known	Nilpotent	
17	M_{32}	Known	Nilpotent	Theorem 8.1
18	D_{32}	Known	Maximal class	Theorem 8.1
19	SD_{32}	Known	Maximal class	Theorem 8.1
20	Q_{32}	Known	Maximal class	Theorem 8.1
21	$C_4^2 \times C_2$	Known	Controlled	[270]
22	$MNA(2,1) \times C_2$	Known	Constrained	Proposition 13.10
23	$(C_4 \rtimes C_4) \times C_2$	Known	Nilpotent	GAP
24	$C_4^2 \rtimes C_2$	Known	Nilpotent	GAP
25	$D_8 \times C_4$	Known		Theorem 9.7
26	$Q_8 \times C_4$	Known		Theorem 9.28
27	$C_2^4 \rtimes C_2$			
28	$(C_4 \times C_2^2) \rtimes C_2$	Known	Constrained	Proposition 13.11
29	$(Q_8 \times C_2) \rtimes C_2$	Known	Constrained	Proposition 13.11
30	$(C_4 \times C_2^2) \rtimes C_2$	Known	Nilpotent	GAP
31	$(C_4 \times C_4) \rtimes C_2$	Known	Nilpotent	GAP
32	$C_2^2.C_2^3$	Known	Nilpotent	GAP
33	$(C_4 \times C_4) \rtimes C_2$	Partly	Controlled	Proposition 13.12
34	$(C_4 \times C_4) \rtimes C_2$	Partly	Controlled	Proposition 13.12
35	$C_4 \rtimes Q_8$	Known	Nilpotent	GAP
36	$C_8 \times C_2^2$	Known	Controlled	[270]
37	$M_{16} \times C_2$	Known	Nilpotent	GAP
38	$D_8 * C_8$	Known		Theorem 9.18
39	$D_{16} \times C_2$	Known		Theorem 9.7
40	$SD_{16} \times C_2$	Known		Theorem 9.37

(continued)

Table 13.2 (continued)

Small group id	Structure	Invariants	Comments	Reference
41	$Q_{16} \times C_2$	Known		Theorem 9.28
42	$D_{16} * C_4$	Known		Theorem 9.18
43	$(D_8 \times C_2) \rtimes C_2$			
44	$(Q_8 \times C_2) \rtimes C_2$			
45	$C_4 \times C_2^3$	Known	Controlled	Theorem 13.9
46	$D_8 \times C_2^2$			
47	$Q_8 \times C_2^2$		Controlled	
48	$(D_8 * C_4) \times C_2$		Controlled	
49	$D_8 * D_8$	Partly	Controlled	[252]
50	$D_8 * Q_8$	Partly	Controlled	[252]
51	C_2^5		Controlled	

Proposition 13.13 *Let D be a 2-group and let $u \in Z(D)$ such that $D/\langle u \rangle$ is isomorphic to one of the following groups*

(i) SmallGroup$(32, q)$ *for $q \in \{11, 22, 28, 29, 33, 34\}$,*
(ii) a group which admits only the nilpotent fusion system.

Then Brauer's $k(B)$-Conjecture holds for every 2-block with defect group D.

Proof This is an application of Theorem 4.10. For the wreath product we refer to [160]. All other cases were handled above.   □

One can use GAP and the previous results to verify Brauer's $k(B)$-Conjecture for 244 of the 267 defect groups of order 64. Here we also use the following elementary observation: Let $z \in Z(D)$ such that every fusion system on $D/\langle z \rangle$ is controlled. If $C_{\mathrm{Aut}(D)}(z)$ is a 2-group, then Brauer's $k(B)$-Conjecture holds for every block with defect group D (cf. Proposition 11.1).

For the purpose of further research we state all indices q such that Brauer's $k(B)$-Conjecture for the defect group SmallGroup$(64, q)$ is *not* known so far:

$$134, 135, 136, 137, 138, 139, 202, 224, 229, 230, 231, 238,$$

$$239, 242, 254, 255, 257, 258, 259, 261, 262, 264, 267.$$

This implies the following corollary.

Corollary 13.14 *Let B be a 2-block with defect group D of order at most 64. If D is generated by two elements, then Brauer's $k(B)$-Conjecture holds for B.*

Corollary 13.15 *Let D be a 2-group containing a cyclic subgroup of index at most 4. Then Brauer's $k(B)$-Conjecture holds for every block with defect group D.*

Proof We may assume that D is not metacyclic. In particular, $|D|/\exp D = 4$. If D is abelian, the result follows from Theorem 13.7. Hence, let us assume that D

is non-abelian. Then D is one of the groups given in Theorem 2 in [205]. We will consider this list of groups case by case and apply the results above. In many cases we get a cyclic central extension of a metacyclic group where Theorem 13.8 applies. We remark that the terms "quasi-dihedral" and "semidihedral" have different meanings in [205].

The group G_1 is metacyclic. For the groups G_2 and G_3 we even know the block invariants precisely. Now consider G_4. Here the element a lies in the center. In particular the group is a cyclic central extension of a group of order 4. The $k(B)$-Conjecture follows. For the group G_5 the element b lies in the center. Moreover, $G_5/\langle b \rangle$ is abelian and has a cyclic subgroup of index 2. Again the claim holds. The groups G_6, G_7, G_8 and G_9 are metacyclic. The groups G_{10} and G_{11} are cyclic central extensions of metacyclic groups. In G_{12} the subgroup $\langle a \rangle$ is normal; in particular $a^{2^{m-3}} \in Z(G_{12})$. Moreover, b is central in $G_{12}/\langle a^{2^{m-3}} \rangle$ and $G_{12}/\langle a^{2^{m-3}} \rangle \cong D_{2^{m-2}} \times C_2$. The claim follows. In G_{13} and G_{14} we see that b is central and the corresponding quotient is certainly metacyclic. Next, $a^{2^{m-3}} \in Z(G_{15})$ and $G_{15}/\langle a^{2^{m-3}} \rangle \cong D_{2^{m-2}} \times C_2$. Exactly the same argument applies to G_{16}. For G_{17} we have $c^{-1}a^2c = abab = a^{2+2^{m-3}}$ and $a^4 \in Z(G_{17})$. Since $G_{17}/\langle a^4 \rangle$ has order 16, the claim follows.

The group G_{18} is slightly more complicated. In general, the core of $\langle a \rangle$ has index at most 8. Thus, $a^{2^{m-3}}$ is always central (in all of these groups). Adjusting notation slightly gives

$$G_{18}/\langle a^{2^{m-3}} \rangle \cong \langle a, b, c \mid a^{2^{m-3}} = b^2 = c^2 = [a, b] = 1, \ cac = a^{-1}b \rangle.$$

We define new elements in this quotient by $\tilde{v} := a^2 b$, $\tilde{x} := bc$ and $\tilde{a} := ac$. Then $\tilde{v}^{2^{m-4}} = 1$, $\tilde{a}^2 = b$ and $\tilde{a}^4 = 1$. Moreover, $cbc = c(acac)c = b$. It follows that $\tilde{x}^2 = 1$ and $\tilde{x}\tilde{v}\tilde{x} = \tilde{v}^{-1}$. Hence, $\langle \tilde{v}, \tilde{x} \rangle \cong D_{2^{m-3}}$. Now $\tilde{a}\tilde{v}\tilde{a}^{-1} = ca^2bc = a^{-2}b = \tilde{v}^{-1}$ and finally $\tilde{a}\tilde{x}\tilde{a}^{-1} = a^2c = \tilde{v}\tilde{x}$. Since $G_{18}/\langle a^{2^{m-3}} \rangle = \langle \tilde{v}, \tilde{x}, \tilde{a} \rangle$, we see that this is precisely the group from Theorem 10.23. The claim follows.

The groups G_{19}, G_{20} and G_{21} are metacyclic. In G_{22} the element a^4 is central and $G_{22}/\langle a^4 \rangle$ has order 16. Let us consider G_{23}. Similarly as above we have

$$G_{23}/\langle a^{2^{m-3}} \rangle \cong \langle a, b, c \mid a^{2^{m-3}} = b^2 = c^2 = [a, b] = 1, \ cac = a^{-1+2^{m-4}}b \rangle$$

(observe that the relation $[b, c] \equiv 1 \pmod{\langle a^{2^{m-3}} \rangle}$ follows from $b \equiv a^{1+2^{m-4}}cac$). Here we define $\tilde{v} := a^{2+2^{m-4}}b$, $\tilde{x} := bc$ and $\tilde{a} := ac$. Again $\langle \tilde{v}, \tilde{x} \rangle \cong D_{2^{m-3}}$. Moreover, $\tilde{a}^2 = a^{2^{m-4}}b$, $\tilde{a}^4 = 1$ and $\tilde{a}\tilde{x}\tilde{a}^{-1} = bca^{-1}cac = a^{2+2^{m-4}}c = \tilde{v}\tilde{x}$. So $G_{23}/\langle a^{2^{m-4}}b \rangle$ is the group from Theorem 10.23. Now it is easy to see that $G_{24}/\langle a^{2^{m-3}} \rangle \cong G_{25}/\langle a^{2^{m-3}} \rangle \cong G_{23}/\langle a^{2^{m-3}} \rangle$. Finally the group G_{26} has order 32; so also here the $k(B)$-Conjecture holds. This completes the proof.   □

For every integer $n \geq 6$ there are exactly 33 groups of order 2^n satisfying the hypothesis of Corollary 13.15. For Olsson's Conjecture we get partial results.

Proposition 13.16 *Let D be a 2-group and $x \in D$ such that $|D : \langle x \rangle| \leq 4$, and suppose that one of the following holds:*

(i) x is conjugate to x^{-5^n} in D for some $n \in \mathbb{Z}$,
(ii) $\langle x \rangle \trianglelefteq D$.

Then Olsson's Conjecture holds for all blocks with defect group D.

Proof Let B be a block with defect group D and fusion system $\mathscr{F}$. We may assume that D is non-metacyclic.

(i) By hypothesis, x is conjugate to x^{-5^n} in $\mathscr{F}$. This condition is preserved if we replace x by an $\mathscr{F}$-conjugate. Hence, we may assume that $\langle x \rangle$ is fully $\mathscr{F}$-normalized. Then x is conjugate to x^{-5^n} in D. In particular, $|C_D(x)/\langle x \rangle| \leq 2$. Hence, b_x dominates a block of $C_G(x)/\langle x \rangle$ with cyclic defect group $C_D(x)/\langle x \rangle$. This shows $l(b_x) = 1$. Now we can apply Theorem 5.3 which gives $k_0(B) \leq 8$. In case $|D : D'| = 4$ a theorem of Taussky (see Satz III.11.9 in [128]) implies that D has maximal class which was excluded.
(ii) We consider the order of $C_D(x)$.

Case (1): $C_D(x) = \langle x \rangle$.

Since D is non-metacyclic, $D/\langle x \rangle$ is non-cyclic. Hence, we are in case (13.16).

Case (2): $x \in Z(D)$.

If D is abelian, the result follows from Corollary 13.15. Thus, we may assume that D is non-abelian. Then every conjugacy class of D has length at most 2. By a result of Knoche (see for example Aufgabe III.24b in [128]) this is equivalent to $|D'| = 2$. Let $y \in D \setminus Z(D)$. Then $C_D(y)$ is non-cyclic. After replacing y by xy if necessary, we have $|\langle x \rangle| = |\langle y \rangle|$. By Proposition 4.3 it suffices to show that $\langle y \rangle$ is fully $\mathscr{F}$-normalized. By Alperin's Fusion Theorem every $\mathscr{F}$-isomorphism on $\langle y \rangle$ is a composition of automorphisms of $\mathscr{F}$-essential subgroups containing y or of D itself. Assume that $E < D$ is $\mathscr{F}$-essential such that $\langle y \rangle \leq E$. Since E is metacyclic and $\mathrm{Aut}(E)$ is not a 2-group, Proposition 10.2 implies $E \cong Q_8$ or $E \cong C_2 \times C_2$. In particular, $|D| \leq 16$. Moreover, Proposition 6.11 implies that D has maximal class. This contradiction shows that there are no $\mathscr{F}$-essential subgroups containing y. Then of course $\langle y \rangle$ is fully $\mathscr{F}$-normalized.

Case (3): $|C_D(x)/\langle x \rangle| = 2$.

Let $y \in C_D(x) \setminus \langle x \rangle$ be of order 2. If $z \in D \setminus C_D(x)$, we may assume that $\langle x, z \rangle$ is a modular 2-group by (13.16). In particular we have $|\langle z \rangle| = 2$ after replacing z by zx^m for some $m \in \mathbb{Z}$ if necessary. Let $|\langle x \rangle| = 2^r$ for some $r \in \mathbb{N}$. Since $\langle x \rangle \trianglelefteq D$, we have $zyz^{-1} \in \{y, yx^{2^{r-1}}\}$. In case $zyz^{-1} = yx^{2^{r-1}}$ it is easy to see that $|D : \langle xy \rangle| = 4$ and $xy \in Z(D)$. Then we are done by case (2). Thus, we may assume that $zyz^{-1} = y$ and $y \in Z(D)$. Then D is given as follows:

$$D = \langle x, z \rangle \times \langle y \rangle \cong M_{2^{r+1}} \times C_2.$$

Now we have $|D'| = 2$ and the claim follows from Proposition 4.3 applied to the subsection (x, b_x). Here observe that $\langle x \rangle$ is fully $\mathscr{F}$-normalized, since $\langle x \rangle \trianglelefteq D$. $\quad\square$

Theorem 13.17 *Olsson's Conjecture holds for all 2-blocks of defect at most 5.*

Proof Let B be a block with defect group D of order 32. Assume first that B is controlled. One can show with GAP that there is always an element $x \in D$ such that $|C_D(x)| = |D : D'|$. If in addition D is abelian, Olsson's Conjecture coincides with Brauer's $k(B)$-Conjecture and we are done. If D is non-abelian, then $|C_D(x)/\langle x \rangle| \le 8$. Thus, an application of Theorems 4.2 and 13.1 gives Olsson's Conjecture.

Now suppose that B is not controlled. Then by Table 13.1, it suffices to consider only the defect groups $D := \texttt{SmallGroup}(32, m)$ where $m \in \{27, 43, 44, 46\}$. Let $\mathscr{F}$ be the fusion system of B. Then we can find (with GAP) an element $u \in D$ such that $|C_D(u)| = |D : D'|$. Moreover, we can choose u such that every element $v \in D$ of the same order also satisfies $|C_D(u)| = |D : D'|$. Hence, the subgroup $\langle u \rangle$ is fully $\mathscr{F}$-centralized. In particular $C_D(u)$ is a defect group of the block b_u. Since $|C_D(u)/\langle u \rangle| \le 8$, the claim follows as before. $\quad\square$

13.3 Minimal Non-metacyclic Defect Groups

In this section we prove a minor result on minimal non-metacyclic defect groups. As usual, minimal non-metacyclic means the whole group is not metacyclic, but all proper subgroups are. Blackburn [30] showed that there are only five minimal non-metacyclic 2-groups. This allows us the give a complete classification of the corresponding blocks. This result appeared in [246].

Theorem 13.18 *Let B be a 2-block with minimal non-metacyclic defect group D. Then one of the following holds:*

(i) *B is nilpotent.*
(ii) *$D \cong C_2^3$. Then $k(B) = k_0(B) = 8$ and $l(B) \in \{3, 5, 7\}$.*
(iii) *$D \cong Q_8 \times C_2$ or $D \cong D_8 * C_4$. Then $k(B) = 14$, $k_0(B) = 8$, $k_1(B) = 6$ and $l(B) = 3$.*

Proof By Theorem 66.1 in [24], D is isomorphic to C_2^3, $Q_8 \times C_2$, $D * C_4$ or to $\texttt{SmallGroup}(32, 32)$. Hence, the result follows from Theorems 13.1, 9.28, 9.18 and Table 13.1. $\quad\square$

Chapter 14
Abelian Defect Groups

14.1 The Brauer-Feit Bound

Let B be a p-block of a finite group G with defect d. Then there is a well-known bound on $k(B)$ proved by Brauer and Feit in 1959.

Theorem 14.1 (Brauer-Feit [42]) *If $d > 2$, then $k(B) < p^{2d-2}$.*

In this chapter (which is an enhanced version of [253]) we are interested in the case where B has an abelian defect group D. Brauer himself already verified the $k(B)$-Conjecture if D is abelian of rank at most 2. For abelian defect groups of rank 3, he obtained $k(B) < p^{5d/3}$ (see for example Theorem VII.10.13 in [81]; observe that $<$ and $\leq$ are mixed up there).

Using a recent result by Halasi and Podoski [101] we substantially improve the Brauer-Feit bound for abelian defect groups.

Theorem 14.2 *Let B be a p-block of a finite group with abelian defect group of order $p^d > p$. Then*

$$k(B) < p^{3d/2-1/2}. \tag{14.1}$$

Proof Let D be a defect group of B. By Corollary 1.2 in [101] there exist elements $x, y \in D$ such that $C_{I(B)}(x) \cap C_{I(B)}(y) = 1$. Without loss of generality, $x \neq 1$. Consider a B-subsection (x, b_x). As usual, b_x dominates a block $\overline{b_x}$ with defect group $\overline{D} := D/\langle x \rangle$ and $I(\overline{b_x}) \cong C_{I(B)}(x)$. We write $\overline{y} := y\langle x \rangle \in \overline{D}$. Choose a $\overline{b_x}$-subsection $(\overline{y}, \beta_{\overline{y}})$ and $\alpha \in I(\beta_{\overline{y}})$. We may regard α as an element of $C_{I(B)}(x)$. Hence, α acts trivially on $\langle x \rangle$ and on $\langle x, y \rangle / \langle x \rangle$. Since α is a p'-element, it must act trivially on $\langle x, y \rangle$ (see for example Theorem 5.3.2 in [94]). This shows $\alpha = 1$ and

© Springer International Publishing Switzerland 2014
B. Sambale, *Blocks of Finite Groups and Their Invariants*, Lecture Notes in Mathematics 2127, DOI 10.1007/978-3-319-12006-5_14

$e(\beta_{\overline{y}}) = 1$. Thus, $\overline{b}_x$ satisfies the $k(B)$-Conjecture. In particular, $l(b_x) = l(\overline{b}_x) < k(\overline{b}_x) \le |\overline{D}| \le p^{d-1}$ (or $l(\overline{b}_x) = k(\overline{b}_x) = 1 < p^{d-1}$). Since B has abelian defect groups, Theorem 7.14 shows $k(B) = k_0(B)$. Now Theorem 4.12 implies

$$k(B) \le p^d \sqrt{l(b_x)} < p^{3d/2-1/2}. \qquad \square$$

Robinson [232, Theorem 2.1(iii)] gave a proof of Eq. (14.1) under the hypothesis that p does not belong to a finite set of primes which depends on the rank of D. For $p = 2$, Theorem 14.2 can be improved further by invoking Theorem 4.13 (see Proposition 14.15). In special situations one may choose $x \in D$ in the proof above such that the order of x is large. We illustrate this by an example. Suppose $D \cong C_{p^n}^m$ for some $n, m \in \mathbb{N}$. Then $I(B)$ acts faithfully on $D/\Phi(D)$. Thus, by Halasi and Podoski [101] we may assume that x has order p^n. Then Eq. (14.1) becomes $k(B) \le p^{3d/2-n/2}$.

14.2   Abelian Groups of Small Rank

Theorem 14.2 already improves Brauer's bound for abelian defect groups of rank 3. We give an even better bound.

Proposition 14.3 *Let B be a p-block of a finite group with abelian defect group of rank 3 and order p^d. Then*

$$k(B) < p^{4d/3}.$$

Proof Let D be a defect group of B, and let $x \in D$ be an element of maximal order p^c. Then for the B-subsection (x, b_x) the block b_x dominates a block $\overline{b}_x$ with defect group $D/\langle x \rangle$ of rank 2. Hence, $l(b_x) = l(\overline{b}_x) < k(\overline{b}_x) \le |D/\langle x \rangle| = p^{d-c}$. Since D has rank 3, it follows that $p^{d-c} \le p^{2d/3}$. By Theorem 7.14, we have $k(B) = k_0(B)$. Thus, Theorem 4.12 implies

$$k(B) \le p^d \sqrt{l(b_x)} < p^{4d/3}. \qquad \square$$

In the following we improve Proposition 14.3 for small primes.

Lemma 14.4 *Let D be an abelian p-group, and let $A \le \mathrm{Aut}(D)$ be a p'-group such that $|A| \le 4$ or $A \cong S_3$. Then for the Cartan matrix $C = (c_{ij})$ of $F[D \rtimes A]$*

there exists a positive definite, integral quadratic form $q = \sum_{1 \leq i \leq j \leq k(A)} q_{ij} x_i x_j$ such that

$$\sum_{1 \leq i \leq j \leq k(A)} q_{ij} c_{ij} \leq |D|.$$

Proof Let $H := D \rtimes A$. After going over to $H/Z(H)$, we may assume that $Z(H) = 1$ and $A \neq 1$. Now we determine the decomposition matrix of FH by discussing the various isomorphism types of A. Assume first that $|A| = 2$. The irreducible Brauer characters of H are just the inflations of $H/D \cong C_2$. Since $D = [D, A] \subseteq H' \subseteq D$ (see Theorem 5.2.3 in [94]), we see that H has just two linear characters. Hence, the character group $\hat{D} := \mathrm{Irr}(D) \cong D$ splits under the action of A into one orbit of length 1 (containing the trivial character) and $(|D| - 1)/2$ orbits of length 2. We compute the irreducible (ordinary) characters of H via induction. The trivial character contributes two rows $(1, 0)$, $(0, 1)$ to the decomposition matrix of H. An orbit of length 2 in $\hat{D}$ gives just one row $(1, 1)$. For $\chi \in \mathrm{Irr}(H)$ we denote the corresponding row in the decomposition matrix by r_χ. Let $q = x_1^2 + x_2^2 - x_1 x_2$ the positive definite quadratic form corresponding to the Dynkin diagram of type A_2. Then we have

$$\sum_{1 \leq i \leq j \leq 2} q_{ij} c_{ij} = \sum_{\chi \in \mathrm{Irr}(H)} q(r_\chi) = k(H) \leq |D|.$$

Here the last inequality holds by the affirmative solution of Brauer's $k(B)$-Conjecture for solvable groups, but one could certainly use more elementary arguments. Exactly the same proof works for $|A| = 3$.

Suppose next that $A \cong C_4$. Here the action of A on $\hat{D}$ gives one orbit of length 1, α orbits of length 2, and β orbits of length 4. As before we get rows of the form $(1, 0, 0, 0)$, $(0, 1, 0, 0)$, $(0, 0, 1, 0)$, $(0, 0, 0, 1)$ and $(1, 1, 1, 1)$ in the decomposition matrix. Let $\chi \in \hat{D}$ be a character in an orbit of length 2. Then χ extends to $D \rtimes \Phi(A)$. Hence, if we arrange the Brauer characters of H suitably, χ contributes two rows $(1, 1, 0, 0)$ and $(0, 0, 1, 1)$ to the decomposition matrix. Again we have $q(r_\chi) = 1$ for all $\chi \in \mathrm{Irr}(H)$ where q is the quadratic form corresponding to the Dynkin diagram of type A_4. The claim follows.

The case $A \cong C_2^2$ is slightly more complicated. First note that $p > 2$. Again $\hat{D}$ splits into one orbit of length 1, α orbits of length 2, and β orbits of length 4. Suppose first that there is an element $1 \neq g \in A$ which acts freely on $\hat{D}$. In this case we may arrange the four irreducible Brauer characters of H in such a way that every row of the decomposition matrix has the form $(1, 0, 0, 0)$, $(0, 1, 0, 0)$, $(0, 0, 1, 0)$, $(0, 0, 0, 1)$, $(1, 1, 0, 0)$, $(0, 0, 1, 1)$, $(1, 0, 0, 1)$, $(0, 1, 1, 0)$ or $(1, 1, 1, 1)$. Let q be the quadratic form corresponding to the positive definite matrix

$$\frac{1}{2} \begin{pmatrix} 2 & -1 & 1 & -1 \\ -1 & 2 & -1 & . \\ 1 & -1 & 2 & -1 \\ -1 & . & -1 & 2 \end{pmatrix}.$$

Then it can be seen that $q(r_\chi) = 1$ for every $\chi \in \mathrm{Irr}(H)$. The claim follows as above. Now we treat the case where every non-trivial element of A has a non-trivial fixed point on $\hat{D}$. We write $A = \{1, g_1, g_2, g_3\}$, $A_i := C_{\hat{D}}(g_i)$ and $\alpha_i := |A_i| > 1$ for $i = 1, 2, 3$. Without loss of generality, $\alpha_1 \le \alpha_2 \le \alpha_3$. Since A acts faithfully on $\hat{D}$, we have $A_2 \cap A_3 = 1$ and $A_2 \times A_3 \le \hat{D}$. Moreover, $\alpha = (\alpha_1 + \alpha_2 + \alpha_3 - 3)/2$ and $\beta = (|D| - \alpha_1 - \alpha_2 - \alpha_3 + 2)/4 \ge (\alpha_2\alpha_3 - \alpha_1 - \alpha_2 - \alpha_3 + 2)/4$. Now the inequality

$$\alpha \le 3(\beta - 1)$$

reduces to $\alpha_1 + \alpha_2 + \alpha_3 \le 3\alpha_3 \le \alpha_2\alpha_3$ which is true since $\alpha_2 \ge p > 2$. We may arrange the irreducible Brauer characters of H such that the decomposition matrix consists of $(\alpha_1 - 1)/2$ pairs of rows $(1, 0, 1, 0)$, $(0, 1, 0, 1)$, $(\alpha_2 - 1)/2$ pairs of the form $(1, 0, 0, 1)$, $(0, 1, 1, 0)$, and $(\alpha_3 - 1)/2$ pairs of the form $(1, 1, 0, 0)$, $(0, 0, 1, 1)$. Let q be the quadratic form corresponding to the Dynkin diagram of type A_4. Then $q(1, 0, 1, 0) = q(0, 1, 0, 1) = q(1, 0, 0, 1) = 2$ and $q(r) = 1$ for all other types of rows r. Since $(\alpha_3 - 1)/2 \ge \alpha/3$ and $(\alpha_1 - 1)/2 \le \alpha/3$, it follows that

$$\sum_{1 \le i \le j \le 4} q_{ij}c_{ij} = \sum_{\chi \in \mathrm{Irr}(H)} q(r_\chi) \le 4 + \frac{2}{3}\alpha + \alpha + \frac{4}{3}\alpha + \beta$$

$$= 4 + 3\alpha + \beta \le 1 + 2\alpha + 4\beta = |\hat{D}| = |D|.$$

Finally assume that $A \cong S_3$. Then $p \ge 5$. We may arrange the three irreducible Brauer characters of H such that their degrees are $(1, 2, 1)$. As above we get three rows in the decomposition matrix $(1, 0, 0)$, $(0, 1, 0)$ and $(0, 0, 1)$. Again we consider the action of A on $\hat{D}$. Let α be the number of orbits of length 2, let β the number of orbits of length 3, and let γ be the number of regular orbits. Then we get α triples of rows $(0, 1, 0)$, $(0, 1, 0)$, $(1, 0, 1)$, β pairs of rows $(1, 1, 0)$, $(0, 1, 1)$, and γ rows of the form $(1, 2, 1)$ in the decomposition matrix of H. Let q be the quadratic form corresponding to the Dynkin diagram of type A_3. We discuss some special cases separately. In case $\alpha = 0$ we obtain with the notation introduced above:

$$\sum_{1 \le i \le j \le 3} q_{ij}c_{ij} = \sum_{\chi \in \mathrm{Irr}(H)} q(r_\chi) = 3 + 2\beta + 2\gamma \le 1 + 3\beta + 6\gamma = |D|.$$

Thus, in the following we suppose that $\alpha > 0$. Let $h \in A$ be an element of order 3 and $A_1 := C_{\hat{D}}(h)$. Obviously, $\alpha = (|A_1| - 1)/2 \ge 2$, since $p \ge 5$. We denote the three involutions in A by g_1, g_2 and g_3. Moreover, let $B_i := C_{\hat{D}}(g_i)$. It is easy to see that h permutes the sets B_1, B_2 and B_3 transitively. In particular, $\beta = |B_i| - 1$. Also, $A_1 \cap B_1 = 1$ and $A_1 \times B_1 \le \hat{D}$. We conclude that

$$\gamma = \frac{|D| - 2\alpha - 3\beta - 1}{6} \ge \frac{(2\alpha + 1)(\beta + 1) - 2\alpha - 3\beta - 1}{6} = \frac{\alpha\beta - \beta}{3}.$$

In case $\beta > 0$ we even have $\beta \geq p - 1 \geq 4$ and $\gamma \geq 2$. Then it follows that $\alpha \leq 3\gamma/\beta + 1 \leq 3\gamma - 2$. For $\beta = 0$ we still have $|D| \geq (2\alpha + 1)p$ and $\gamma \geq 2(2\alpha + 1)/3$. So in any case the inequality

$$\alpha \leq 3\gamma - 2$$

holds. Now we change the ordering of the Brauer characters such that their degrees are $(1, 1, 2)$. Then as above

$$\sum_{1 \leq i \leq j \leq 3} q_{ij} c_{ij} = \sum_{\chi \in \mathrm{Irr}(H)} q(r_\chi) = 3 + 3\alpha + 3\beta + 3\gamma \leq 1 + 2\alpha + 3\beta + 6\gamma = |D|.$$

This finishes the proof.   □

By Sect. 4.1 it is known that Lemma 14.4 fails for example for $A \cong C_3^2$. Our next lemma is quite technical, but powerful.

Lemma 14.5 *Let B be a p-block of a finite group with defect group D. If there exists an element $x \in Z(D)$ such that $D/\langle x \rangle$ is abelian, and $|C_{I(B)}(x)| \leq 4$ or $C_{I(B)}(x) \cong S_3$, then Brauer's $k(B)$-Conjecture holds for B.*

Proof We consider a B-subsection (x, b_x). The aim of the proof is to apply Theorem 4.2 in connection with Lemma 14.4. Let C be the Cartan matrix of b_x. As usual, b_x dominates a block $\overline{b_x}$ with abelian defect group $\overline{D} := D/\langle x \rangle$, Cartan matrix $\overline{C} := \frac{1}{|\langle x \rangle|} C = (c_{ij})$, and $I(\overline{b_x}) \cong C_{I(B)}(x)$. By work of Usami and Puig [226, 227, 270, 271] there exists a perfect isometry between $\overline{b_x}$ and its Brauer correspondent with normal defect group. By Theorem 4.11 in [48] the Cartan matrices are preserved under perfect isometries up to basic sets. Thus, we may assume that $\overline{b_x}$ has normal defect group $\overline{D}$. By Theorem 1.19, $\overline{b_x}$ is Morita equivalent to the group algebra $F[\overline{D} \rtimes I(\overline{b_x})]$ except possibly if $I(\overline{b_x}) \cong C_2^2$ (which has non-trivial Schur multiplier $H^2(C_2^2, F^\times) \cong C_2$). Let us first handle this exceptional case. Here $\overline{b_x}$ is Morita equivalent to a (non-trivial) twisted group algebra $F_\gamma[\overline{D} \rtimes C_2^2]$ where the 2-cocycle γ is uniquely determined. By Proposition 1.20, the Cartan matrix of $\overline{b_x}$ is the same as the Cartan matrix of a non-principal block of a group of type $\overline{D} \rtimes D_8$ (note that D_8 is a covering group of C_2^2). The group algebra of $\overline{D} \rtimes D_8$ has $k(D_8) = 5$ irreducible Brauer characters. Four of them lie in the principal block. Therefore, the Cartan matrix of $\overline{b_x}$ has dimension $5 - 4 = 1$. Hence, we are done in the exceptional case.

Now assume that $\overline{b_x}$ is Morita equivalent to FH where $H := \overline{D} \rtimes I(\overline{b_x})$. Then by Lemma 14.4 there is a positive definite quadratic form $q = \sum_{1 \leq i \leq j \leq k(\overline{b_x})} q_{ij} x_i x_j$ such that

$$\sum_{1 \leq i \leq j \leq k(\overline{b_x})} q_{ij} c_{ij} \leq |\overline{D}|.$$

The result follows easily by Theorem 4.2.   □

The following lemma generalizes Corollary 1.2(ii) in [232].

Lemma 14.6 *Let B be a block of a finite group with abelian defect group D. If $I(B)$ contains an abelian subgroup of index at most 4, then Brauer's $k(B)$-Conjecture holds for B.*

Proof Let $A \leq I(B)$ be abelian such that $|I(B) : A| \leq 4$. It is well-known that A has a regular orbit on D, i.e. there exists an element $x \in D$ such that $C_A(x) = 1$. Hence, $|C_{I(B)}(x)| \leq 4$, and the claim follows from Lemma 14.5.   □

We also have a dual version.

Lemma 14.7 *Let B be a block of a finite group with abelian defect group. If $|I(B)'| \leq 4$, then Brauer's $k(B)$-Conjecture holds for B.*

Proof By [143, Theorem 1.1] there exists an element $u \in D$ such that $|C_{I(B)}(u)| \leq 4$. Now the claim follows from Lemma 14.5.   □

We remark that Lemma 14.6 also holds under the more general hypothesis that $I(B)$ contains a subgroup R of index at most 4 such that R has a regular orbit on D. Since many non-abelian groups also guarantee regular orbits, it is worthwhile to study small groups with this property in detail. We begin with a special case.

Proposition 14.8 *Let $A \cong D_{2n}$ with $n \geq 3$ and let p be a prime such that $p \nmid 2n$. Suppose that for any $d \mid n$, $d - 1$ is not a non-trivial p-power (this is always true if n is odd). Then any faithful action of A on an elementary abelian p-group provides regular orbits.*

Proof Let V be an absolutely irreducible $\mathbb{F}_q A$-module where $q = p^m$ for some $m \in \mathbb{N}$. Then by Lemma 3.I in [82] it suffices to show that $\overline{A} := A / C_A(V)$ has a regular orbit on V. Since A has an abelian subgroup of index 2, we have $\dim V \in \{1, 2\}$. We may assume that $\dim V = 2$. Then $\overline{A}$ is non-abelian of order $2d$ for some $d \mid n$. Write $\overline{A} = \langle \sigma \rangle \rtimes \langle \tau \rangle$ such that $|\langle \sigma \rangle| = d > 2$. By way of contradiction suppose that $\overline{A}$ does not have a regular orbit on V. Let $\mathcal{M}$ be the set of subgroups of $\overline{A}$ of prime order. Then

$$V = \bigcup_{H \in \mathcal{M}} C_V(H).$$

Since V is not a union of q proper subspaces, we have $q < |\mathcal{M}| \leq 2d - 1$. Let $M \in GL(2, q)$ be the matrix which describes the action of σ on V. Let λ be an eigenvalue of M in the algebraic closure of $\mathbb{F}_q$. Since $M^d = 1$, λ is a d-th root of unity. Since M is diagonalizable in the algebraic closure of $\mathbb{F}_q$, we may even assume that λ is a primitive d-th root of unity (recall that $\overline{A}$ acts faithfully). Since M is conjugate to its inverse, also $\lambda^{-1} \neq \lambda$ is an eigenvalue of M. In particular, the characteristic polynomial has the form $(X - \lambda)(X - \lambda^{-1}) = X^2 - (\lambda + \lambda^{-1})X + 1 \in \mathbb{F}_q[X]$. Hence, $\lambda + \lambda^{-1} \in \mathbb{F}_q$ and $\lambda^q + \lambda^{-q} = (\lambda + \lambda^{-1})^q = \lambda + \lambda^{-1}$. This shows that $q \equiv \pm 1 \pmod{d}$. Suppose first that $d \mid q + 1$. Since $q < 2d - 1$, we obtain $d = q + 1$.

Table 14.1 Small groups without regular orbits

Size	id	Size	id	Size	id	Size	id	Size	id	Size	id	Size	id
8	3	40	6	64	41	64	253	80	50	96	111	108	23
12	4	40	8	64	52	64	254	80	51	96	117	108	24
16	7	40	10	64	95	64	258	81	7	96	121	108	28
16	8	40	12	64	101	64	261	84	12	96	135	108	42
16	11	40	13	64	115	64	263	84	13	96	179	112	4
16	13	48	5	64	118	64	265	84	14	96	186	112	5
20	4	48	6	64	119	72	5	88	5	96	189	112	15
21	1	48	7	64	124	72	8	88	7	96	192	112	25
24	5	48	25	64	129	72	17	88	9	96	200	112	28
24	6	48	29	64	131	72	20	93	1	96	206	112	29
24	8	48	33	64	134	72	25	96	6	96	207	112	30
24	14	48	35	64	137	72	27	96	7	96	208	112	31
28	3	48	36	64	138	72	28	96	12	96	209	112	38
32	9	48	37	64	141	72	30	96	27	96	210	112	40
32	11	48	43	64	142	72	46	96	28	96	212	112	42
32	19	48	47	64	146	72	48	96	34	96	213	120	18
32	25	48	48	64	152	72	49	96	44	96	215	120	20
32	27	48	51	64	157	76	3	96	54	96	219	120	23
32	28	52	4	64	173	80	4	96	62	96	223	120	25
32	34	56	4	64	187	80	6	96	64	96	226	120	27
32	39	56	5	64	189	80	16	96	67	96	230	120	28
32	40	56	9	64	196	80	25	96	68	100	4	120	30
32	42	56	12	64	198	80	26	96	78	100	14	120	46
32	43	60	12	64	202	80	29	96	80	104	5	124	3
32	46	63	3	64	203	80	31	96	87	104	8		
32	48	64	6	64	211	80	36	96	98	104	10		
32	50	64	12	64	226	80	37	96	106	104	12		
36	4	64	32	64	230	80	39	96	107	104	13		
36	12	64	34	64	250	80	44	96	109	105	1		
40	5	64	38	64	251	80	46	96	110	108	4		

However, this contradicts our hypothesis. Thus, we have $d = q - 1$ and $\lambda \in \mathbb{F}_q$. Therefore, we may assume $M = \left(\begin{smallmatrix} \lambda & 0 \\ 0 & \lambda^{-1} \end{smallmatrix} \right)$. Let T be the matrix which describes the action of τ. Since $T^2 = 1$ and $TMT = M^{-1}$, we may assume $T = \left(\begin{smallmatrix} 0 & 1 \\ 1 & 0 \end{smallmatrix} \right)$. Then $C_{\overline{A}}(1, 0) = 1$, and we have a contradiction. □

Proposition 14.9 *Let A be a group of order less than 128. Then there is a finite p-group P such that $A \le \text{Aut}(P)$, $p \nmid |A|$ and A does not have regular orbits on P if and only if A is isomorphic to* SmallGroup(n, i) *where (n, i) is one of the pairs given in Table 14.1.*

Proof The proof is computer assisted. Suppose that A does not have a regular orbit on P. By Lemma 2.6.2 in [106], we may assume that P is an elementary abelian p-group, i.e. a vector space over $\mathbb{F}_p$. Let $\mathcal{M}$ be the set of subgroups of A of prime order. Then

$$P = \bigcup_{H \in \mathcal{M}} C_P(H).$$

Since P cannot be the union of p proper subgroups, we get $p < |\mathcal{M}| < |A|$. Hence, p is bounded in terms of A.

By Maschke's Theorem, P decomposes into irreducible A-invariant subgroups $P = P_1 \oplus \ldots \oplus P_n$. Suppose that P_{n-1} is isomorphic to P_n as $\mathbb{F}_p A$-module. Then A still acts faithfully on $P_1 \oplus \ldots \oplus P_{n-1}$ and there is still no regular orbit. Thus, we may assume that the P_i are pairwise non-isomorphic. In particular, there are only finitely many possibilities for P up to isomorphism. In order to make the computation efficient, we need some more details.

If A is abelian, then it is well-known that A always has regular orbits. More generally, Yang [290] proved that a nilpotent group A has regular orbits provided the following holds: A does not involve D_8 and if $p = 2$, then A does not involve $C_r \wr C_r$ for any Mersenne prime r. Therefore, we do not need to consider these cases.

While building combinations of the P_i, we can certainly leave out the trivial representation. Suppose that A acts faithfully on $P = P_1 \oplus \ldots \oplus P_n$, but not faithfully on any proper subsum $P_{i_1} \oplus \ldots \oplus P_{i_k}$. Let

$$K_i := C_A(P_1) \cap \ldots \cap C_A(P_{i-1}) \cap C_A(P_{i+1}) \cap \ldots \cap C_A(P_n)$$

for $i = 1, \ldots, n$. Since $K_i \cap K_j = 1$ for $i \neq j$, every K_i contains a minimal normal subgroup N_i and $N_i \neq N_j$ for $i \neq j$. In particular, n is bounded by the number of minimal normal subgroups of A. Moreover, every P_i contains at least $n - 1$ distinct minimal normal subgroups.

Let us consider the (faithful) action of $A_i := A / C_A(P_i)$ on P_i. Suppose we have already found regular orbits of A_i on P_i for all i. Then there exist $x_i \in P_i$ such that $C_A(x_i) = C_A(P_i)$. Then $C_A(x_1 \ldots x_n) = 1$ and we are done. Hence, in order to find actions without regular orbits it suffices to consider sums $P_1 \oplus \ldots \oplus P_n$ such that at least one A_i has no regular orbit on P_i. This allows us to apply induction on $|A|$.

Now we consider the opposite situation. Assume that A is a direct product $A = A_1 \times A_2$ such that A_1 acts faithfully without regular orbits on an elementary abelian p-group P_1. Suppose further that $p \nmid |A_2|$. Then we may choose any faithful $\mathbb{F}_p A_2$-module P_2. It is easy to see that A has no regular orbit on the inflation $P_1 \oplus P_2$.

Another interesting inductive condition is the following. Suppose that we have found a subgroup $A_1 \leq A$ such that A_1 always has regular orbits and $A_2 \cap A_1 \neq 1$ for all $1 \neq A_2 \leq A$. Then for $x \in P$ such that $C_{A_1}(x) = 1$ we also have $C_A(x) = 1$, i.e. A has a regular orbit. This applies for example to quaternion groups A with $A_1 = Z(A)$.

We also need to discuss the question, how to check for regular orbits efficiently. We pick elements $x \in P$ randomly and check if $C_A(x) = 1$. This usually works quite well if $|P|$ is large. However, if we did not find regular orbits among the first, say 30, random choices, we compute all the orbits sizes. Since there are usually many regular orbits, we only have to compute all the orbits sizes in small cases.

While working through the list of groups A of order less than 128 in GAP, it turns out that certain irreducible representations are not available. This concerns the dihedral groups $A \cong D_m$ where

$$m \in \{46, 50, 58, 74, 82, 86, 92, 94, 98, 102, 106, 110, 116, 118, 122, 124\}$$

and the group $A \cong C_{37} \rtimes C_3$. Proposition 14.8 works for all dihedral groups above except the last one $A \cong D_{124}$. But here, GAP shows that there is in fact an irreducible, faithful representation on $\mathbb{F}_{61}^2$ without regular orbits. Now we handle the group $A \cong C_{37} \rtimes C_3$ by hand. Let $S \in \mathrm{Syl}_{37}(A)$ and $\mathrm{Syl}_3(A) = \{T_1, \dots, T_{37}\}$. Assume that A acts faithfully on the elementary abelian p-group P without regular orbits. Then

$$P = C_P(S) \cup \bigcup_{i=1}^{37} C_P(T_i).$$

Since S has a regular orbit on P, we have $C_P(T_i) \neq 1$ for some i. Since A acts transitively on $C_P(T_1), \dots, C_P(T_{37})$, we also have $|C_P(T_1)| = \dots = |C_P(T_{37})| =: p^b$. Let $|C_P(S)| =: p^a$. Since $C_P(S) \cap C_P(T_i) = C_P(\langle S, T_i \rangle) = C_P(A) = 1$ and $C_P(T_i) \cap C_P(T_j) = 1$ for $i \neq j$, we obtain

$$0 \equiv |P| = p^a + 37(p^b - 1) \equiv p^a - 37 \pmod{p}.$$

This implies $a = 0$ (because $p \neq 37$). Thus, $p \mid 36$ and $p = 2$. Since $1 + 37(2 - 1) = 38$ and $1 + 37(4 - 1) = 112$ are not 2-powers, we have $b \geq 3$. However, then $0 \equiv |P| \equiv -36 \pmod 8$. A contradiction.

Our algorithm takes very long for the group $D_8 \times C_2^2 \times C_3$. We will also give a theoretical argument here. If a group A has regular orbits on any elementary abelian p-group for a prime p, then A also has regular orbits on any finite-dimensional vector space over $\mathbb{F}_{p^n}$ for any $n \in \mathbb{N}$ (since $\mathrm{GL}(m, \mathbb{F}_{p^n}) \leq \mathrm{Aut}(C_p^{mn})$). Our algorithm shows that $D_8 \times C_2^2$ has regular orbits for all $p \geq 5$ (however not for $p = 3$). Now Theorem 5.I in [82] shows that $D_8 \times C_2^2 \times C_3$ has regular orbits for all $p \geq 5$ and we are done. $\square$

One can show that 84 % of the groups of order less than 128 provide regular orbits in the situation above (for this reason we list the complementary set in Table 14.1). Proposition 14.9 will be applied later in Proposition 14.13, but we need to settle a special case for $p = 2$ first.

Lemma 14.10 *Let A be a p'-automorphism group of an abelian p-group $P \cong \prod_{i=1}^{n} C_{p^i}^{m_i}$. Then A is isomorphic to a subgroup of*

$$\prod_{i=1}^{n} GL(m_i, p)$$

where $GL(0, p) := 1$.

Proof As a p'-group, A acts faithfully on $P/\Phi(P)$. Hence, the canonical homomorphism

$$A \longrightarrow \prod_{i=1}^{n} \mathrm{Aut}(\Omega_{n-i+1}(P)\Phi(P)/\Omega_{n-i}(P)\Phi(P)) \tag{14.2}$$

is injective. Since $\Omega_i(P)\Phi(P)/\Omega_{i-1}(P)\Phi(P)$ is elementary abelian of rank m_i for $i = 1, \ldots, n$, the claim follows. $\square$

Combining Lemmas 14.6 and 14.10 gives the following result which is probably not new.

Corollary 14.11 *Let B be a p-block of a finite group with abelian defect group $D \cong \prod_{i=1}^{n} C_{p^i}^{m_i}$ such that $m_i \leq 1$ for $i = 1, \ldots, n$. Then Brauer's $k(B)$-Conjecture holds for B.*

Now we turn to abelian p-groups with homocyclic factors. Here it is necessary to restrict p.

Theorem 14.12 *Let B be a 2-block of a finite group with abelian defect group $D \cong \prod_{i=1}^{n} C_{2^i}^{m_i}$. Assume that one of the following holds:*

(i) For some $i \in \{1, \ldots, n\}$ we have $m_i \leq 4$ and $m_j \leq 2$ for all $j \neq i$.
(ii) D has rank 5.

Then Brauer's $k(B)$-Conjecture holds for B.

Proof

(i) For each $k \in \{1, \ldots, n\}$ we define A_k to be the image of the canonical map

$$I(B) \longrightarrow \mathrm{Aut}(\Omega_{n-k+1}(D)\Phi(D)/\Omega_{n-k}(D)\Phi(D)) \cong GL(m_k, p).$$

Then we can refine the monomorphism from Eq. (14.2) to $I(B) \to \prod_{k=1}^{n} A_k$. Since $GL(2, 2) \cong S_3$, we have $A_j \leq C_3$ for $j \neq i$. In order to apply Lemma 14.6, it suffices to show that $A_i \leq GL(4, 2)$ contains an abelian subgroup of index at most 4. Since A_i has odd order, we have $|A_i| \mid (2^4-1)(2^3-1)(2^2-1) = 3^2 \cdot 5 \cdot 7$. It can be seen further that $|A_i| \in \{1, 3, 5, 7, 9, 15, 21\}$. The claim follows.

(ii) Now assume that D has rank 5. The case $|D| = 32$ was already handled in Theorem 13.7. Thus, by part (i) we may assume that $C_4^5 \leq D$ and $I(B) \leq$ GL$(5, 2)$. As usual, $e(B)$ is a divisor of $3^2 \cdot 5 \cdot 7 \cdot 31$. Suppose first that $31 \mid e(B)$. One can show that every group whose order divides $3^2 \cdot 5 \cdot 7 \cdot 31$ has a normal Sylow 31-subgroup. Therefore $I(B)$ lies in the normalizer of a Sylow 31-subgroup of GL$(5, 2)$. Thus, we may assume $e(B) = 31 \cdot 5$. Here Lemma 14.6 does not apply. However, we can still show the existence of a regular orbit. Obviously, $I(B)$ cannot have a regular orbit on $D/\Phi(D) \cong C_2^5$. However, using GAP one can show that $I(B)$ has a regular orbit on $\Omega_2(D) \cong C_4^5$. So we can find a subsection (u, b_u) such that $l(b_u) = 1$. The claim follows in this case. Now we can assume that $31 \nmid e(B)$. In case $7 \mid e(B)$ we see again that $I(B)$ has a normal Sylow 7-subgroup and $e(B) = 3^2 \cdot 7$ without loss of generality. It is easy to see that every group of order $3^2 \cdot 7$ has an abelian subgroup of index 3. Thus, we may finally suppose that $7 \nmid e(B)$. Then $I(B)$ is abelian itself. This completes the proof. $\square$

Theorem 14.12 improves an unpublished result by Robinson [229]. In the next proposition we investigate how far we can go only by restricting the inertial index.

Proposition 14.13 *Let B be a block of a finite group with abelian defect group and $e(B) \leq 255$. Then the $k(B)$-Conjecture is satisfied for B.*

Proof Let $I(B)$ be an arbitrary group of order at most 255, and let D be a defect group of B. We compute with GAP the set $\mathscr{L}$ of subgroups of $I(B)$ which have order less than 128 and are not on the list in Table 14.1. For every $H \in \mathscr{L}$ we check the following condition:

$$\forall L \leq I(B) : L \cap H = 1 \Longrightarrow |L| \leq 4 \vee L \cong S_3. \tag{14.3}$$

By Proposition 14.9 there is an $x \in D$ such that $\mathrm{C}_{I(B)}(x) \cap H = \mathrm{C}_H(x) = 1$. Hence, if Condition (14.3) is true for some $H \in \mathscr{L}$, we get $|\mathrm{C}_{I(B)}(x)| \leq 4$ or $\mathrm{C}_{I(B)}(x) \cong S_3$. Then the $k(B)$-Conjecture follows from Lemma 14.5. It turns out that (14.3) is false for only a few groups which will be handled case by case.

For $I(B) \cong C_{31} \rtimes C_5$ one can show that we have a regular orbit unless $p = 2$. Thus, let $p = 2$. We study the (faithful) action of $I(B)$ on $\Omega(D)$. By Theorem 14.12 we may assume $|\Omega(D)| \geq 2^6$. A GAP calculation shows that $I(B)$ has eight irreducible representations over $\mathbb{F}_2$ and their degrees are $1, 4, 5, \ldots, 5$. Moreover, the image of the second representation has order 5 while the last six representations are faithful. In particular the action of $I(B)$ on $\Omega(D)$ is not irreducible. So we decompose $\Omega(D) = V_1 \times \ldots \times V_n$ into irreducible $I(B)$-invariant subgroups V_i. Without loss of generality, V_1 is faithful. Hence, we find an element $v_1 \in V_1$ such that $\mathrm{C}_{I(B)}(v_1)$ has order 5. If there is at least one more non-trivial summand, say V_2, we find another element $v_2 \in V_2$ such that $\mathrm{C}_{I(B)}(v_1) \nsubseteq \mathrm{C}_{I(B)}(v_2)$. It follows that $\mathrm{C}_{I(B)}(v) = 1$ for $v := v_1 v_2$. Therefore, we may assume that $I(B)$ acts trivially on $V_2 \times \ldots \times V_n$. By Theorem 5.2.3 in [94], also D decomposes as $D = \mathrm{C}_D(I(B)) \times [D, I(B)]$. It follows that $[D, I(B)] \cong C_{2^a}^5$ for some $a \geq 1$.

In case $a \geq 2$ we have seen in the proof of Theorem 14.12 that $I(B)$ has a regular orbit on $[D, I(B)]$. Hence, $[D, I(B)]$ is elementary abelian of order 32. Define $|C_D(I(B))| =: 2^k$. Then B has 2^{k+1} subsections up to conjugation. Half of them have inertial index 155 while the other half have inertial index 5. Let (u, b_u) be one of the B-subsections with $I(b_u) \cong I(B)$. In order to determine $l(b_u)$ we may suppose that $C_D(I(B)) = 1$ by Theorem 1.39 (applied inductively). Now take a non-trivial b_u-subsection (v, β_v). Then the Cartan matrix of β_v is given by $2(3 + \delta_{ij})_{1 \leq i,j \leq 5}$ up to basic sets (see proof of Theorem 13.7). Theorem 4.2 gives $k(b_u) \leq 16$. Since (v, β_v) is the only non-trivial b_u-subsection up to conjugation, we obtain $l(b_u) \leq 11$. Similarly we can show that $l(b_u) \leq 5$ if (u, b_u) is a B-subsection such that $e(b_u) = 5$. Now we get $k(B) \leq 2^k \cdot 11 + 2^k \cdot 5 = 2^{k+4} \leq |D|$, because $k(B)$ is the sum over the numbers $l(b_u)$ (see Theorem 1.35). This completes the case $e(B) = 155$.

The next exceptional group is $I(B) \cong \mathtt{SmallGroup}(160, 199)$. Here $Z(I(B))$ is the unique minimal normal subgroup of $I(B)$. In particular every faithful representation contains a faithful, irreducible representation as a direct summand. Using GAP we show that only the prime $p = 3$ is "interesting". If $I(B)$ acts faithfully and irreducibly on D, then one can find an element $x \in D$ such that $|C_{I(B)}(x)| \leq 2$. Therefore, the $k(B)$-Conjecture follows from Lemma 14.5.

We continue with $I(B) \cong GL(3, 2)$. Here the algorithm of Proposition 14.9 shows that $I(B)$ has regular orbits. Finally, we have the following exceptions: $I(B) \in \{C_{29} \times C_7, C_{41} \rtimes C_5, C_{23} \rtimes C_{11}\}$. Here the arguments for $C_{37} \rtimes C_3$ from the proof of Proposition 14.9 show that there are always regular orbits. We omit the details.   □

For $e(B) = 256$ the arguments in Proposition 14.13 fail as one can see by the following example. There is a subgroup $A \leq GL(4, 3)$ of order 256 such that C_3^4 splits under the action of A into orbits of lengths 1, 16, 32 and 32. Hence, the corresponding stabilizers have order at least 8.

As an application of various results we present two other propositions on 2-blocks.

Proposition 14.14 *Let B be a 2-block with abelian defect group of order* 64. *Then* $k(B) \leq 3 \cdot 64$.

Proof Let D be a defect group of B. By Theorem 14.12 we may assume that D is elementary abelian. Moreover, by Proposition 14.13 we may assume that $e(B) \geq 256$. As usual, $I(B)$ is a subgroup of $H := GL(6, 2)$. Since $I(B)$ has odd order, $I(B)$ is solvable. In particular, there exists a prime p such that $O_p(I(B)) \neq 1$. Hence, $I(B) \leq N_H(O_p(I(B)))$. Now we can use GAP to run through the local subgroups of H. It turns out that $I(B) \cong (C_7 \rtimes C_3)^2$. Since C_7^2 has a regular orbit on D, there exists a B-subsection (u, b_u) such that $I(b_u) \cong C_3^2$. We consider the block $\overline{b_u}$ of $C_G(u)/\langle u \rangle$ with defect group C_2^5 dominated by b_u. Since C_3^2 has a non-trivial fixed point v on C_2^5, Theorem 1.39 implies $l(b_u) = l(\overline{b_u}) = l(\beta_v)$ where (v, β_v) is a $\overline{b_u}$-subsection. Again β_v dominates a block $\overline{\beta_v}$ with defect group

C_2^4. Thus, Theorem 13.2 shows $l(b_u) = l(\overline{\beta_v}) \le 9$. Now the claim follows from Theorems 4.12 and 7.14. $\square$

Proposition 14.15 *Let B be a 2-block with abelian defect groups and odd defect $d > 1$. Then*

$$k(B) \le 2^d (2^{\frac{d-1}{2}} - 1).$$

Proof As in Theorem 14.2 we find a subsection (u, b_u) such that $l(b_u) < 2^{d-1}$. Since $\lfloor \sqrt{2^{d-1} - 1} \rfloor = 2^{\frac{d-1}{2}} - 1$ is odd, the claim follows from Theorems 4.13 and 7.14. $\square$

A corresponding result for even defects would be a bit confusing.

The next theorem handles the $k(B)$-Conjecture for 3-blocks with abelian defect groups of rank at most 3 as a special case.

Theorem 14.16 *Let B be a 3-block of a finite group with defect group $D \cong \prod_{i=1}^{n} C_{3^i}^{m_i}$ such that for two $i, j \in \{1, \ldots, n\}$ we have $m_i, m_j \le 3$, and $m_k \le 1$ for all $i \ne k \ne j$. Then Brauer's $k(B)$-Conjecture holds for B.*

Proof As in the proof of Theorem 14.12 we may assume that $I(B) \le \mathrm{GL}(3, 3) \times \mathrm{GL}(3, 3)$. By Lemma 14.6, it suffices to show that every $3'$-subgroup of $\mathrm{GL}(3, 3)$ has an abelian subgroup of index at most 2. In order to do so, we may assume $I(B) \le \mathrm{GL}(3, 3)$. Then $e(B)$ is a divisor of $(3^3 - 1)(3^2 - 1)(3 - 1) = 2^5 \cdot 13$. In case $13 \mid e(B)$, Sylow's Theorem shows that $I(B)$ has a normal Sylow 13-subgroup. Hence, $I(B)$ lies in the normalizer of the Sylow 13-subgroup in $\mathrm{GL}(3, 3)$. Thus, $e(B) = 2 \cdot 13$ without loss of generality. The claim holds. Suppose next that $I(B)$ is a 2-group. It can be shown that a Sylow 2-subgroup of $\mathrm{GL}(3, 3)$ is isomorphic to $SD_{16} \times C_2$; so it contains an abelian maximal subgroup. Obviously the same holds for $I(B)$ and the claim follows. $\square$

For $p = 5$ it is necessary to restrict the rank of the defect group.

Theorem 14.17 *Let B be a 5-block of a finite group with abelian defect group of rank 3. Then Brauer's $k(B)$-Conjecture holds for B.*

Proof We consider the (faithful) action of $I(B)$ on $\Omega(D) \cong C_5^3$. In particular, $I(B) \le \mathrm{GL}(3, 5)$. Fortunately, GAP is able to compute a set of representatives for the conjugacy classes of $5'$-subgroups of $\mathrm{GL}(3, 5)$. In particular we obtain $e(B) \mid 2^7 \cdot 3$ or $e(B) \mid 2^2 \cdot 3 \cdot 31$. A further analysis shows that there is an element $x \in \Omega(D)$ such that $|C_{I(B)}(x)| \le 4$ or $C_{I(B)}(x) \cong S_3$. The claim follows by Lemma 14.5. $\square$

For the defect group C_7^3 the proof above would not work. More precisely, it is possible here that $I(B)$ has order 6^4, the largest orbit on D has length 6^3 and the corresponding stabilizer is isomorphic to C_6. Hence, the existence of a perfect isometry for b_x is unknown.

Chapter 15
Blocks with Few Characters

In the previous chapters we investigated the numerical invariants of a block B for a given defect group D. In this chapter we consider the opposite situation, i.e. we determine D if $k(B)$ is given. In general, this is a difficult task. Problem 21 on Brauer's list [36] asks whether there are finitely many choices for D if $k(B)$ is fixed. This is known to be true provided the Alperin-McKay Conjecture holds (see [170]).

For small values of $k(B)$, the following things are known:

 (i) We have $k(B) = 1$ if and only if $D = 1$.
 (ii) We have $k(B) = 2$ if and only if $|D| = 2$ (see [32]).
 (iii) If $k(B) \leq 4$ and $l(B) = 1$, then $|D| = k(B)$ (see [162]).
 (iv) If $k(B) = 5$ and $l(B) = 1$, then $D \in \{C_5, D_8, Q_8\}$ (see [56]).

In this chapter we study the case $k(B) = 3$ and $l(B) = 2$. Most of the results come from [167]. We will show under additional hypotheses that $|D| = 3$, and it is conjectured that this holds in general.

We start with the classification of the transitive linear groups. Here

$$\Gamma L(1, p^n) := \mathbb{F}_{p^n}^\times \rtimes \mathrm{Aut}(\mathbb{F}_{p^n}) \cong C_{p^n-1} \rtimes C_n$$

denotes the semilinear group of degree 1. Moreover, 2_-^{1+4} is the extraspecial group $D_8 * Q_8$ of order 32.

Theorem 15.1 (Hering) *Let $G \leq \mathrm{GL}(n, p)$ act (naturally and) transitively on $\mathbb{F}_p^n \setminus \{0\}$. Then $n = km$ and one of the following holds:*

1. $G \leq \Gamma L(1, p^n)$,
2. $k \geq 2$ *and* $\mathrm{SL}(k, p^m) \trianglelefteq G$,
3. $k \geq 4$ *is even and* $\mathrm{Sp}(k, p^m)' \trianglelefteq G$,
4. $k = 6$, $p = 2$ *and* $G_2(2^m)' \trianglelefteq G$,
5. $n = 2$, $p \in \{5, 7, 11, 19, 23, 29, 59\}$ *and G is given in Table 15.1,*

© Springer International Publishing Switzerland 2014
B. Sambale, *Blocks of Finite Groups and Their Invariants*, Lecture Notes in Mathematics 2127, DOI 10.1007/978-3-319-12006-5_15

Table 15.1 Sporadic transitive linear groups

Degree	Order	Structure	Small group id	Primitive group id
5^2	24	$SL(2,3)$	3	15
	48	$SL(2,3) \rtimes C_2$	33	18
	96	$SL(2,3) \rtimes C_4$	67	19
7^2	48	$SL(2,3).C_2$	28	25
	144	$(SL(2,3).C_2) \times C_3$	121	29
11^2	120	$SL(2,3) \times C_5$	15	39
	240	$GL(2,3) \times C_5$	103	42
	120	$SL(2,5)$	5	56
	600	$SL(2,5) \times C_5$	54	57
19^2	1080	$SL(2,5) \times C_9$	63	86
23^2	528	$(SL(2,3).C_2) \times C_{11}$	87	59
29^2	840	$SL(2,5) \times C_7$	13	106
	1680	$(SL(2,5).C_2) \times C_7$	408	110
59^2	3480	$SL(2,5) \times C_{29}$	–	–
3^4	160	$2^{1+4}_- \rtimes C_5$	199	71
	320	$2^{1+4}_-.D_{10}$	1581	90
	640	$2^{1+4}_-.(C_5 \rtimes C_4)$	21454	99
	1920	$2^{1+4}_-.A_5$	241003	130
	3840	$2^{1+4}_-.S_5$	–	129
	240	$SL(2,5).C_2$	89	124
	480	$SL(2,5).C_4$	221	126
	480	$(SL(2,5).C_2) \rtimes C_2$	947	127
	960	$(SL(2,5).C_4) \rtimes C_2$	5688	128

6. $n = 4$, $p = 2$ and $G \cong A_7$,
7. $n = 4$, $p = 3$ and G is given in Table 15.1,
8. $n = 6$, $p = 3$ and $G \cong SL(2,13)$.

Proof In Sect. 5 of Hering's paper [110] which is quoted in Remark XII.7.5 in [130] the classification appeared in a slightly inaccurate form. For example part IV (part (4) in [130]) states for $n = 2$ and $p = 23$ that G contains a normal subgroup $N \cong Q_8$ such that $C_G(N) = Z(N)$. Then $|G| \leq 48$ and G cannot act transitively on a set with $23^2 - 1 = 528$ elements.

The classification we use here is from Theorem 69.7 in [141]. Observe that $G_2(2)' \cong PSU(3,3)$ (and $G_2(2^m)' \cong G_2(2^m)$ for $m \geq 2$). Hence, we do not need case E5 in [110]. Moreover, the exceptional case $G \cong A_6$ for $p^n = 2^4$ in both references is unnecessary, since $A_6 \cong Sp(4,2)'$. On the other hand, $Sp(k, p^m)' \cong Sp(k, p^m)$ for $k \geq 6$ or $p^m \geq 3$ (see Propositions 3.7–3.9 in [99]). Thus, we do not weaken the statement by replacing $Sp(k, p^m)$ with $Sp(k, p^m)'$.

Presentations of the solvable exceptional groups are given in Huppert [127]. The groups where $p^n = 3^2$ are already included in case (ii). In order to find all exceptions, we do the following. The group $H := \mathbb{F}_p^n \rtimes G$ acts 2-transitively and thus primitively on $\mathbb{F}_p^n$. Hence, we can run through the library of primitive permutation groups (of degree less than 2500) in GAP. In almost each case we give the id number of H in this list and the id number of G in the Small Groups Library. In case $n = 2$ and $p = 59$ these numbers are not available. Instead we can access the subgroups of $\mathrm{GL}(2, 59)$ directly. In this way we obtain Table 15.1 which confirms most statements on the Wikipedia page [286].   □

Although Theorem 15.1 depends on the classification of the finite simple groups, the following result only uses Passman's classification [220] of the p-solvable transitive linear groups which is CFSG-free. It was developed mostly by Külshammer and already announced in the introduction of [165] without proof.

Proposition 15.2 *Let B be a block of a finite group G with normal defect group D, and suppose that $k(B) = 3$. Then $|D| = 3$.*

Proof By results of Fong and Reynolds, we may assume that D is a Sylow p-subgroup of G, and that $Z := O_{p'}(G)$ is cyclic and central in G. By the remark above, we may also assume that $l(B) = 2$. By Theorem 1.35 we know that B has only two subsections. In particular, G acts transitively on $D \setminus \{1\}$ by conjugation. Hence, D is elementary abelian. We write $|D| = p^d$. By the Hall-Higman Lemma, the kernel of the action of G on D is ZD. Observe that G/ZD is a p'-group. Hence by Theorem 15.1, apart from finitely many exceptions, G/ZD is isomorphic to a subgroup of $\Gamma\mathrm{L}(1, p^d)$. In particular, G/ZD has a cyclic normal subgroup H/ZD whose order s divides $p^d - 1$ such that G/H is cyclic of order t dividing d. Since G/ZD acts transitively on $D \setminus \{1\}$, we also have $(p^d - 1) \mid |G : ZD| = st$.

It is well-known that $\mathrm{IBr}(B) = \mathrm{IBr}(G|\zeta) := \{\chi \in \mathrm{IBr}(G) : (\chi_{|Z}, \zeta) \neq 0\}$ for some $\zeta \in \mathrm{IBr}(Z)$. Let us consider $\mathrm{IBr}(H|\zeta)$. On the one hand, $|\mathrm{IBr}(G|\zeta)| = |\mathrm{IBr}(B)| = l(B) = 2$ implies that G has at most two orbits on $\mathrm{IBr}(H|\zeta)$. Moreover, each of these orbits has length at most $|G : H| = t$. Thus, $|\mathrm{IBr}(H|\zeta)| \leq 2t \leq 2d$.

On the other hand, we have $ZD/D \leq Z(H/D)$. Since H/ZD is cyclic, H/D has to be abelian. In particular we have $|\mathrm{IBr}(H|\zeta)| = |H : ZD| = s$. Thus, $s = |\mathrm{IBr}(H|\zeta)| \leq 2d$, and $p^d - 1 \leq |G : ZD| \leq st \leq 2d^2$.

If $p = 2$, then our result follows easily since $k_0(B) \equiv 0 \pmod{4}$ for $d \geq 2$. Thus, we may assume that $p \geq 3$. If $d = 1$, then the claim follows easily from Theorem 8.6. Hence, we may assume that $d \geq 2$ and $p \geq 3$. If $d = 2$, then $p^2 \leq 1 + 8 = 9$, i.e. $p = 3$. This case leads to a contradiction by making use of the results in [154]. Therefore, we may assume that $d \geq 3$ and $p \geq 3$, so that $3^d \leq p^d \leq 1 + d^2$. However, this is impossible.

It remains to deal with the exceptional cases in Theorem 15.1; so we may assume that

$$|D| \in \{5^2, 7^2, 11^2, 19^2, 23^2, 29^2, 59^2, 3^4\}.$$

Suppose first that $d = 2$, and choose a non-trivial B-subsection (u, b_u). Then b_u dominates a unique block $\overline{b_u}$ of $C_G(u)/\langle u \rangle$, and $\overline{b_u}$ has defect 1. Since $1 = l(b_u) = l(\overline{b_u})$ we conclude that $\overline{b_u}$ has inertial index 1. Thus, b_u has inertial index

1 as well, and G/ZD acts regularly on $D \setminus \{1\}$. Hence, G/Z is a Frobenius group with Frobenius kernel ZD/Z and Frobenius complement G/ZD. In particular the Sylow subgroups of G/ZD are cyclic or (generalized) quaternion. Thus, the Schur multiplier of G/ZD is trivial. Hence, we may assume that $Z = 1$. But then B is the only p-block of G, so that G has class number 3. This implies that $|G| \leq 6$, a contradiction.

We are left with the case $|D| = 3^4$. By Table 15.1, $|G/Z| = 2^k 3^4 5$ with $k \in \{5, 6, 7\}$. Since as above b_u does not have inertial index 2, only $k \in \{6, 7\}$ is admissible. Hence, $G/ZD \cong$ SmallGroup$(320, 1581)$ or $G/ZD \cong$ SmallGroup$(640, 21454)$. In the latter case the Schur multiplier of G/ZD is trivial again. Hence, let $|G/ZD| = 320$. Here GAP shows that the Schur multiplier has order 2. Thus, we may assume that $|Z| = 2$ and $G/D \cong$ SmallGroup$(640, 19095)$ (a Schur covering group). Moreover, B is not the principal block of G (see Proposition IV.5.32 in [19]). By Brauer's First Main Theorem (and its extensions) one can see that $\mathcal{O}G$ consists of just two blocks. The whole group algebra has $k(G/D) = 22$ simple modules while the principal block has $k(G/ZD) = 14$ simples modules. This gives the contradiction $l(B) = k(G/D) - k(G/ZD) = 8$.   □

Now we can carry over the proof in [170] to our situation.

Theorem 15.3 *Let B be a block of a finite group G with defect group D such that $k(B) = 3$. Suppose that the Alperin-McKay Conjecture holds for B. Then $|D| = 3$.*

Proof Let b be the Brauer correspondent of B in $N_G(D)$. Then b dominates a unique block $\overline{b}$ of $N_G(D)/\Phi(D)$ (see Corollary 7 in [230]), and $\overline{b}$ has defect group $\overline{D} := D/\Phi(D)$ which is abelian and normal in $N_G(D)/\Phi(D)$. Moreover, we have

$$k(\overline{b}) = k_0(\overline{b}) \leq k_0(b) = k_0(B) \leq k(B) = 3.$$

If we assume that $k(\overline{b}) \leq 2$, then we get $|\overline{D}| \leq 2$. Thus, D is a cyclic 2-group which is impossible. This shows that we must have $k(\overline{b}) = 3$.

Since $\overline{D}$ is normal in $N_G(D)/\Phi(D)$, Proposition 15.2 implies that $|\overline{D}| = 3$. Thus, D is cyclic, and Theorem 8.6 yields the result.   □

Next we turn to blocks with non-exotic fusion systems. This leads to a question about finite groups which is answered by the following strong result. This is also related to the classification of fusion systems on extraspecial groups mentioned on page 164.

Theorem 15.4 *Let p be a prime, and let G be a finite group in which any two non-trivial cyclic p-subgroups are conjugate. Then one of the following holds:*

1. *The Sylow p-subgroups of G are elementary abelian.*
2. *$p = 3$ and $O^{p'}(G/O_{p'}(G))$ is isomorphic to Ru, J_4 or $^2F_4(q)'$ with $q = 2^{6b \pm 1}$ and $b \geq 0$.*
3. *$p = 5$ and $G/O_{p'}(G)$ is isomorphic to Th.*

In cases (ii) and (iii) the Sylow p-subgroups of G are of type p_+^{1+2}.

The proof of Theorem 15.4 relies heavily on a paper by Navarro and Tiep [202] (see Theorem 15.11 below) and also on the classification of the finite simple groups. We omit the details.

Proposition 15.5 *Let B be a p-block of a finite group G with $k(B) - l(B) = 1$. Suppose that the fusion system of B is non-exotic (for instance if B is the principal block or if G is p-solvable). Then the defect groups of B are elementary abelian.*

Proof Assume that a defect group D of B is non-abelian. Let $\mathcal{F}$ be the fusion system of B, and let H be a finite group such that $D \in \mathrm{Syl}_p(H)$ and $\mathcal{F} = \mathcal{F}_D(H)$. By Theorem 1.35, one can see that $\mathcal{F}$ has exactly two conjugacy classes. In particular, H satisfies the hypothesis of Theorem 15.4. It follows that $p \in \{3, 5\}$ and D is of type p_+^{1+2}.

Suppose first that $p = 5$. Then $\mathcal{F}$ is the fusion system of Th on one of its Sylow 5-subgroups. Moreover, $|\mathrm{Out}_{\mathcal{F}}(D)| = 96$ by Ruiz and Viruel [241]. Proposition 11.8 shows that B is Morita equivalent to the principal 5-block B_0 of Th. In particular, we have $k(B_0) - l(B_0) = 1$. Let (u, b_u) be a non-trivial B_0-subsection. Then b_u is the principal 5-block of $C_{Th}(u)$, and $l(b_u) = 1$. Thus, $C_{Th}(u)$ is 5-nilpotent by [129, Theorem VII.14.9]. However, the fusion system of b_u is not nilpotent, since the kernel of the canonical map $\mathrm{Out}_{\mathcal{F}}(D) \to \mathrm{Aut}_{\mathcal{F}}(Z(D))$ cannot be trivial. Contradiction.

It remains to consider the case $p = 3$. Let (u, b_u) denote a non-trivial B-subsection, and denote by $\overline{b_u}$ the unique 3-block of $C_G(u)/\langle u \rangle$ dominated by b_u. Then $1 = l(b_u) = l(\overline{b_u})$, and $\overline{b_u}$ has an elementary abelian defect group of order 9. By Theorem 15.4 and [241], we may assume that $\mathcal{F}$ is the fusion system of ${}^2F_4(2)'$ or J_4 on one of its Sylow 3-subgroups. Thus, the inertial quotient of B is isomorphic to D_8 or SD_{16} respectively, by the results in [241] (cf. [198]). It follows easily that the inertial quotient of $\overline{b_u}$ is isomorphic to C_4 or Q_8 respectively. However, if $I(\overline{b_u}) \cong C_4$, then the results of [154] lead to the contradiction $l(\overline{b_u}) = 4$.

Thus, we may assume that the inertial quotient of $\overline{b_u}$ is isomorphic to Q_8. Then the arguments in [154] show that there are only two $\overline{b_u}$-subsections, and we obtain $k(\overline{b_u}) = 2$. However, then the defect groups of $\overline{b_u}$ have order 2, a contradiction.   □

In a recent article [112], we have shown that the non-exoticness condition in Proposition 15.5 is superfluous. We will not go into the details here. The result applies for example to blocks with multiplicity 1 introduced by Michler [191].

Corollary 15.6 *Let B be a block with non-exotic fusion system and $k(B) = 3$. Then the defect groups of B are elementary abelian.*

We remind the reader that it is not known if there are any blocks with exotic fusion systems. Nevertheless, it seems difficult to conclude $|D| = 3$ in the situation of Corollary 15.6. Even in the case $D \cong C_3^2$ and $I(B) \cong C_8$ it is not known if $k(B) = 3$ can occur. Using generalized decomposition numbers one can see that $|D|$ is a sum of three non-zero squares provided $k(B) = 3$. Hence, $|D| \neq 25$. Moreover, if $p \equiv -1 \pmod 8$, then $|D| = p^{2k}$ for some $k \geq 1$. By Proposition 1.46, the Cartan matrix of B has determinant $|D|$.

For the principal block we can say slightly more. Here we also give a CFSG-free argument for the fact that the defect group is abelian. This proposition did not appear so far.

Proposition 15.7 *Let B be the principal p-block of a finite group with defect d and $k(B) = 3$. Then B has elementary abelian defect groups, d is odd, and $p = 3$ or $p \equiv 11 \pmod{24}$.*

Proof We may assume that $p > 2$ and $l(B) = 2$. Let D be a defect group of B. By the remark above, the determinant of the Cartan matrix of B is p^d. Hence, in the language of [149], the stable Grothendieck group of B is cyclic. Thus, Theorem 3.1 in [149] shows that the stable center $\overline{Z}(B)$ is a symmetric algebra (in fact the condition on the stable Grothendieck group in this theorem is superfluous). Now it follows from Theorem 1.1 of the same paper that D is abelian.

Let Q be the generalized decomposition matrix of B. Then

$$Q = \begin{pmatrix} 1 & 0 & 1 \\ * & * & a \\ * & * & b \end{pmatrix}$$

where the first row corresponds to the trivial character. By the orthogonality relations, $p^d - 1 = a^2 + b^2$ and $\gcd(a, b) = 1$. Hence, $p^d \equiv 3 \pmod 8$. It follows that $p \equiv 3 \pmod 8$ and $d \equiv 1 \pmod 2$. Moreover, $p = 3$ or $p \equiv p^d \equiv -1 \pmod 3$. The claim follows.   □

In the situation of Proposition 15.7 it is further known that $p^d - 1$ has no divisors $q \equiv 3 \pmod 4$. The smallest example for $p \neq 3$ is $p^d = 11^5 = 1 + 153^2 + 371^2$ (observe that $p^d = 11$ is excluded by Theorem 8.6).

In the following we consider slightly more general questions.

Proposition 15.8 *Let B be a p-block of a finite group G with constrained fusion system $\mathscr{F}$ (for example if G is p-solvable). Then all B-subsections are major if and only if B has abelian defect groups.*

Proof Let D be a defect group of B. If D is abelian, then it is well-known that all B-subsections are major. Now assume conversely that all B-subsections are major. Then every element $x \in D$ is $\mathscr{F}$-conjugate to an element $y \in Z(D) \subseteq C_D(O_p(\mathscr{F})) \subseteq O_p(\mathscr{F})$. It follows that $\mathscr{F}$ is controlled and $D = Z(D)$.   □

For $p = 2$ we can drop the constrained condition on $\mathscr{F}$ by a recent result of Henke [109].

Proposition 15.9 (Henke) *Let $\mathscr{F}$ be a fusion system on a finite 2-group P such that every element in P is conjugate to an element in $Z(P)$. Then P is abelian.*

As a consequence we obtain an old result by Camina and Herzog [53].

Corollary 15.10 (Camina-Herzog) *Let G be a finite group such that $|G : C_G(x)|$ is odd for every 2-element $x \in G$. Then G has abelian Sylow 2-subgroups.*

The original proof of Corollary 15.10 uses Walter's classification of the finite simple groups with abelian Sylow 2-subgroups. In contrast, the proof of Henke's result is fairly elementary. The Camina-Herzog Theorem was generalized by Navarro and Tiep [202].

Theorem 15.11 (Navarro-Tiep) *Let $p \notin \{3, 5\}$ be a prime, and let G be a finite group such that $|G : C_G(x)| \not\equiv 0 \pmod{p}$ for every p-element $x \in G$. Then G has abelian Sylow p-subgroups.*

After all these results we propose the following question.

Question B Let $\mathscr{F}$ be an exotic fusion system on a finite p-group P such that any two non-trivial elements of P are conjugate in $\mathscr{F}$. Does it follow that $P \cong 7_+^{1+2}$?

Observe that there are precisely three exotic fusion systems on 7_+^{1+2} with the desired property (see [241]). In the proof of Theorem 11.9 we have already used the fact that these fusion systems cannot occur for blocks. Note also that fusion systems on abelian groups are controlled and thus non-exotic.

We give some evidence for Question B which has not been published.

Lemma 15.12 *Let $\mathscr{F}$ be an exotic fusion system on a p-group P such that any two non-trivial elements of P are conjugate in $\mathscr{F}$. Then the following holds:*

1. $\exp(P) = p > 2$.
2. $\mathrm{Out}_{\mathscr{F}}(P)$ *acts transitively on* $Z(P) \setminus \{1\}$. *In particular,* $Z(P) \subseteq P' = \Phi(P)$.
3. $C_{P/P'}(\mathrm{Out}_{\mathscr{F}}(P)) = 1$.
4. *For every element* $x \in P \setminus Z(P)$, *the subgroup* $C_P(x)$ *is contained in an* $\mathscr{F}$-*essential subgroup. In particular, every maximal subgroup* $M < P$ *such that* $Z(P) < Z(M)$ *is* $\mathscr{F}$-*essential.*

Proof

 (i) Obviously, $\exp(P) = p$. Since groups of exponent 2 are abelian, we have $p > 2$.
 (ii) By Burnside's Theorem for fusion systems (a mild extension of Theorem A.8 in [19]), any two non-trivial elements in $Z(P)$ are conjugate under $\mathrm{Out}_{\mathscr{F}}(P)$. Since P is non-abelian, we have $1 \neq P' \cap Z(P)$. As a characteristic subgroup we must have $Z(P) = P' \cap Z(P) \subseteq P'$. Finally, P/P' also has exponent p, so we get $P' = \Phi(P)$.
 (iii) Let $\mathscr{N} := N_{\mathscr{F}}(P)$. By Yoshida's Transfer Theorem for fusion systems (Theorem Y in [66]) and Alperin's Fusion Theorem we have

$$\langle x^f x^{-1} : x \in P, \ f \in \mathrm{Aut}_{\mathscr{F}}(P)\rangle = \langle x^f x^{-1} : x \in Q \leq P, \ f \in \mathrm{Aut}_{\mathscr{N}}(Q)\rangle$$

$$= \langle x^f x^{-1} : x \in Q \leq P, \ f \in \mathrm{Aut}_{\mathscr{F}}(Q)\rangle$$

$$= \langle xy^{-1} : x \text{ and } y \text{ are } \mathscr{F}\text{-conjugate}\rangle = P.$$

Now the claim follows from 8.4.2 in [159].

(iv) Choose a morphism $\varphi : \langle x \rangle \to P$ such that $\varphi(x) \in Z(P)$. Then the extension axiom for fusion systems (see Proposition I.2.5 in [19]) implies that φ can be extended to $\psi : C_P(x) \to P$. If there is no $\mathscr{F}$-essential subgroup $Q \leq P$ such that $C_P(x) \leq Q$, then Alperin's Fusion Theorem would show that ψ is in fact induced from $\mathrm{Aut}(P)$. But this is impossible, since $x \notin Z(P)$. For the second claim choose $x \in Z(M) \setminus Z(P)$. $\square$

Proposition 15.13 Let $\mathscr{F}$ be an exotic fusion system on a 3-group P such that any two non-trivial elements of P are conjugate in $\mathscr{F}$. Then $|P| \geq 3^7$.

Proof We may assume that $|P| \geq 3^4$. It is easy to compute the groups of order at most 3^6 and exponent 3 in GAP. Using Lemma 15.12, there are no candidates for $|P| = 3^4$. Now let $|P| = 3^5$. The extraspecial group $P \cong 3_+^{1+4}$ is excluded by Theorem 5.3 in [265]. It follows that P is uniquely determined and of the form $C_3^4 \rtimes C_3$. Moreover, $P' = \Phi(P) = Z(P) \cong C_3^2$. Hence, all $\mathscr{F}$-essential subgroups must be maximal by Proposition 6.12. Moreover, P has only one abelian maximal subgroup. By Alperin's Fusion Theorem, there is at least one $\mathscr{F}$-essential subgroup $Q \cong C_3 \times 3_+^{1+2}$. Since $\mathrm{Out}_{\mathscr{F}}(Q)$ does not have a normal 3-subgroup, the canonical map $\mathrm{Out}_{\mathscr{F}}(Q) \to \mathrm{Aut}(Q/Z(Q)) \times \mathrm{Aut}(Z(Q)/\Phi(Q))$ must be injective. However, P/Q acts trivially on $Q/Z(Q) = Q/P'$ and on $Z(Q)/\Phi(Q) \cong C_3$. Contradiction.

Finally, let $|P| = 3^6$. Then GAP shows that only two possibilities for P are feasible, namely $\mathtt{SmallGroup}(3^6, i)$ where $i \in \{122, 469\}$. The possibility $i = 469$ leads to the Sylow 3-subgroup of $\mathrm{SL}(3, 9)$. Here Theorem 4.5.1 of [57] shows that $\mathscr{F}$ is non-exotic. Now let $P \cong \mathtt{SmallGroup}(3^6, 122)$. Then $Z(P) = P' = \Phi(P) \cong C_3^3$ and $\mathrm{Out}(P)/O_3(\mathrm{Out}(P)) \cong \mathrm{GL}(3, 3)$. By Lemma 15.12(ii), 13 divides $|\mathrm{Out}_{\mathscr{F}}(P)|$. Since $\mathrm{Out}_{\mathscr{F}}(P)$ is a $3'$-group, we have $|\mathrm{Out}_{\mathscr{F}}(P)| \mid 2^5 \cdot 13$. Hence, Sylow's Theorem gives $\mathrm{Out}_{\mathscr{F}}(P) \leq N_{\mathrm{GL}(3,3)}(P_{13})$ for some $P_{13} \in \mathrm{Syl}_{13}(\mathrm{GL}(3, 3))$. This shows $\mathrm{Out}_{\mathscr{F}}(P) \cong C_{26}$. However, it can be verified with GAP that then $\mathrm{Out}_{\mathscr{F}}(P)$ does not act transitively on $Z(P) \setminus \{1\}$. $\square$

It follows from results on the Burnside Problem that in the situation of Proposition 15.13 the group P has nilpotency class at most 3 and $P'' = 1$.

Proposition 15.14 Let $\mathscr{F}$ be an exotic fusion system on a p-group P such that any two non-trivial elements of P are conjugate in $\mathscr{F}$. Then $|P| \neq p^4$.

Proof By Proposition 15.13, we may assume that $p \geq 5$. The non-abelian groups of order p^4 and exponent p are given in Lemma 3.2 in [157]: $C_p \times p_+^{1+2}$ and

$$Q := \langle a, b, c, d \mid [a, b] = [a, c] = [a, d] = [b, c] = 1, \ [b, d] = a, \ [c, d] = b \rangle$$

where p-powers of generators and not mentioned commutator relations between generators are defined to be trivial. The group $C_p \times p_+^{1+2}$ is excluded by Lemma 15.12(ii). Hence, assume $P \cong Q$. Then $C_p^3 \cong A := \langle a, b, c \rangle \leq P$. Moreover, $Z(P) = \langle a \rangle$. By Lemma 15.12(iv), A is $\mathscr{F}$-essential and every element $x \in A \setminus Z(P)$ is $\mathrm{Aut}_{\mathscr{F}}(A)$-conjugate to an element in $Z(P)$. Suppose for the moment that there is another abelian maximal subgroup $A_1 \neq A$. Then

$C_p^2 \cong A \cap A_1 \subseteq Z(AA_1) = Z(P)$. A contradiction. Thus, $\mathrm{Out}_{\mathscr{F}}(P)$ acts on A and $\mathrm{Aut}_{\mathscr{F}}(A)$ acts transitively on $Z(P) \setminus \{1\}$. This shows that $\mathrm{Aut}_{\mathscr{F}}(A)$ even acts transitively on $A \setminus \{1\}$; so it is a transitive linear group of degree p^3. Moreover, the order of $\mathrm{Aut}_{\mathscr{F}}(A)$ is divisible by p exactly once. However, by Theorem 15.1 there is no such transitive linear group. $\qquad\square$

References

1. Alperin, J.L.: The main problem of block theory. In: Proceedings of the Conference on Finite Groups (University of Utah, Park City, UT, 1975), pp. 341–356. Academic, New York (1976)
2. Alperin, J.L.: Weights for finite groups. In: The Arcata Conference on Representations of Finite Groups (Arcata, CA, 1986), Proceedings of the Symposium on Pure Mathematics , vol. 47.1, pp. 369–379. American Mathematical Society, Providence (1987)
3. Alperin, J.L., Brauer, R., Gorenstein, D.: Finite simple groups of 2-rank two. Scripta Math. **29**(3–4), 191–214 (1973). Collection of articles dedicated to the memory of Abraham Adrian Albert
4. Alperin, J.L., Broué, M.: Local methods in block theory. Ann. Math. (2) **110**(1), 143–157 (1979)
5. An, J.: Dade's conjecture for the Tits group. New Zealand J. Math. **25**(2), 107–131 (1996)
6. An, J.: The Alperin and Dade conjectures for Ree groups $^2F_4(q^2)$ in non-defining characteristics. J. Algebra **203**(1), 30–49 (1998)
7. An, J.: Controlled blocks of the finite quasisimple groups for odd primes. Adv. Math. **227**(3), 1165–1194 (2011)
8. An, J., Dietrich, H.: The AWC-goodness and essential rank of sporadic simple groups. J. Algebra **356**, 325–354 (2012)
9. An, J., Dietrich, H.: The essential rank of fusion systems of blocks of symmetric groups. Int. J. Algebra Comput. **22**(1), 1250002, 15 (2012)
10. An, J., Dietrich, H.: The essential rank of Brauer categories for finite groups of lie type. Bull. Lond. Math. Soc. **45**(2), 363–369 (2013)
11. An, J., Eaton, C.W.: Modular representation theory of blocks with trivial intersection defect groups. Algebr. Represent. Theory **8**(3), 427–448 (2005)
12. An, J., Eaton, C.W.: Blocks with extraspecial defect groups of finite quasisimple groups. J. Algebra **328**, 301–321 (2011)
13. An, J., Eaton, C.W.: Nilpotent blocks of quasisimple groups for odd primes. J. Reine Angew. Math. **656**, 131–177 (2011)
14. An, J., Eaton, C.W.: Nilpotent blocks of quasisimple groups for the prime two. Algebr. Represent. Theory **16**(1), 1–28 (2013)
15. An, J., O'Brien, E.A.: The Alperin and Dade conjectures for the O'Nan and Rudvalis simple groups. Commun. Algebra **30**(3), 1305–1348 (2002)
16. An, J., O'Brien, E.A., Wilson, R.A.: The Alperin weight conjecture and Dade's conjecture for the simple group J_4. LMS J. Comput. Math. **6**, 119–140 (2003)
17. Andersen, K.K.S., Oliver, B., Ventura, J.: Reduced, tame and exotic fusion systems. Proc. Lond. Math. Soc. (3) **105**(1), 87–152 (2012)

© Springer International Publishing Switzerland 2014
B. Sambale, *Blocks of Finite Groups and Their Invariants*, Lecture Notes in Mathematics 2127, DOI 10.1007/978-3-319-12006-5

18. Aschbacher, M.: Simple connectivity of p-group complexes. Israel J. Math. **82**(1–3), 1–43 (1993)
19. Aschbacher, M., Kessar, R., Oliver, B.: Fusion Systems in Algebra and Topology. London Mathematical Society Lecture Note Series, vol. 391. Cambridge University Press, Cambridge (2011)
20. Barker, L.: On p-soluble groups and the number of simple modules associated with a given Brauer pair. Q. J. Math. Oxford Ser. (2) **48**(190), 133–160 (1997)
21. Barnes, E.S.: Minkowski's fundamental inequality for reduced positive quadratic forms. J. Aust. Math. Soc. Ser. A **26**(1), 46–52 (1978)
22. Bender, H.: Transitive Gruppen gerader Ordnung, in denen jede Involution genau einen Punkt festläßt. J. Algebra **17**, 527–554 (1971)
23. Berkovich, Y.: Groups of Prime Power Order, vol. 1. de Gruyter Expositions in Mathematics, vol. 46. Walter de Gruyter GmbH & Co. KG, Berlin (2008)
24. Berkovich, Y., Janko, Z.: Groups of Prime Power Order, vol. 2. de Gruyter Expositions in Mathematics, vol. 47. Walter de Gruyter GmbH & Co. KG, Berlin (2008)
25. Berkovich, Y., Janko, Z.: Groups of Prime Power Order, vol. 3. de Gruyter Expositions in Mathematics, vol. 56. Walter de Gruyter GmbH & Co. KG, Berlin (2011)
26. Bertels, J.: Blöcke mit der 2-sylowgruppe von PSU(3, 4) als defektgruppe. Diplomarbeit, Jena (2012)
27. Bessenrodt, C., Olsson, J.B.: Spin representations and powers of 2. Algebr. Represent. Theory **3**(3), 289–300 (2000)
28. Blackburn, N.: On a special class of p-groups. Acta Math. **100**, 45–92 (1958)
29. Blackburn, N.: Über das Produkt von zwei zyklischen 2-Gruppen. Math. Z. **68**, 422–427 (1958)
30. Blackburn, N.: Generalizations of certain elementary theorems on p-groups. Proc. Lond. Math. Soc. (3) **11**, 1–22 (1961)
31. Brandt, H., Intrau, O.: Tabellen reduzierter positiver ternärer quadratischer Formen. Abh. Sächs. Akad. Wiss. Math.-Nat. Kl. **45**(4), 1–261 (1958)
32. Brandt, J.: A lower bound for the number of irreducible characters in a block. J. Algebra **74**(2), 509–515 (1982)
33. Brauer, R.: Investigations on group characters. Ann. Math. (2) **42**, 936–958 (1941)
34. Brauer, R.: Number theoretical investigations on groups of finite order. In: Proceedings of the International Symposium on Algebraic Number Theory, Tokyo and Nikko, 1955, pp. 55–62. Science Council of Japan, Tokyo (1956)
35. Brauer, R.: On the structure of groups of finite order. In: Proceedings of the International Congress of Mathematicians, Amsterdam, 1954, vol. 1, pp. 209–217. Erven P. Noordhoff N.V., Groningen (1957)
36. Brauer, R.: Representations of finite groups. In: Lectures on Modern Mathematics, vol. I, pp. 133–175. Wiley, New York (1963)
37. Brauer, R.: Some applications of the theory of blocks of characters of finite groups. I. J. Algebra **1**, 152–167 (1964)
38. Brauer, R.: Some applications of the theory of blocks of characters of finite groups. II. J. Algebra **1**, 307–334 (1964)
39. Brauer, R.: On blocks and sections in finite groups. II. Am. J. Math. **90**, 895–925 (1968)
40. Brauer, R.: Defect groups in the theory of representations of finite groups. Ill. J. Math. **13**, 53–73 (1969)
41. Brauer, R.: On 2-blocks with dihedral defect groups. In: Symposia Mathematica, vol. XIII (Convegno di Gruppi e loro Rappresentazioni, INDAM, Rome, 1972), pp. 367–393. Academic, London (1974)
42. Brauer, R., Feit, W.: On the number of irreducible characters of finite groups in a given block. Proc. Natl. Acad. Sci. USA **45**, 361–365 (1959)
43. Brauer, R., Nesbitt, C.: On the modular characters of groups. Ann. Math. (2) **42**, 556–590 (1941)

44. Broué, M.: On characters of height zero. In: The Santa Cruz Conference on Finite Groups (University of California, Santa Cruz, CA, 1979), Proceedings of the Symposium on Pure Mathematics, vol. 37, pp. 393–396. American Mathematical Society, Providence (1980)
45. Broué, M.: Les l-blocs des groups $GL(n, q)$ et $U(n, q^2)$ et leurs structures locales. Astérisque **133–134**, 159–188 (1986). Seminar Bourbaki, vol. 1984/85
46. Broué, M.: Isométries de caractères et équivalences de Morita ou dérivées. Inst. Hautes Études Sci. Publ. Math. **71**, 45–63 (1990)
47. Broué, M.: Isométries parfaites, types de blocs, catégories dérivées. Astérisque **181–182**, 61–92 (1990)
48. Broué, M.: Equivalences of blocks of group algebras. In: Finite-Dimensional Algebras and Related Topics (Ottawa, ON, 1992), NATO Adv. Sci. Inst. Ser. C Math. Phys. Sci., vol. 424, pp. 1–26. Kluwer Academic Publishers, Dordrecht (1994)
49. Broué, M., Olsson, J.B.: Subpair multiplicities in finite groups. J. Reine Angew. Math. **371**, 125–143 (1986)
50. Brunat, O., Gramain, J.B.: Perfect isometries and murnaghan-nakayama rules (2013). arXiv: 1305.7449v1
51. Cabanes, M., Enguehard, M.: Representation Theory of Finite Reductive Groups. New Mathematical Monographs, vol. 1. Cambridge University Press, Cambridge (2004)
52. Cabanes, M., Picaronny, C.: Types of blocks with dihedral or quaternion defect groups. J. Fac. Sci. Univ. Tokyo Sect. IA Math. **39**(1), 141–161 (1992). Revised version: http://www.math.jussieu.fr/~cabanes/type99.pdf
53. Camina, A.R., Herzog, M.: Character tables determine abelian Sylow 2-subgroups. Proc. Am. Math. Soc. **80**(3), 533–535 (1980)
54. Carter, R.W.: Centralizers of semisimple elements in the finite classical groups. Proc. Lond. Math. Soc. (3) **42**(1), 1–41 (1981)
55. Carter, R.W.: Finite Groups of Lie Type. Pure and Applied Mathematics (New York). Wiley, New York (1985). Conjugacy classes and complex characters, A Wiley-Interscience Publication
56. Chlebowitz, M., Külshammer, B.: Symmetric local algebras with 5-dimensional center. Trans. Am. Math. Soc. **329**(2), 715–731 (1992)
57. Clelland, M.R.: Saturated fusion systems and finite groups. Ph.D. thesis, Birmingham (2006)
58. Conlon, S.B.: Twisted group algebras and their representations. J. Aust. Math. Soc. **4**, 152–173 (1964)
59. Conway, J.H., Curtis, R.T., Norton, S.P., Parker, R.A., Wilson, R.A.: ATLAS of finite groups. Oxford University Press, Eynsham (1985). Maximal subgroups and ordinary characters for simple groups, With computational assistance from J.G. Thackray
60. Craven, D.A.: Control of fusion and solubility in fusion systems. J. Algebra **323**(9), 2429–2448 (2010)
61. Craven, D.A.: The Theory of Fusion Systems. Cambridge Studies in Advanced Mathematics, vol. 131. Cambridge University Press, Cambridge (2011). An algebraic approach
62. Craven, D.A., Eaton, C.W., Kessar, R., Linckelmann, M.: The structure of blocks with a Klein four defect group. Math. Z. **268**(1–2), 441–476 (2011)
63. Craven, D.A., Glesser, A.: Fusion systems on small p-groups. Trans. Am. Math. Soc. **364**(11), 5945–5967 (2012)
64. Curtis, C.W., Reiner, I.: Methods of Representation Theory. Vol. I. Wiley Classics Library. Wiley, New York (1990). With applications to finite groups and orders, Reprint of the 1981 original, A Wiley-Interscience Publication
65. Dade, E.C.: Blocks with cyclic defect groups. Ann. Math. (2) **84**, 20–48 (1966)
66. Díaz, A., Glesser, A., Park, S., Stancu, R.: Tate's and Yoshida's theorems on control of transfer for fusion systems. J. Lond. Math. Soc. (2) **84**(2), 475–494 (2011)
67. Díaz, A., Ruiz, A., Viruel, A.: All p-local finite groups of rank two for odd prime p. Trans. Am. Math. Soc. **359**(4), 1725–1764 (2007)
68. Dietz, J.: Stable splittings of classifying spaces of metacyclic p-groups, p odd. J. Pure Appl. Algebra **90**(2), 115–136 (1993)

69. Donovan, P.W.: Dihedral defect groups. J. Algebra **56**(1), 184–206 (1979)
70. Du, S., Jones, G., Kwak, J.H., Nedela, R., Škoviera, M.: 2-Groups that factorise as products of cyclic groups, and regular embeddings of complete bipartite graphs. Ars Math. Contemp. **6**(1), 155–170 (2013)
71. Düvel, O.: On Donovan's conjecture. J. Algebra **272**(1), 1–26 (2004)
72. Eaton, C.W.: Generalisations of conjectures of Brauer and Olsson. Arch. Math. (Basel) **81**(6), 621–626 (2003)
73. Eaton, C.W.: The equivalence of some conjectures of Dade and Robinson. J. Algebra **271**(2), 638–651 (2004)
74. Eaton, C.W., Kessar, R., Külshammer, B., Sambale, B.: 2-Blocks with abelian defect groups. Adv. Math. **254**, 706–735 (2014)
75. Eaton, C.W., Külshammer, B., Sambale, B.: 2-Blocks with minimal nonabelian defect groups ii. J. Group Theory **15**, 311–321 (2012)
76. Eaton, C.W., Moretó, A.: Extending Brauer's Height Zero Conjecture to blocks with nonabelian defect groups. Int. Math. Res. Not. **2014**(20), 5581–5601 (2014)
77. Enguehard, M.: Isométries parfaites entre blocs de groupes symétriques. Astérisque **181–182**, 157–171 (1990)
78. Enguehard, M.: Vers une décomposition de Jordan des blocs des groupes réductifs finis. J. Algebra **319**(3), 1035–1115 (2008)
79. Erdmann, K.: Blocks whose defect groups are Klein four groups: a correction. J. Algebra **76**(2), 505–518 (1982)
80. Erdmann, K.: Blocks of Tame Representation Type and Related Algebras. Lecture Notes in Mathematics, vol. 1428. Springer, Berlin (1990)
81. Feit, W.: The Representation Theory of Finite Groups. North-Holland Mathematical Library, vol. 25. North-Holland Publishing, Amsterdam (1982)
82. Fleischmann, P.: Finite groups with regular orbits on vector spaces. J. Algebra **103**(1), 211–215 (1986)
83. Fong, P.: On the characters of p-solvable groups. Trans. Am. Math. Soc. **98**, 263–284 (1961)
84. Fong, P., Harris, M.E.: On perfect isometries and isotypies in finite groups. Invent. Math. **114**(1), 139–191 (1993)
85. Fong, P., Srinivasan, B.: The blocks of finite general linear and unitary groups. Invent. Math. **69**(1), 109–153 (1982)
86. Fujii, M.: On determinants of Cartan matrices of p-blocks. Proc. Jpn. Acad. Ser. A Math. Sci. **56**(8), 401–403 (1980)
87. Gantmacher, F.R.: Matrizentheorie. Hochschulbücher für Mathematik, vol. 86. VEB Deutscher Verlag der Wissenschaften, Berlin (1986)
88. Gao, S.: On Brauer's $k(B)$-problem for blocks with metacyclic defect groups of odd order. Arch. Math. (Basel) **96**(6), 507–512 (2011)
89. Gao, S.: Blocks of full defect with nonabelian metacyclic defect groups. Arch. Math. (Basel) **98**, 1–12 (2012)
90. Gao, S., Zeng, J.: On the number of ordinary irreducible characters in a p-block with a minimal nonabelian defect group. Commun. Algebra **39**(9), 3278–3297 (2011)
91. Glesser, A.: Sparse fusion systems. Proc. Edinb. Math. Soc. (2) **56**(1), 135–150 (2013)
92. Gluck, D.: Rational defect groups and 2-rational characters. J. Group Theory **14**(3), 401–412 (2011)
93. Gluck, D., Wolf, T.R.: Brauer's height conjecture for p-solvable groups. Trans. Am. Math. Soc. **282**(1), 137–152 (1984)
94. Gorenstein, D.: Finite groups. Harper & Row Publishers, New York (1968)
95. Gorenstein, D., Harada, K.: Finite Groups Whose 2-Subgroups are Generated by at Most 4 Elements. American Mathematical Society, Providence (1974). Memoirs of the American Mathematical Society, No. 147
96. Gorenstein, D., Lyons, R.: The local structure of finite groups of characteristic 2 type. Mem. Am. Math. Soc. **42**(276), 1–731 (1983)

97. Gorenstein, D., Lyons, R., Solomon, R.: The Classification of the Finite Simple Groups. Number 3. Part I. Chapter A. Mathematical Surveys and Monographs, vol. 40. American Mathematical Society, Providence (1998). Almost simple K-groups

98. Gorenstein, D., Walter, J.H.: The characterization of finite groups with dihedral Sylow 2-subgroups. I. J. Algebra **2**, 85–151 (1965)

99. Grove, L.C.: Classical Groups and Geometric Algebra. Graduate Studies in Mathematics, vol. 39. American Mathematical Society, Providence (2002)

100. Guralnick, R.M.: Commutators and commutator subgroups. Adv. Math. **45**(3), 319–330 (1982)

101. Halasi, Z., Podoski, K.: Every coprime linear group admits a base of size two (2013). arXiv: 1212.0199v2

102. Hall Jr., M.: The Theory of Groups. The Macmillan Co., New York (1959)

103. Hall, M., Jr., Senior, J.K.: The Groups of Order 2^n ($n \le 6$). The Macmillan Co., New York (1964)

104. Harada, K.: Groups with a certain type of Sylow 2-subgroups. J. Math. Soc. Jpn. **19**, 203–307 (1967)

105. Harris, M.E., Knörr, R.: Brauer correspondence for covering blocks of finite groups. Commun. Algebra **13**(5), 1213–1218 (1985)

106. Hartley, B., Turull, A.: On characters of coprime operator groups and the Glauberman character correspondence. J. Reine Angew. Math. **451**, 175–219 (1994)

107. Hendren, S.: Extra special defect groups of order p^3 and exponent p^2. J. Algebra **291**(2), 457–491 (2005)

108. Hendren, S.: Extra special defect groups of order p^3 and exponent p. J. Algebra **313**(2), 724–760 (2007)

109. Henke, E.: A characterization of saturated fusion systems over abelian 2-groups. Adv. Math. **257**, 1–5 (2014)

110. Hering, C.: Transitive linear groups and linear groups which contain irreducible subgroups of prime order. Geometriae Dedicata **2**, 425–460 (1974)

111. Hermann, P.Z.: On finite p-groups with isomorphic maximal subgroups. J. Aust. Math. Soc. Ser. A **48**(2), 199–213 (1990)

112. Héthelyi, L., Kessar, R., Külshammer, B., Sambale, B.: Blocks with transitive fusion systems. (2014, submitted)

113. Héthelyi, L., Külshammer, B.: Characters, conjugacy classes and centrally large subgroups of p-groups of small rank. J. Algebra **340**, 199–210 (2011)

114. Héthelyi, L., Külshammer, B., Sambale, B.: A note on olsson's conjecture. J. Algebra **398**, 364–385 (2014)

115. Higman, G.: Suzuki 2-groups. Ill. J. Math. **7**, 79–96 (1963)

116. Hiss, G.: Morita equivalences between blocks of finite chevalley groups. In: Proceedings of Representation Theory of Finite and Algebraic Groups, Osaka University, pp. 128–136 (2000)

117. Hiss, G., Kessar, R.: Scopes reduction and Morita equivalence classes of blocks in finite classical groups. J. Algebra **230**(2), 378–423 (2000)

118. Hiss, G., Kessar, R.: Scopes reduction and Morita equivalence classes of blocks in finite classical groups. II. J. Algebra **283**(2), 522–563 (2005)

119. Hiss, G., Lux, K.: Brauer Trees of Sporadic Groups. Oxford Science Publications. The Clarendon Press Oxford University Press, New York (1989)

120. Holloway, M., Koshitani, S., Kunugi, N.: Blocks with nonabelian defect groups which have cyclic subgroups of index p. Arch. Math. (Basel) **94**(2), 101–116 (2010)

121. Holm, T.: Blocks of tame representation type and related algebras: derived equivalences and hochschild cohomology. Habilitationsschrift, Magdeburg (2001)

122. Holm, T.: Notes on donovan's conjecture for blocks of tame representation type (2014). http://www.iazd.uni-hannover.de/~tholm/ARTIKEL/donovan.ps

123. Holm, T., Kessar, R., Linckelmann, M.: Blocks with a quaternion defect group over a 2-adic ring: the case $\tilde{A}_4$. Glasg. Math. J. **49**(1), 29–43 (2007)

124. Horimoto, H., Watanabe, A.: On a perfect isometry between principal p-blocks of finite groups with cyclic p-hyperfocal subgroups (2012). Preprint
125. Humphreys, J.E.: Defect groups for finite groups of Lie type. Math. Z. **119**, 149–152 (1971)
126. Huppert, B.: Über das Produkt von paarweise vertauschbaren zyklischen Gruppen. Math. Z. **58**, 243–264 (1953)
127. Huppert, B.: Zweifach transitive, auflösbare Permutationsgruppen. Math. Z. **68**, 126–150 (1957)
128. Huppert, B.: Endliche Gruppen. I. Die Grundlehren der Mathematischen Wissenschaften, Band 134. Springer, Berlin (1967)
129. Huppert, B., Blackburn, N.: Finite Groups. II. Grundlehren der Mathematischen Wissenschaften, vol. 242. Springer, Berlin (1982)
130. Huppert, B., Blackburn, N.: Finite Groups. III. Grundlehren der Mathematischen Wissenschaften, vol. 243. Springer, Berlin (1982)
131. Isaacs, I.M.: Blocks with just two irreducible Brauer characters in solvable groups. J. Algebra **170**(2), 487–503 (1994)
132. Isaacs, I.M.: Finite Group Theory. Graduate Studies in Mathematics, vol. 92. American Mathematical Society, Providence (2008)
133. Isaacs, I.M., Navarro, G.: Characters of p'-degree of p-solvable groups. J. Algebra **246**(1), 394–413 (2001)
134. Isaacs, I.M., Navarro, G.: New refinements of the McKay conjecture for arbitrary finite groups. Ann. Math. (2) **156**(1), 333–344 (2002)
135. Itô, N.: Über das Produkt von zwei zyklischen 2-Gruppen. Publ. Math. Debrecen **4**, 517–520 (1956)
136. Itô, N., Ôhara, A.: Sur les groupes factorisables par deux 2-groupes cycliques. I. Cas où leur groupe des commutateurs est cyclique. Proc. Jpn. Acad. **32**, 736–740 (1956)
137. Itô, N., Ôhara, A.: Sur les groupes factorisables par deux 2-groupes cycliques. II. Cas où leur groupe des commutateurs n'est pas cyclique. Proc. Jpn. Acad. **32**, 741–743 (1956)
138. Jacobsen, M.W.: Block fusion systems of the alternating groups (2012). arXiv:1204.2702v1
139. James, G., Kerber, A.: The Representation Theory of the Symmetric Group. Encyclopedia of Mathematics and its Applications, vol. 16. Addison-Wesley Publishing, Reading (1981). With a foreword by P.M. Cohn, With an introduction by Gilbert de B. Robinson
140. Janko, Z.: Finite 2-groups with exactly one nonmetacyclic maximal subgroup. Israel J. Math. **166**, 313–347 (2008)
141. Johnson, N.L., Jha, V., Biliotti, M.: Handbook of Finite Translation Planes. Pure and Applied Mathematics (Boca Raton), vol. 289. Chapman & Hall/CRC, Boca Raton (2007)
142. Karpilovsky, G.: The Schur Multiplier. London Mathematical Society Monographs. New Series, vol. 2. The Clarendon Press Oxford University Press, New York (1987)
143. Keller, T.M., Yang, Y.: Abelian quotients and orbit sizes of finite groups (2014). arXiv:1407. 6436v1
144. Kemper, G., Lübeck, F., Magaard, K.: Matrix generators for the Ree groups $^2G_2(q)$. Commun. Algebra **29**(1), 407–413 (2001)
145. Kessar, R.: Blocks and source algebras for the double covers of the symmetric and alternating groups. J. Algebra **186**(3), 872–933 (1996)
146. Kessar, R.: Scopes reduction for blocks of finite alternating groups. Q. J. Math. **53**(4), 443–454 (2002)
147. Kessar, R.: Introduction to block theory. In: Group Representation Theory, pp. 47–77. EPFL Press, Lausanne (2007)
148. Kessar, R., Koshitani, S., Linckelmann, M.: Conjectures of Alperin and Broué for 2-blocks with elementary abelian defect groups of order 8. J. Reine Angew. Math. **671**, 85–130 (2012)
149. Kessar, R., Linckelmann, M.: On blocks with Frobenius inertial quotient. J. Algebra **249**(1), 127–146 (2002)
150. Kessar, R., Linckelmann, M.: On perfect isometries for tame blocks. Bull. Lond. Math. Soc. **34**(1), 46–54 (2002)
151. Kessar, R., Linckelmann, M., Navarro, G.: A characterisation of nilpotent blocks (2014). arXiv:1402.5871v1

152. Kessar, R., Malle, G.: Quasi-isolated blocks and Brauer's height zero conjecture. Ann. Math. (2) **178**(1), 321–384 (2013)
153. Kessar, R., Stancu, R.: A reduction theorem for fusion systems of blocks. J. Algebra **319**(2), 806–823 (2008)
154. Kiyota, M.: On 3-blocks with an elementary abelian defect group of order 9. J. Fac. Sci. Univ. Tokyo Sect. IA Math. **31**(1), 33–58 (1984)
155. Klein, A.A.: On Fermat's theorem for matrices and the periodic identities of $M_n(GF(q))$. Arch. Math. (Basel) **34**(5), 399–402 (1980)
156. Koshitani, S.: Conjectures of Donovan and Puig for principal 3-blocks with abelian defect groups. Commun. Algebra **31**(5), 2229–2243 (2003)
157. Koshitani, S., Külshammer, B., Sambale, B.: On loewy lengths of blocks. Math. Proc. Camb. Philos. Soc. **156**(3), 555–570 (2014)
158. Koshitani, S., Miyachi, H.: Donovan conjecture and Loewy length for principal 3-blocks of finite groups with elementary abelian Sylow 3-subgroup of order 9. Commun. Algebra **29**(10), 4509–4522 (2001)
159. Kurzweil, H., Stellmacher, B.: The Theory of Finite Groups. Universitext. Springer, New York (2004)
160. Külshammer, B.: On 2-blocks with wreathed defect groups. J. Algebra **64**, 529–555 (1980)
161. Külshammer, B.: On p-blocks of p-solvable groups. Commun. Algebra **9**(17), 1763–1785 (1981)
162. Külshammer, B.: Symmetric local algebras and small blocks of finite groups. J. Algebra **88**(1), 190–195 (1984)
163. Külshammer, B.: Crossed products and blocks with normal defect groups. Commun. Algebra **13**(1), 147–168 (1985)
164. Külshammer, B.: A remark on conjectures in modular representation theory. Arch. Math. (Basel) **49**(5), 396–399 (1987)
165. Külshammer, B.: Landau's theorem for p-blocks of p-solvable groups. J. Reine Angew. Math. **404**, 189–191 (1990)
166. Külshammer, B.: Donovan's conjecture, crossed products and algebraic group actions. Israel J. Math. **92**(1–3), 295–306 (1995)
167. Külshammer, B., Navarro, G., Sambale, B., Tiep, P.H.: On finite groups with two conjugacy classes of p-elements and related questions for p-blocks. Bull. Lond. Math. Soc. **46**(2), 305–314 (2014)
168. Külshammer, B., Okuyama, T.: On centrally controlled blocks of finite groups. (unpublished)
169. Külshammer, B., Puig, L.: Extensions of nilpotent blocks. Invent. Math. **102**(1), 17–71 (1990)
170. Külshammer, B., Robinson, G.R.: Alperin-McKay implies Brauer's problem 21. J. Algebra **180**(1), 208–210 (1996)
171. Külshammer, B., Sambale, B.: The 2-blocks of defect 4. Represent. Theory **17**, 226–236 (2013)
172. Külshammer, B., Wada, T.: Some inequalities between invariants of blocks. Arch. Math. (Basel) **79**(2), 81–86 (2002)
173. Landrock, P.: A counterexample to a conjecture on the Cartan invariants of a group algebra. Bull. Lond. Math. Soc. **5**, 223–224 (1973)
174. Landrock, P.: Finite groups with a quasisimple component of type $PSU(3, 2^n)$ on elementary abelian form. Ill. J. Math. **19**, 198–230 (1975)
175. Landrock, P.: The non-principal 2-blocks of sporadic simple groups. Commun. Algebra **6**(18), 1865–1891 (1978)
176. Landrock, P.: On the number of irreducible characters in a 2-block. J. Algebra **68**(2), 426–442 (1981)
177. Leedham-Green, C.R., Plesken, W.: Some remarks on Sylow subgroups of general linear groups. Math. Z. **191**(4), 529–535 (1986)
178. Liebeck, H.: The location of the minimum of a positive definite integral quadratic form. J. Lond. Math. Soc. (2) **3**, 477–484 (1971)
179. Liedahl, S.: Enumeration of metacyclic p-groups. J. Algebra **186**(2), 436–446 (1996)

180. Linckelmann, M.: Derived equivalence for cyclic blocks over a P-adic ring. Math. Z. **207**(2), 293–304 (1991)

181. Linckelmann, M.: A derived equivalence for blocks with dihedral defect groups. J. Algebra **164**(1), 244–255 (1994)

182. Linckelmann, M.: The source algebras of blocks with a Klein four defect group. J. Algebra **167**(3), 821–854 (1994)

183. Linckelmann, M.: Fusion category algebras. J. Algebra **277**(1), 222–235 (2004)

184. Linckelmann, M.: Introduction to fusion systems. In: Group Representation Theory, pp. 79–113. EPFL Press, Lausanne (2007). Revised version: http://web.mat.bham.ac.uk/C.W.Parker/Fusion/fusion-intro.pdf

185. Lux, K., Pahlings, H.: Representations of Groups. Cambridge Studies in Advanced Mathematics, vol. 124. Cambridge University Press, Cambridge (2010). A computational approach

186. Malle, G., Navarro, G.: Inequalities for some blocks of finite groups. Arch. Math. (Basel) **87**(5), 390–399 (2006)

187. Malle, G., Navarro, G.: Blocks with equal height zero degrees. Trans. Am. Math. Soc. **363**(12), 6647–6669 (2011)

188. Mann, A.: On p-groups whose maximal subgroups are isomorphic. J. Aust. Math. Soc. Ser. A **59**(2), 143–147 (1995)

189. Maplesoft, a division of Waterloo Maple Inc.: Maple 16 (2012). http://www.maplesoft.com/products/Maple/

190. Mazurov, V.D.: Finite groups with metacyclic Sylow 2-subgroups. Sibirsk. Mat. Ž. **8**, 966–982 (1967)

191. Michler, G.O.: On blocks with multiplicity one. In: Representations of Algebras (Puebla, 1980). Lecture Notes in Mathematics, vol. 903, pp. 242–256. Springer, Berlin (1981)

192. Moretó, A., Navarro, G.: Heights of characters in blocks of p-solvable groups. Bull. Lond. Math. Soc. **37**(3), 373–380 (2005)

193. Murai, M.: On subsections of blocks and Brauer pairs. Osaka J. Math. **37**(3), 719–733 (2000)

194. Naehrig, M.: Die Brauer-Bäume des Monsters M in Charakteristik 29. Diplomarbeit, Aachen (2002)

195. Nagao, H.: On a conjecture of Brauer for p-solvable groups. J. Math. Osaka City Univ. **13**, 35–38 (1962)

196. Nagao, H., Tsushima, Y.: Representations of finite groups. Academic, Boston (1989). Translated from the Japanese

197. Naik, V.: Groups of order 32 (2013). http://groupprops.subwiki.org/wiki/Groups_of_order_32

198. Narasaki, R., Uno, K.: Isometries and extra special Sylow groups of order p^3. J. Algebra **322**(6), 2027–2068 (2009)

199. Navarro, G.: The McKay conjecture and Galois automorphisms. Ann. Math. (2) **160**(3), 1129–1140 (2004)

200. Navarro, G., Späth, B.: On Brauer's height zero conjecture. J. Eur. Math. Soc. **16**, 695–747 (2014)

201. Navarro, G., Tiep, P.H.: Brauer's height zero conjecture for the 2-blocks of maximal defect. J. Reine Angew. Math. **669**, 225–247 (2012)

202. Navarro, G., Tiep, P.H.: Abelian Sylow subgroups in a finite group. J. Algebra **398**, 519–526 (2014)

203. Nebe, G., Sloane, N.: A catalogue of lattices (2014). http://www.math.rwth-aachen.de/~Gabriele.Nebe/LATTICES/

204. Neukirch, J.: Algebraische Zahlentheorie. Springer, Berlin (1992)

205. Ninomiya, Y.: Finite p-groups with cyclic subgroups of index p^2. Math. J. Okayama Univ. **36**, 1–21 (1994)

206. Ninomiya, Y., Wada, T.: Cartan matrices for blocks of finite p-solvable groups with two simple modules. J. Algebra **143**(2), 315–333 (1991)

207. Nipp, G.L.: Quaternary Quadratic Forms. Springer, New York (1991). Computer generated tables, With a 3.5″ IBM PC floppy disk

208. Noeske, F.: Adgc for sporadic groups (2014). http://www.math.rwth-aachen.de/~Felix. Noeske/tabular.pdf
209. O'Brien, E.A.: Hall-senior number vs small group id (2014). http://permalink.gmane.org/gmane.comp.mathematics.gap.user/2426
210. Okuyama, T., Wajima, M.: Character correspondence and p-blocks of p-solvable groups. Osaka J. Math. **17**(3), 801–806 (1980)
211. Oliver, B., Ventura, J.: Saturated fusion systems over 2-groups. Trans. Am. Math. Soc. **361**(12), 6661–6728 (2009)
212. Olsson, J.B.: On 2-blocks with quaternion and quasidihedral defect groups. J. Algebra **36**(2), 212–241 (1975)
213. Olsson, J.B.: McKay numbers and heights of characters. Math. Scand. **38**(1), 25–42 (1976)
214. Olsson, J.B.: On the subsections for certain 2-blocks. J. Algebra **46**(2), 497–510 (1977)
215. Olsson, J.B.: Lower defect groups. Commun. Algebra **8**(3), 261–288 (1980)
216. Olsson, J.B.: Inequalities for block-theoretic invariants. In: Representations of Algebras (Puebla, 1980). Lecture Notes in Mathematics, vol. 903, pp. 270–284. Springer, Berlin (1981)
217. Olsson, J.B.: On subpairs and modular representation theory. J. Algebra **76**(1), 261–279 (1982)
218. Olsson, J.B.: Combinatorics and Representations of Finite Groups. Vorlesungen aus dem Fachbereich Mathematik der Universität GH Essen, Essen, vol. 20 (1993)
219. Park, S.: The gluing problem for some block fusion systems. J. Algebra **323**(6), 1690–1697 (2010)
220. Passman, D.S.: p-Solvable doubly transitive permutation groups. Pac. J. Math. **26**, 555–577 (1968)
221. Puig, L.: Nilpotent blocks and their source algebras. Invent. Math. **93**(1), 77–116 (1988)
222. Puig, L.: Pointed groups and construction of modules. J. Algebra **116**(1), 7–129 (1988)
223. Puig, L.: The hyperfocal subalgebra of a block. Invent. Math. **141**(2), 365–397 (2000)
224. Puig, L.: Frobenius Categories Versus Brauer Blocks. Progress in Mathematics, vol. 274. Birkhäuser Verlag, Basel (2009). The Grothendieck group of the Frobenius category of a Brauer block
225. Puig, L.: Nilpotent extensions of blocks. Math. Z. **269**(1–2), 115–136 (2011)
226. Puig, L., Usami, Y.: Perfect isometries for blocks with abelian defect groups and Klein four inertial quotients. J. Algebra **160**(1), 192–225 (1993)
227. Puig, L., Usami, Y.: Perfect isometries for blocks with abelian defect groups and cyclic inertial quotients of order 4. J. Algebra **172**(1), 205–213 (1995)
228. Rickard, J.: Derived categories and stable equivalence. J. Pure Appl. Algebra **61**(3), 303–317 (1989)
229. Robinson, G.R.: On the number of characters in a block and the Brauer-Feit matrix (unpublished)
230. Robinson, G.R.: The number of blocks with a given defect group. J. Algebra **84**(2), 493–502 (1983)
231. Robinson, G.R.: On the number of characters in a block. J. Algebra **138**(2), 515–521 (1991)
232. Robinson, G.R.: On Brauer's $k(B)$ problem. J. Algebra **147**(2), 450–455 (1992)
233. Robinson, G.R.: Local structure, vertices and Alperin's conjecture. Proc. Lond. Math. Soc. (3) **72**(2), 312–330 (1996)
234. Robinson, G.R.: Dade's projective conjecture for p-solvable groups. J. Algebra **229**(1), 234–248 (2000)
235. Robinson, G.R.: Weight conjectures for ordinary characters. J. Algebra **276**(2), 761–775 (2004)
236. Robinson, G.R.: Amalgams, blocks, weights, fusion systems and finite simple groups. J. Algebra **314**(2), 912–923 (2007)
237. Robinson, G.R.: Large character heights, $Qd(p)$, and the ordinary weight conjecture. J. Algebra **319**(2), 657–679 (2008)
238. Robinson, G.R.: On the focal defect group of a block, characters of height zero, and lower defect group multiplicities. J. Algebra **320**(6), 2624–2628 (2008)

239. Roitman, M.: On Zsigmondy primes. Proc. Am. Math. Soc. **125**(7), 1913–1919 (1997)
240. Rouquier, R.: The derived category of blocks with cyclic defect groups. In: Derived Equivalences for Group Rings. Lecture Notes in Mathematics, vol. 1685, pp. 199–220. Springer, Berlin (1998)
241. Ruiz, A., Viruel, A.: The classification of p-local finite groups over the extraspecial group of order p^3 and exponent p. Math. Z. **248**(1), 45–65 (2004)
242. Rédei, L.: Das "schiefe Produkt" in der Gruppentheorie. Comment. Math. Helv. **20**, 225–264 (1947)
243. Sambale, B.: 2-Blocks with minimal nonabelian defect groups. J. Algebra **337**, 261–284 (2011)
244. Sambale, B.: 2-blöcke mit metazyklischen und minimal nichtabelschen defektgruppen. Dissertation, Südwestdeutscher Verlag für Hochschulschriften, Saarbrücken (2011)
245. Sambale, B.: Cartan matrices and Brauer's $k(B)$-conjecture. J. Algebra **331**, 416–427 (2011)
246. Sambale, B.: Cartan matrices and Brauer's $k(B)$-conjecture II. J. Algebra **337**, 345–362 (2011)
247. Sambale, B.: Blocks with defect group $D_{2^n} \times C_{2^m}$. J. Pure Appl. Algebra **216**, 119–125 (2012)
248. Sambale, B.: Fusion systems on metacyclic 2-groups. Osaka J. Math. **49**, 325–329 (2012)
249. Sambale, B.: Blocks with central product defect group $D_{2^n} * C_{2^m}$. Proc. Am. Math. Soc. **141**(12), 4057–4069 (2013)
250. Sambale, B.: Blocks with defect group $Q_{2^n} \times C_{2^m}$ and $SD_{2^n} \times C_{2^m}$. Algebr. Represent. Theory **16**(6), 1717–1732 (2013)
251. Sambale, B.: Brauer's Height Zero Conjecture for metacyclic defect groups. Pac. J. Math. **262**(2), 481–507 (2013)
252. Sambale, B.: Further evidence for conjectures in block theory. Algebra Number Theory **7**(9), 2241–2273 (2013)
253. Sambale, B.: On the Brauer-Feit bound for abelian defect groups. Math. Z. **276**, 785–797 (2014)
254. Sambale, B.: The Alperin-McKay Conjecture for metacyclic, minimal non-abelian defect groups. Proc. Am. Math. Soc. (2014, to appear). arXiv:1403.5153v1
255. Sambale, B.: Cartan matrices and Brauer's $k(B)$-Conjecture III. Manuscripta Math. (2014, to appear)
256. Sambale, B.: Cartan matrices of blocks of finite groups (2014). http://www.minet.uni-jena. de/algebra/personen/sambale/matrices.pdf
257. Sambale, B.: Fusion systems on bicyclic 2-groups. Proc. Edinb. Math. Soc. (to appear)
258. Sawabe, M.: A note on finite simple groups with abelian Sylow p-subgroups. Tokyo J. Math. **30**(2), 293–304 (2007)
259. Sawabe, M., Watanabe, A.: On the principal blocks of finite groups with abelian Sylow p-subgroups. J. Algebra **237**(2), 719–734 (2001)
260. Schmid, P.: The Solution of the $k(GV)$ Problem. ICP Advanced Texts in Mathematics, vol. 4. Imperial College Press, London (2007)
261. Schulz, N.: Über p-blöcke endlicher p-auflösbarer Gruppen. Dissertation, Universität Dortmund (1980)
262. Scopes, J.: Cartan matrices and Morita equivalence for blocks of the symmetric groups. J. Algebra **142**(2), 441–455 (1991)
263. Späth, B.: A reduction theorem for the Alperin-McKay conjecture. J. Reine Angew. Math. **680**, 153–189 (2013)
264. Späth, B.: A reduction theorem for the blockwise Alperin weight conjecture. J. Group Theory **16**(2), 159–220 (2013)
265. Stancu, R.: Control of fusion in fusion systems. J. Algebra Appl. **5**(6), 817–837 (2006)
266. The GAP Group: GAP – Groups, Algorithms, and Programming, Version 4.6.5 (2013). http://www.gap-system.org
267. Thomas, R.M.: On 2-groups of small rank admitting an automorphism of order 3. J. Algebra **125**(1), 27–35 (1989)
268. Uno, K.: Dade's conjecture for tame blocks. Osaka J. Math. **31**(4), 747–772 (1994)

269. Uno, K.: Conjectures on character degrees for the simple Thompson group. Osaka J. Math. **41**(1), 11–36 (2004)

270. Usami, Y.: On p-blocks with abelian defect groups and inertial index 2 or 3. I. J. Algebra **119**(1), 123–146 (1988)

271. Usami, Y.: Perfect isometries for blocks with abelian defect groups and dihedral inertial quotients of order 6. J. Algebra **172**(1), 113–125 (1995)

272. Usami, Y.: Perfect isometries and isotypies for blocks with abelian defect groups and the inertial quotients isomorphic to $Z_3 \times Z_3$. J. Algebra **182**(1), 140–164 (1996)

273. Usami, Y.: Perfect isometries and isotypies for blocks with abelian defect groups and the inertial quotients isomorphic to $Z_4 \times Z_2$. J. Algebra **181**(3), 727–759 (1996)

274. Usami, Y.: Perfect isometries for principal blocks with abelian defect groups and elementary abelian 2-inertial quotients. J. Algebra **196**(2), 646–681 (1997)

275. van der Waall, R.W.: On p-nilpotent forcing groups. Indag. Math. (N.S.) **2**(3), 367–384 (1991)

276. van der Waerden, B.L., Gross, H.: Studien zur Theorie der quadratischen Formen. Lehrbücher und Monographien aus dem Gebiete der exakten Wissenschaften, Mathematische Reihe, Band 34. Birkhäuser Verlag, Basel (1968)

277. Waldmüller, R.: Untersuchungen zu donovans vermutung für klassische gruppen. Dissertation, Aachen (2005)

278. Wang, B.: Modular representations of direct products. MM Res. Preprints **22**, 256–263 (2003)

279. Watanabe, A.: p-Blocks and p-regular classes in a finite group. Kumamoto J. Sci. (Math.) **15**(1), 33–38 (1982)

280. Watanabe, A.: Notes on p-blocks of characters of finite groups. J. Algebra **136**(1), 109–116 (1991)

281. Watanabe, A.: On perfect isometries for blocks with abelian defect groups and cyclic hyperfocal subgroups. Kumamoto J. Math. **18**, 85–92 (2005)

282. Watanabe, A.: Appendix on blocks with elementary abelian defect group of order 9. In: Representation Theory of Finite Groups and Algebras, and Related Topics (Kyoto, 2008), pp. 9–17. Kyoto University Research Institute for Mathematical Sciences, Kyoto (2010)

283. Watanabe, A.: The number of irreducible brauer characters in a p-block of a finite group with cyclic hyperfocal subgroup. J. Algebra **416**, 167–183 (2014)

284. Webb, P.: An introduction to the representations and cohomology of categories. In: Group Representation Theory, pp. 149–173. EPFL Press, Lausanne (2007)

285. Weir, A.J.: Sylow p-subgroups of the classical groups over finite fields with characteristic prime to p. Proc. Am. Math. Soc. **6**, 529–533 (1955)

286. Wikipedia: List of transitive finite linear groups (2014). http://en.wikipedia.org/wiki/List_of_transitive_finite_linear_groups

287. Yang, S.: On Olsson's conjecture for blocks with metacyclic defect groups of odd order. Arch. Math. (Basel) **96**, 401–408 (2011)

288. Yang, S.: 3-Blocks with Abelian defect groups isomorphic to $Z_{3^m} \times Z_{3^n}$. Acta Math. Sin. (Engl. Ser.) **29**(12), 2245–2250 (2013)

289. Yang, S., Gao, S.: On the control of fusion in the local category for the p-block with a minimal nonabelian defect group. Sci. China Math. **54**, 325–340 (2011)

290. Yang, Y.: Regular orbits of nilpotent subgroups of solvable linear groups. J. Algebra **325**, 56–69 (2011)

Index

© Springer International Publishing Switzerland 2014
B. Sambale, *Blocks of Finite Groups and Their Invariants*, Lecture Notes
in Mathematics 2127, DOI 10.1007/978-3-319-12006-5

LECTURE NOTES IN MATHEMATICS ⚇ Springer

Edited by J.-M. Morel, B. Teissier; P.K. Maini

Editorial Policy (for the publication of monographs)

1. Lecture Notes aim to report new developments in all areas of mathematics and their applications - quickly, informally and at a high level. Mathematical texts analysing new developments in modelling and numerical simulation are welcome.

   Monograph manuscripts should be reasonably self-contained and rounded off. Thus they may, and often will, present not only results of the author but also related work by other people. They may be based on specialised lecture courses. Furthermore, the manuscripts should provide sufficient motivation, examples and applications. This clearly distinguishes Lecture Notes from journal articles or technical reports which normally are very concise. Articles intended for a journal but too long to be accepted by most journals, usually do not have this "lecture notes" character. For similar reasons it is unusual for doctoral theses to be accepted for the Lecture Notes series, though habilitation theses may be appropriate.

2. Manuscripts should be submitted either online at www.editorialmanager.com/lnm to Springer's mathematics editorial in Heidelberg, or to one of the series editors. In general, manuscripts will be sent out to 2 external referees for evaluation. If a decision cannot yet be reached on the basis of the first 2 reports, further referees may be contacted: The author will be informed of this. A final decision to publish can be made only on the basis of the complete manuscript, however a refereeing process leading to a preliminary decision can be based on a pre-final or incomplete manuscript. The strict minimum amount of material that will be considered should include a detailed outline describing the planned contents of each chapter, a bibliography and several sample chapters.

   Authors should be aware that incomplete or insufficiently close to final manuscripts almost always result in longer refereeing times and nevertheless unclear referees' recommendations, making further refereeing of a final draft necessary.

   Authors should also be aware that parallel submission of their manuscript to another publisher while under consideration for LNM will in general lead to immediate rejection.

3. Manuscripts should in general be submitted in English. Final manuscripts should contain at least 100 pages of mathematical text and should always include

   – a table of contents;
   – an informative introduction, with adequate motivation and perhaps some historical remarks: it should be accessible to a reader not intimately familiar with the topic treated;
   – a subject index: as a rule this is genuinely helpful for the reader.

   For evaluation purposes, manuscripts may be submitted in print or electronic form (print form is still preferred by most referees), in the latter case preferably as pdf- or zipped ps-files. Lecture Notes volumes are, as a rule, printed digitally from the authors' files. To ensure best results, authors are asked to use the LaTeX2e style files available from Springer's web-server at:

   ftp://ftp.springer.de/pub/tex/latex/svmonot1/ (for monographs) and
   ftp://ftp.springer.de/pub/tex/latex/svmultt1/ (for summer schools/tutorials).

Additional technical instructions, if necessary, are available on request from lnm@springer.com.

4. Careful preparation of the manuscripts will help keep production time short besides ensuring satisfactory appearance of the finished book in print and online. After acceptance of the manuscript authors will be asked to prepare the final LaTeX source files and also the corresponding dvi-, pdf- or zipped ps-file. The LaTeX source files are essential for producing the full-text online version of the book (see http://www.springerlink.com/openurl.asp?genre=journal&issn=0075-8434 for the existing online volumes of LNM). The actual production of a Lecture Notes volume takes approximately 12 weeks.

5. Authors receive a total of 50 free copies of their volume, but no royalties. They are entitled to a discount of 33.3 % on the price of Springer books purchased for their personal use, if ordering directly from Springer.

6. Commitment to publish is made by letter of intent rather than by signing a formal contract. Springer-Verlag secures the copyright for each volume. Authors are free to reuse material contained in their LNM volumes in later publications: a brief written (or e-mail) request for formal permission is sufficient.

Addresses:

Professor J.-M. Morel, CMLA,
École Normale Supérieure de Cachan,
61 Avenue du Président Wilson, 94235 Cachan Cedex, France
E-mail: morel@cmla.ens-cachan.fr

Professor B. Teissier, Institut Mathématique de Jussieu,
UMR 7586 du CNRS, Équipe "Géométrie et Dynamique",
175 rue du Chevaleret
75013 Paris, France
E-mail: teissier@math.jussieu.fr

For the "Mathematical Biosciences Subseries" of LNM:

Professor P. K. Maini, Center for Mathematical Biology,
Mathematical Institute, 24-29 St Giles,
Oxford OX1 3LP, UK
E-mail: maini@maths.ox.ac.uk

Springer, Mathematics Editorial, Tiergartenstr. 17,
69121 Heidelberg, Germany,
Tel.: +49 (6221) 4876-8259

Fax: +49 (6221) 4876-8259
E-mail: lnm@springer.com